COOPERATIVES IN AGRICULTURE

DAVID W. COBIA

Editor

Prentice Hall Career & Technology
Englewood Cliffs, New Jersey 07632

Library of Congress Cataloging-in-Publication Data

Cooperatives in agriculture.

Includes index.
1. Agriculture, Cooperative. I. Cobia, David W.
HD1491.A3C67 1989   334'.683   88-15132

Editorial/production supervision and
  interior design: Lillian Glennon
Cover design: Edsal Enterprizes
Manufacturing buyer: Robert Anderson

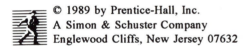
Printed in the United States of America
10 9 8 7 6 5 4

ISBN   0-13-172461-4

PRENTICE-HALL INTERNATIONAL (UK) LIMITED, *London*.
PRENTICE-HALL OF AUSTRALIA PTY. LIMITED, *Sydney*
PRENTICE-HALL CANADA INC., *Toronto*
PRENTICE-HALL HISPANOAMERICANA, S.A., *Mexico*
PRENTICE-HALL OF INDIA PRIVATE LIMITED, *New Delhi*
PRENTICE-HALL OF JAPAN, INC., *Tokyo*
SIMON & SCHUSTER ASIA PTE. LTD., *Singapore*
EDITORA PRENTICE-HALL DO BRASIL, LTDA., *Rio de Janeiro*

Contents

Part VI
Management
17

Part VII
Structural Dynamics
20

Preface

PURPOSE AND AUDIENCE

Our primary motivation has been to develop a textbook for university-level courses focused primarily on agricultural cooperatives. As with most such books, the authors hope it will be used as the key reference on the subject by instructors who include cooperative education in other courses and by professionals in government and industry. We believe that it can also be used to develop training materials for other audiences, such as employees of cooperatives and trade and high school students.

Many cooperatives are struggling through a wrenching adjustment. Traditional organizational linkages have been destroyed and cooperative principles are being challenged. A new generation of participants is shaping the role of cooperatives. It is hoped that this book will play a significant part in helping you, the reader, understand and even assist in the adjustment process. The objective of the adjustment process is to create and maintain cooperatives that will contribute to the economic well-being of their members.

The primary objective behind gaining a competent understanding of cooperatives is effective management and communication by those who deal with cooperatives in any capacity. This includes those who direct, manage, work for, control (by electing directors), own, patronize, receive benefits from, compete with, regulate, or study cooperatives, as well as those who expect to be involved in any of these activities in the future.

Many cooperative members, employees, and cooperative educators consider cooperatives to be a unique, superior form of business. Many others exhibit a high level of misunderstanding and demonstrate a strong bias against cooperatives. The reasons for positive and negative attitudes and good and poor understanding are based on beliefs, feelings, interests, experience, and education. This is why the term "cooperative" and many of its associated terms and concepts have so many different meanings to different people. Even those closely associated with cooperatives often disagree on key concepts such as the definition and role of a cooperative.

Our purpose is to provide a strong educational foundation that is both logically consistent and realistic. Areas of strong disagreements that exist are recognized. As a result, your knowledge of cooperatives will be improved, and your attitudes will be based on facts, not fiction. More important, you will be more effective in your dealings with cooperatives, whatever your present or future role.

FEATURES

There are four distinguishing characteristics of this book. First, we have made an attempt to create a balanced approach, giving equal treatment to topics of equal importance. A few novel tactics are included to acknowledge creativity and to give readers a launching pad for further innovation. We also have sought a geographic and commodity balance in our use of illustrative examples.

Second, we have focused attention on unique aspects of practices, management, and theory of agricultural cooperatives, and excluded concepts that apply to business firms in general. However, we also have referred readers to books that authors suggest are among the best available on these topics. A third distinguishing feature of this book that will exist for a short time is that it is current. The fourth is that we have woven legal topics (except those about antitrust laws) into related topics throughout the book. Most of these sections were written by James Baarda.

CONTENT AND SEQUENCE OF CHAPTERS

In choosing the content of this book, we concluded a full quarter or semester course would take all our available time and space just to cover the key unique features of cooperatives. Thus we excluded topics that are generally applicable to any business, resulting in a necessarily minimal treatment of basic principles of marketing, finance, and management. Although the setting and focus is on agricultural cooperatives, the principles and issues apply to cooperatives in other industries as well.

Authors press for their own ideas on the proper sequence of topics. The sequence used in this text is not entirely satisfactory to any author, but in the editors opinion it represents the best compromise.

Part I is intended to define cooperatives, contrast them with other forms of business, discuss key features, and explore principles that are used as guides, organizations and conduct. Part II, which completes the stage setting, includes an explanation of the taxonomy, scope, structure, and historical framework of cooperatives. In Part III the economic justification for cooperatives and the theoretical constructs on which other topics (especially marketing and, to a lesser extent, finance and management) are established. Marketing is the focus of Part IV. The marketing section immediately follows the theory section because concepts developed in the theory section are applied in marketing more than they are in any other section. The marketing section is followed by Part V, on finance, because it builds on definitions and concepts covered in marketing. Part VI, on management, follows the marketing and finance sections. Some have argued that the order should be reversed to management, finance, and then marketing. Each order has its advantages. The management section, in a strictly topical approach, would perhaps be placed more appropriately before marketing and finance, because both of these topics are logically subtopics of management. However, management is placed fifth because of the link between theory and marketing and between marketing and finance. Further, many authors argue that the relative importance of management issues can more readily be appreciated if the reader has been exposed to the earlier topics. Finally, communications, the last chapter in the management section, provides a transition to some topics in Part VII on structural dynamics.

There is a healthy variety in the sequence that instructors can use to present concepts to students. For example, one instructor places heavy emphasis on market structure and public policy while presenting most of the course. Another instructor presents business management concepts simultaneously with topics on cooperatives. A third uses the chapter about starting a cooperative to review economic justification, theory, and rationale for different organization forms. Yet another uses the section on tax policy to introduce financial concepts.

Some instructors may wish to modify the way they present the materials in light of the organization of this book while others will want to continue with their past practices. We believe the book will lend itself to many approaches.

TERMINOLOGY

No comprehensive taxonomy of cooperatives has been developed or widely accepted. Therefore, we have given some common words a special or specific meaning so as to describe a concept clearly and accurately. We realize that it is frustrating when the meaning one has always given a common or familiar term is modified. Even long-time students of cooperatives will note that some special cooperative terms have been given a slightly different meaning than the one with which they may be familiar.

To some it will appear we are much like the character in *Alice in Wonderland* who said,: "When I use a word it means just what I choose it to mean—neither more nor less." However, in our stubbornness we have a higher purpose than simply insisting on one point of view based on personal biases or convenience. We need a logically consistent jargon with which to communicate clearly, accurately, and unambiguously the nature of cooperatives. In the past, cooperative terminology has been plagued with lack of rigor in this regard. Some terminology has been inadequate, confusing, misleading, or even incorrect in succinctly describing the distinctive features of cooperatives. For example, cooperatives have used *savings* in place of *net income* and *dividends* when the correct term was *patronage refunds*.

We hope to correct many of these deficiencies in this book. Undoubtedly many traditionalists will complain. The authors of the various chapters have and continue to press for their own ideal terms, definitions, and descriptions. The editor has reconciled these as much as possible. Cooperative thought, including terminology, is in a state of transition. This is frustrating to the student and educator alike. Nevertheless, we believe that the student deserves and the educator needs a level of rigor and precision in this book comparable to that found in popular economics and management college-level textbooks.

CRITIQUES WELCOMED

We anticipate the need to publish a revised version. New concepts and a new organizational framework have been introduced here for the first time. The book will undoubtedly require reworking as environmental developments take place and as we continue the dialogue on appropriate terminology, concepts, and analytical framework. Therefore, we encourage readers to share their insights. We have found that when given an opportunity, many industry participants and students assume with enthusiasm the task of commenting on what is right and wrong with a textbook. We welcome such feedback. Students, instructors, industry participants, trade association observers, and others are invited to send their candid comments and suggestions to David W. Cobia, Department of Agricultural Economics, North Dakota State University, Fargo, ND 58105.

Acknowledgments

This book is empirical evidence that people involved with cooperatives cooperate! There are 15 contributing authors. Several of them also reviewed other chapters. The reviewers who unselfishly shared their time and effort are DeeVon Bailey, Calvin Berry, Thomas Brewer, Terry Centner, Gail Cramer, Reynold Dahl, Ron Deiter, Stanley Dreyer, Donald A. Fredrick, John Hagen, James Haskell, Robert Jacobson, Richard King, Larry Mack, William Nelson, Brice Ratchford, Jeffrey Royer, Bernard Sanders, Brian Schmiesing, Richard Sexton, Lyle Solverson, David Thompson, Mike Turner, and Roger A. Wissman. These people have made their contributions as an additional burden to regular professional pressures.

In an act of faith, several instructors volunteered to use early versions of this book as a text in their classes. They were DeeVon Bailey, David Cobia, Reynold Dahl, Ron Deiter, Kenneth D. Duft, Frank Groves, John Hagen, Robert Jacobson, Brice Ratchford, Lee Schrader, and Hank Wallace.

Four other persons not professionally involved with cooperatives provided invaluable assistance. Dr. Gordon Erlandson, an agricultural economics professor emeritus, carefully examined the manuscript and drawing on a professional life as a devoted teacher, provided editorial and pedagogical suggestions from what he perceived to be the student's point of view. Patricia Cobia has spent considerable time consolidating reviewer comments, typing, and making editorial changes on the computer. Kathy Cobia and Darla Christensen also aided by entering revisions into the manuscript. Denise Jamsa edited the manuscript for grammatical correctness.

Out-of-pocket expenses were covered by a grant from the Cooperative Foundation, an organization created and funded by MSI Insurance. Thomas F. Ellerbe, Sr. (deceased) served as the foundation contact person. The American Institute of Cooperation (AIC) organized the Blue Ribbon Committee for the Development of a Textbook on Cooperatives in 1985, which provided a sounding board. In addition, AIC covered some travel expenses. Active members of the committee were David G. Barton (chairman), Bruce L. Anderson, William E. Black, David W. Cobia, Kenneth D. Duft, Gene Ingalsbe, Brice Ratchford, Brian H Schmiesing, Michael Turner, and Richard Vilstrup. David Simpson and Owen Hallberg served as AIC staff consultants.

This effort was launched while the editor was on a developmental leave from North Dakota State University at Brigham Young University (BYU). The Department of Agricultural Economics at BYU provided the editor with an office, secretarial assistance, and a pleasant and friendly atmosphere. The editor is also most grateful to North Dakota State University for providing him with a developmental leave salary and time to complete the task.

David W. Cobia, Editor
David G. Barton, Chairman, AIC Committee
May 27, 1988

Part I
Introduction

1

What Is a Cooperative?

David Barton,
Kansas State University

OVERVIEW

A cooperative is a user-owned and user-controlled business that distributes benefits on the basis of use. More specifically, it is distinguished from other businesses by three concepts or principles: First, the user-owner principle. Persons who own and finance the cooperative are those that use it. Second, the user-control principle. Control of the cooperative is by those who use the cooperative. Third, the user-benefits principle. Benefits of the cooperative are distributed to its users on the basis of their use.[1] The user-benefits principle is often stated as business-at-cost. In this chapter we enlarge on the definition of a cooperative, compare cooperatives with other types of businesses, define selected key terms, and touch on methods of putting the above-mentioned principles into action.

Cooperatives account for a significant share of business throughout the entire world, especially in agricultural markets. There are nearly 6,000 farm marketing and supply cooperatives in the United States with almost

[1]This definition and statement of principles is paraphrased from the U.S. Department of Agriculture's report to the Senate Agricultural Appropriations Subcommittee.

5 million members. In 1985, U.S. agricultural cooperatives supplied farmers with 26% of their inputs and marketed 28% of their products. U.S. cooperatives provided 22% of the money borrowed to finance non-real estate operations and 43% of the money borrowed to finance real estate purchases by farmers in 1984. Cooperatives also provide rural residents with electricity, water, telephone service, and insurance.

The reason for cooperatives' popularity among farmers is simple: through cooperatives farmers can pool their financial resources and carry out business activities they could not independently perform as efficiently. In some past cases, existing businesses have not provided them with the goods and services they have desired. In other cases, existing businesses have followed monopolistic practices, thereby extracting monopolistic profits to the farmers' disadvantage. Therefore, farmers have had significant economic incentives to join together and form cooperatives that operate at cost, thus enabling them to enjoy greater profits from acquiring inputs, from receiving services, and from marketing outputs.

These benefits occur in the form of patronage refunds, more favorable prices, services that would otherwise be unavailable, and access to markets and assured sources of supplies. Patronage refunds are price adjustments made after a cooperative's accounting period. For example, a cooperative that generates $200,000 in revenue and pays $180,000 in expenses has $20,000 left over as net income, which it returns to patrons on the basis of patronage. A patron that conducts $4,000 worth of business with the cooperative receives as a refund a proportionate share of the net income (4,000/200,000 or 2%) totaling $400. This is how most cooperatives achieve business at cost. Other cooperatives price their services so that net income is zero. Cooperatives can also allocate net income to patrons as dividends on equity, but these dividends must usually be limited.

Favorable prices are lower prices for supplies and services and higher prices of commodities produced by farmers. Cooperatives provide services when other companies are sometimes unwilling to do so in low-density rural areas. Cooperatives also provide access to markets for farmers who would otherwise have been denied such access when other companies have withdrawn from the market. Cooperatives also prove to be a reliable source of supplies during periods of shortages.

Users are generally called patrons. Patrons eligible to vote in the affairs of the cooperative are called members.

A person becomes a member of a cooperative by meeting its qualifications, including any requirements to purchase equity certificates or stock. Membership gives members the right to vote for members of the board of directors, who then select their own officers and hire a manager or chief executive officer (CEO) to manage the cooperative. Members vote on proposed policies regarding key issues. Control of a cooperative is typically democratic, meaning that each person has only one vote regardless of the amount that the person has invested in the cooperative or volume of business transacted. Where control is not democratic, return on investment is limited to ensure that most benefits are returned to patrons in proportion to use. Voting in cooperatives may also be proportional to use or to equity investment.

Terminology

Listing several terms may be confusing to some readers. However, being aware of other terms, although we discourage their use, will facilitate communication with people of different backgrounds. Reviewing the rationale for terms used in this book will make discussions more precise, avoid misunderstanding, and help clear up confusion over terminology. The definition and rationale for specific terms are given when the relevant topics are first discussed in detail. A glossary is provided at the end of the book.

The terminology used to describe cooperatives and other firms differs widely. Cooperatives are also commonly called *nonprofit corporations* or *patron-owned corporations*. We will use the term *cooperative*[2] because it does not have the wrong connotation of the first term and is more succinct than the second term.

Firms other than cooperatives are called *noncooperatives, investor-owned* or *proprietary firms,* and *profit, private, ordinary, standard,* and *other corporations* or just *corporations*. There are valid reasons for not using any of these terms. They are misleading because cooperatives are also proprietary, private, and investor-owned organizations. They also seek to increase the profits or economic well-being of their members. Even "noncooperative" has several unintended negative connotations and could refer to other institutions, such as churches and government agencies.

The distinction between cooperatives and other businesses is that cooperatives return net income to users or to patrons, while other business firms return net income on the basis of investment. Thus in this book we use the term *investor-oriented firm* or *IOF* to represent business firms other than cooperatives or user-oriented firms. Although these designations are not widely recognized, they avoid erroneous concepts and connotations associated with more common terms.

Our definition of a cooperative is tailored to cooperatives owned by customers or patrons, since very few are controlled or owned by employees. Thus, except for a section in Chapter 4 on employee cooperatives, most future references to cooperatives will be about cooperatives oriented to providing economic benefits to members who are patrons. A *patron* refers to any person, business, or other institution that conducts

---

[2]Current usage in the United States is to spell "cooperative" without a hyphen. Older usage in the United States and current usage in Great Britain and its commonwealth countries is to hyphenate the spelling (co-operative). A short informal form commonly used worldwide is "co-op" in place of "cooperative." A correct abbreviation is "coop."; an incorrect abbreviation leaves off the period (a coop is a building where chickens live).

business with a cooperative or other type of business. Cooperatives are classified by the type of patron: consumers (households), nonprofit and government agencies, or businesses (including farmers). A consumer patron purchases goods and services for final consumption. Nonprofit and government agencies purchase goods and services from cooperatives to improve the quality and quantity of their services. A business patron purchases goods and services from cooperatives to be used in the production of other goods and services or markets its output through cooperatives.

We have used some traditional cooperative terms, such as "patron," in place of the conventional business terms, such as "customer," in most discussions. However, to facilitate comparing cooperatives with IOF business and to provide those of you with an interest in business management, we will describe how some common cooperatives and IOF terms compare. Finally, we are especially interested in cooperatives whose members are agricultural producers. At this point, however, we must first review some features of business in general to better understand the unique function of the agricultural cooperative in today's economy.

Stakeholders

Each business has a set of key stakeholders (Pearce and Robinson, p. 28), including (1) *owners* or investors; (2) *employees*, who work for the business in some capacity, such as in labor or management; (3) *patrons* or *customers*, who patronize the business by using it as a source of supplies or services needed in their household or business or as an outlet for the products they produce; and (4) *other* stakeholders (such as creditors and suppliers), who have vital interests in the business (Figure 1.1).[3] Owners are of two types: *controlling owners*, who have the power to vote or to control the business, and *noncontrolling owners*, who do not have voting rights. The first three stakeholders we have mentioned are sufficient for defining a cooperative and differentiating it from other legal types of business.

Thus far we have used the terms "patron" and "customer" interchangeably. Patrons of cooperatives may sell to or buy from the cooperative. For example, a farmer may sell wheat and buy fertilizer. In both cases the farmer is a patron or customer in a general or informal sense. This is similar to your relationship with a bank. If you deposit or sell money to the bank, or if you borrow or buy money from the bank, you are thought of as a customer by the bank.

Any person or business can be all four types of stakeholders. For example, one person can simultaneously be a customer of IBM, work for IBM, own voting or common stock in IBM, and own nonvoting stock in IBM; there is no direct or necessary relationship between these key stakeholder roles. However, in a cooperative these roles are very closely connected. In

[3] See Pearce and Robinson (p. 28) for additional information on this concept.

Figure 1.1 Key stakeholders in a business.

this chapter we show that this relationship is one of the most fundamental distinguishing features of a cooperative.

LEGAL FORMS

It is also important to understand where cooperatives fit in among the three legal forms of business organization: proprietorships, partnerships, and corporations (Figure 1.2).

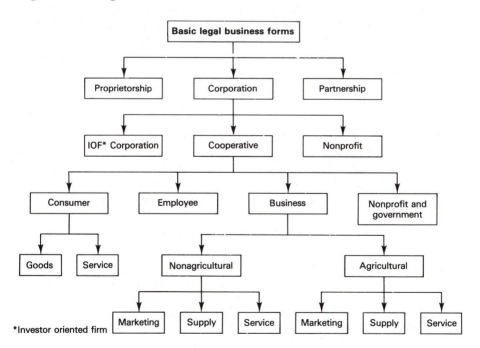

Figure 1.2 Basic Taxonomy of legal forms of organization emphasizing cooperatives.

A *proprietorship* is a business owned and controlled by one entrepreneur who provides the equity capital, makes the key management decisions, and bears unlimited liability for debts and losses in exchange for the opportunity to receive profits. Income taxes are paid by the owner, not by the business, and the proprietorship ceases when the owner dies. Obviously, this organization is in direct contrast to that of the cooperative, whose main objective is to share all business income and responsibilities among several member-patron owners. Proprietorships are the oldest, simplest, most numerous, and smallest type of business (Table 1.1). In the United States over 12 million proprietorships comprise 74% of the businesses by number but only about 7% of total business sales dollars. Proprietorships are usually small businesses, although any size of business can be a proprietorship.

A *partnership* is similar to a proprietorship, except that it is owned and controlled by two or more people or other businesses. A partnership is the simplest business form for group action, including that of a cooperative. However, it is not as prevalent as other business types. In the United States there are 1.5 million partnerships, which account for only about 3% of the total business volume.

There are two kinds of partnerships: general partnerships and limited partnerships. Our reference will be to general partnerships. The primary difference is that a limited partnership has one or more general partners and one or more limited partners. The limited partners contribute equity capital but have limited liability for debts and losses or other claims against the business and generally do not participate in management.

A *corporation* is a legally chartered institution formed by a group of people or other businesses who are granted, as a body of one, certain of the legal powers, rights, and privileges of a person, distinct from those of the individuals making up the group. In other words, the corporation is much like an artificial person. It owns assets, has the right to net income, and is liable for debts, losses, and other claims. It may also vote in certain instances, for example, if it owns stock in another corporation. The corporation is an artificial person with a perpetual existence. For example, if

TABLE 1.1   Numbers and sales of legal forms of business organization, 1981

Business type	Numbers (Thousands)	Total sales (Billions)	Average sales (Thousands)
Proprietorship	12,185	$ 523	$ 43
Partnershi p	1,461	272	186
Corporation	2,812	7,026	2,499

Source: Statistical Abstract of the United States, 1985.

an owner dies or sells equity to another owner or back to the corporation itself, the owner's disassociation or absence does not cause the dissolution of the corporation as it would in a proprietorship or often in a partnership.

A corporation is a more complex form of organization and accounts for most business activity in the United States. There are about 3 million corporations in the United States; these account for 17% of the country's business numbers but 90% of its business volume. The largest businesses are almost all corporations. However, many small businesses, including many one-person businesses, are also corporations; most corporations, in fact, are small. Although U.S. corporations average $2.5 million per year in sales, about 85% of them have assets and annual sales of less than $1 million. By comparison, almost all agricultural cooperatives have sales exceeding $1 million and are therefore in the top 15% of corporations by size.

Nearly all cooperatives are corporations. However, cooperatives are only one type among many types of corporations, including C corporations, nonprofit corporations, and S corporations. C and S corporations are distinctions made in the tax codes. Cooperatives are identified as subchapter T corporations. It is important to understand what distinguishes cooperatives from these other corporations.

A corporation is generally chartered under state laws and must fulfill the organizational and operational requirements of those laws. The laws may vary from state to state. In addition to the general incorporation statutes applying to C corporations, other types of corporations, including cooperatives, rely on statutes that differentiate that type of business.

Ownership is evidenced by certificates of stock in a stock corporation or by other ownership instruments in a nonstock corporation. Voting owners or members maintain control by electing a board of directors. The board then selects its own presiding officers and hires the chief executive officer (CEO) and sometimes other executive officers; it also sets and maintains major policies of the business, and it monitors the operation of the business in behalf of the voting owners or members. Cooperatives have traditionally used the title *president* for the presiding officer of the board and the title of *manager* for the CEO. However, there is a trend among cooperatives, especially among the larger ones, to use the terms more commonly by IOFs, of *chairman* and *president*, respectively.

All owners have limited liability equal to their equity investment for losses, debts, and other financial claims on the business. If the corporation becomes insolvent, creditors cannot legally require owners to supply additional capital beyond what they had already invested.

All C corporations are subject to taxation of their taxable income at corporate rates *before* net income is distributed. In addition, owners are required to report as personal taxable income dividends they received from corporate after-tax income. Therefore, the distribution of corporate earnings as dividends results in double taxation of those earnings, first at the corporate level and again at the individual level. However, special rules apply to cooperatives, to nonprofit corporations, and to S corporations. If cooperative and S corporations comply with certain requirements and distribute net income according to certain regulations, their income is taxable only at the individual owner level. The result is that their net income, like that of proprietorships and partnerships, is taxed only once instead of twice.

Table 1.2 shows the cooperative's relationship to three other legal forms of business. For each of the three decision factors—control, ownership, and benefits—we have asked questions to focus on important characteristics. Study this table carefully to gain a better understanding of the similarities and differences between cooperatives and other types of business.

KEY CONCEPTS

We use the following six concepts to enhance your basic understanding of a cooperative and its place in our economy:

1. The primary purpose is economic benefits for members.
2. Members are usually patrons.
3. Members own and control the cooperative.
4. Qualifying patrons receive distribution of benefits.
5. Cooperatives are private organizations.
6. Public policy establishes the institutional framework.

Primary Purpose

Our study of cooperatives will focus on cooperative businesses whose primary purpose is economic or commercial and whose members therefore join the cooperative primarily for economic reasons. However, cooperatives may pursue some noneconomic objectives as well.

Benefits of social value include all noneconomic results or outcomes of major interest or importance to stakeholders, including the satisfaction many of them experience through the association, unity, and involvement characteristic of member-controlled organizations. Some members like being involved with others to achieve a common purpose. Some members also like electing or serving as directors (although others sometimes object to the amount of involvement required or expected of cooperative members).

These benefits are important. However, one must keep in mind that if cooperative businesses do not satisfactorily fulfill their economic pur-

TABLE 1.2 Comparison of four legal forms of doing business in the United States

Business decision factors	Form of business			
	Proprietorship	Partnership	IOF corporation	Cooperative corporation
1. Control				
a. Who is eligible to vote on selecting board & major policies?	Proprietor	Partners	Common stockholders	Members
b. What are requirements to become a voter?	Be sole owner of the business	Be a partial owner of the business, usually with the permission of other voting partners	Be an owner of common stock of the business	Must meet qualifications of membership (type of business, investment, etc.); most co-ops have open rather than closed or limited membership
c. How many votes each for eligible voters?	One	Usually apportioned by each partner's equity proportion	One vote per share of common stock	Usually democratic control: one-member, one-vote basis
d. Who manages the business?	Proprietor or someone selected by proprietor	Managing partner or chief executive officer (CEO) selected by the voting partners	Board of directors elected directly or indirectly by stockholders; board selects CEO	Board of directors elected directly or indirectly by members; board selects CEO
2. Ownership				
a. Who is eligible to be an owner?	Any person able to provide sufficient equity capital for a one-owner business	Any person able to provide sufficient equity capital and acceptable to other partners	Any person able to buy one or more shares of common stock; closely held corporations may have additional restrictions	Generally restricted to those qualifying for membership but may include any person able to buy minimum investment

TABLE 1.2   (cont'd)

Business decision factors	Form of business			
	Proprietorship	Partnership	IOF corporation	Cooperative corporation
b. How much ownership is required by individual owner?	As proprietor chooses	As partners agree	No limits except minimum cost of one share of stock; closely held corporations may vary	Usually the cost of one share of common stock or membership fee; desirable that ownership be in the hands of current patrons in proportion to use
c. How is ownership transferred?	Privately negotiated or purchase and sale	Privately negotiated or purchase and sale with approval of partners	Privately negotiated or purchase and sale, may require corporate approval	Equity transfers highly restricted
3. Benefits				
a. Why is net income generated?	For distribution to the sole owner as a return on equity investment	For distribution to the owners (partners) as a return on their equity investment	For distribution to the owners (stockholders) as a return on their equity investment	For distribution to patrons based on patronage
b. Who can be patrons?	No restriction	No restriction	No restriction	Generally no restriction but may be restricted to persons who qualify for membership
c. How is net income distributed?				
(1) To owners in proportion to investment as dividends	Unlimited since all net income belongs to proprietor	Unlimited since all net income belongs to partners	Unlimited split between dividends and reserves highly variable in practice; most pay something	Limited by law in most states, usually to 8%; most co-ops pay none

TABLE 1.2 (cont'd)

Business decision factors	Form of business			
	Proprietorship	Partnership	IOF corporation	Cooperative corporation
(2) To patrons as patronage refunds	Permissible, but very unusual	Permissible, but very unusual	Permissible, but very unusual	Very common and often required by law or bylaws; the most unique characteristic of co-ops
(3) To unallocated company reserves as retained earnings	Permissible but all reserves allocated to sole owners	Permissible but reserves are in essence allocated to partners	Permissible and very common; split between dividends and reserves highly variable	Permissible; most co-ops use only nonpatron, nonmember earnings, which are usually relatively small
d. What is income tax obligation?	As proprietor income at individual rates	As partner income at individual rates	Corporation pays taxes at ordinary corporate rates after deductions and adjustments; stockholder pays taxes at individual rates on dividends, causing double taxation on these earnings	Same as IOF corporation except that patronage refunds are subject to single taxation

pose, in the long run they will be unable to fulfill even noneconomic purposes. The end result is often liquidation or reorganization because of economic necessity.

Members have always had one or more economic purposes for joining together to form cooperatives. For example, farmers in a local area have joined together to integrate vertically in the forward direction to market their farm products, such as grain and milk. Farmers in a local area have also joined together to integrate vertically in the backward direction to purchase their farm supplies, such as petroleum products, feed, and fertilizer. In general, members have pursued a form of group action called economic integration; the most common forms being horizontal and vertical integration.

The reasons members have joined together include (1) to obtain a fair or efficient price (i.e., to correct market failure); (2) to reduce costs through economies of size and coordination; (3) to provide markets, supplies, and services that are missing or in danger of being lost; (4) to pool risk; (5) to capture profits from another level; and (6) to benefit from increased market power. A discussion of these motivations and justifications is presented in chap. 8.

Members and Patrons

Cooperatives are controlled by members. Members elect the board of directors and vote on major policies. Members must meet certain qualifications. For example, members of a local agricultural cooperative must usually be agricultural producers. In cooperative businesses, members are usually patrons. For example, Farmland Industries, Agway, Associated Milk Producers, Union Equity, Harvest States, Land O'Lakes, Sunkist Growers, and Ocean Spray are controlled by patrons who are voting members. Patrons are those who do business with a cooperative whether by selling inputs or by purchasing goods and services. This orientation is unusual when compared to that of IOFs, which are investor oriented and investor controlled.

Although patrons or customers are important in IOF businesses such as Cargill and IBM, these businesses do not give patrons formal control. Rather, patron influence is expressed in the market system through purchases and sales. Cooperatives, on the other hand, are controlled by their member-patrons, and net income is returned to patrons as patronage refunds. Patrons may or may not be members. Patronage refunds may be given only to members or may also be given to nonmembers.

The relationships between ownership, control, and distribution of benefits, on the one hand, and patron subgroups of members and nonmembers, participating and nonparticipating patrons, on the other, are complex in cooperatives. The relationships are better understood by examining the

TABLE 1.3   Matrix of cooperative member and patron classifications

Control: Is patron eligible to vote?	Member status	Benefits: Is patron eligible to receive patronage refunds?	
		Yes	No
Yes	Member	Type 1: Member–participating patron Vote —yes Refund —yes Ownership —yes	Type 2: Member–nonparticipating patron Vote —yes[a] Refund — no Ownership — yes
No	Nonmember	Type 3: Nonmember-participating patron Vote —no Refund —yes Ownership —Yes	Type 4: Nonmember-nonparticipating patron Vote — no Refund — no Ownership —no

[a]May not be eligible to vote in some states.

combinations of membership and eligibility to receive patronage refunds (Table 1.3). Cooperative businesses can be classified by the type of stakeholder permitted to be the member and therefore authorized to control the cooperative. Therefore, we have listed four types of stakeholders. These four types are related to the four stakeholders in Figure 1.1 in the following paragraphs.

The owner or investor category alone is not usually a membership category unless it includes investors who are also patrons or employees. It is conceivable but unlikely that a cooperative may be investor oriented. If the only affiliation a member has to have with the cooperative is investment, then control and ownership is no different than in an IOF business. It is unlikely that investor members would authorize a substantial portion of cooperative net income to be distributed to employees or patrons who are not also members. Therefore, cooperatives will nearly always be user oriented. The closest cooperative-like business that is both investor and user oriented is an investment club because members' return and patronage is proportional to investment in the club.

In addition to qualifying as patrons, members must also meet certain additional requirements. Not all patrons of a cooperative are necessarily eligible or admitted to membership. For example, most agricultural cooperatives require that members must be agricultural producers or other cooperatives. In fact, some cooperatives' best patrons are denied membership on these grounds, even though they reside near the cooperatives' operations and purchase fuel, fertilizer, chemicals, and other products from the cooperative. Futhermore, not all producers apply or are admitted to membership. Since not all patrons are granted a voice or vote in control of the cooperative, its patron category may be subdivided into members and nonmembers.

Four basic types of patrons can be defined by the possible combinations of members, nonmembers, and participating and nonparticipating patrons. (Table 1.3). Type 1, member-participating patrons, are the most common. These are patrons owning voting stock and providing most of the patronage. Legal statutes generally require that a minimum of 50% of the patronage come from this group. Type 2, member-nonparticipating patrons, are relatively rare. They are members that do not qualify for patronage refunds because substantial quantity discounts have been given. Type 3 and 4 patrons typically do not qualify for membership because they are not agricultural producers or they are agricultural producers but do not choose to be members. Type 3, nonmember-participating patrons, qualify for patronage refunds but not for membership. Like members, patrons must meet specific requirements and be formally approved by the board of directors to be eligible to receive patronage refunds. Type 4, nonmember-nonparticipating patrons, do not qualify for membership or patronage refunds. Some nonmember patrons are not given patronage refunds because of insufficient amount and frequency of their business or because of their classification, such as a competitor or government organization. The classification is generally made by the board of directors.

Cooperatives may be classified by the kind of patron (Figure 1.2). Those owned and controlled by businesses (including farmers) may be classified as agricultural and nonagricultural. Other categories of cooperatives in this scheme include those owned by consumers, employees, and nonprofit organizations and governmental agencies. Agricultural cooperatives are discussed in Chapter 3 and other types of cooperatives in Chapter 4.

Ownership and Control

Ownership and control are closely related business decision factors. Ownership is represented by equity investment and control is represented by voting rights. However, control often requires at least a minimum of ownership. The major equity decisions concern (1) how much equity should be needed, (2) from whom and how equity is acquired for investment in the business, and (3) from whom and how equity is repurchased or redeemed. When making these decisions, cooperatives have many alternatives not available to most other businesses. There are also restrictions on the use of some alternatives, such as publicly traded stock. These alternatives are described in Part V.

Equity that belongs to individual persons or businesses and is represented by stock certificates or other instruments is *allocated equity*.

Public Policy Regarding Cooperatives

State and federal legislation establishes the institutional framework in which corporations, including cooperatives, operate. These statutes dictate the manner of organization, rights bestowed upon them, method of taxation, and set certain limits on the conduct of their business.

State law Almost all corporations are chartered under state incorporation statutes. Agricultural cooperatives in most states incorporate under a special cooperative statute but are also governed by the general corporation statute. Some organizations have chosen to call themselves cooperatives when, in fact, they have not been properly incorporated. Cooperative statutes in most states prohibit using the word "cooperative" in a business name, such as Farmers Cooperative Grain Company, unless the business is legitimately incorporated as a cooperative.

Federal law The belief that the dispersion of economic power among many competing businesses will yield the greatest benefit to society as a whole has been held by many leaders in the United States since the inception of our country. This concept was formalized by the Sherman Antitrust Act of 1890, which declared monopoly illegal. That law and others that refined it held that two or more parties could not get together to agree on prices and other restrictive marketing practices.

In the late nineteenth and early twentieth centuries many agricultural producers found many farm input and product markets to be inadequate or unfair. Operating individually, they were unable to get needed supplies or sell their products at what they considered reasonable prices. Consequently, they organized farm supply cooperatives through which to buy farm inputs such as fertilizer, feed, fuel, and chemicals, and organized marketing cooperatives to help them market farm products such as grain, milk, cotton, vegetables, and fruit at fair prices.

However, the very act of farmers getting together to organize a cooperative to create a more balanced competitive environment was sometimes held to be in violation of antitrust statutes. Therefore, Congress passed laws that protected cooperatives (and other sectors such as labor) from antitrust litigation. The cornerstone act, popularly known as the magna carta of marketing cooperatives, is the Capper-Volstead Act passed in 1922. It simply provides that farmers can organize marketing cooperatives without violating antitrust laws, as long as members are agricultural producers, no member has more than one vote or dividends on equity are less than 8%, nonmember business is less than 50%, and prices of products marketed are not unduly enhanced. Legislation has also been passed which provides for the single-tax treatment of patronage-based income, thus avoiding the double-tax treatment of IOF corporations.

SUMMARY

Cooperatives are distinctive businesses (compared to proprietorships, partnerships, and other corporations) that operate under three principles of user ownership, user control, and user benefits (generally referred to as business-at-cost). Cooperatives are one of three primary types of corporations. The other two are investor-oriented corporations and nonprofit corporations.

A cooperative is a business controlled by voting members, usually on a democratic (one-person, one-vote) basis. It is owned by its patrons, who obtain most of their ownership by doing business with the cooperative. Two methods of investment are common: (1) a distribution of net income to the patron in the form of an equity investment called a retained patronage refund, or (2) the retention of a portion of the transaction price as an equity investment, called a per-unit capital retain. Direct investment, the most common method in other businesses, is also used but to a much lesser extent.

The financial benefits to patrons of a cooperative have both similarities and differences when compared to other businesses. Products, services, and pricing are usually similar to competing businesses. However, the net income earned from patron business is distributed differently in some respects.

The most distinguishing financial benefit is the distribution of net income, usually the larger part, in the form of patronage refunds. In other words, most net income is distributed to patrons or customers on the basis of business volume, not to owners on the basis of investment. Some net income may be retained as retained earnings in a manner similar to other businesses. Some may also be distributed to owners in the form of dividends on their investment. However, cooperatives limit themselves to paying a maximum amount, usually 8%. In practice, most cooperatives pay no dividends.

Public policy has generally been favorable toward the creation and operation of cooperatives. State laws generally include special incorporation statutes for cooperatives, both agricultural and nonagricultural. Federal laws include the Capper-Volstead Act, which permits group action to form agricultural marketing cooperatives without being in violation of antitrust laws. Tax codes include provisions, which permit cooperative income, classified as patronage refunds, to be taxed only once, either at the cooperative level or the patron level.

Cooperatives are no panacea for meeting the economic needs of a particular group. They are a complex, and to most, an unfamiliar form of organization. Why do we have cooperatives? Where are they prevalent in agriculture? What makes them successful in some cases? How are they justified in our free-market, capitalistic economy?

In this chapter we have laid a foundation to help you answer some of these questions. In the next chapter we describe cooperative principles, another important building block in understanding cooperatives.

DISCUSSION QUESTIONS

1-1.   What is a cooperative?

1-2.   What is its purpose?

1-3.   What are the three principles associated with the definition of a cooperative? What is their common element?

1-4.   What is the difference between a cooperative and other legal forms of business?

1-5.   Why are cooperatives capitalistic?

1-6.   Why were agricultural cooperatives organized by farmers?

1-7.   What is the difference between a participating patron and a nonparticipating patron of a cooperative?

1-8.   What is the difference between a member and a nonmember patron of a cooperative?

1-9.   What kind of benefits are available to member-patrons of a cooperative?

1-10.   What does it mean when we say that cooperatives are private organizations?

1-11.   What kind of public policy is associated with cooperatives? Why?

REFERENCES

ABRAHAMSEN, MARTIN A., *Cooperative Business Enterprise*. New York: McGraw-Hill, 1976.

BAUMOL, WILLIAM J., and ALAN S. BLINDER, *Economics: Principles and Policy*, 4th ed. New York: Harcourt Brace Jovanovich, 1988.

MCBRIDE, GLYNN, *Agricultural Cooperatives*. Westport, CN: AVE, 1986.

PEARCE, JOHN A., II, and RICHARD B. ROBINSON, Jr., *Strategic Management*. Homewood, IL: Richard D. Irwin, 1982.

ROY, EWELL PAUL, *Cooperatives: Development, Principles, and Management*, 4th ed. Danville, IL: The Interstate Printers & Publishers, Inc., 1981.

SCHAARS, MARVIN A., *Cooperatives, Principles and Practices*. Univ. Center for Cooperatives A1457, Univ. of Wisconsin-Madison, 1971.

U.S. DEPARTMENT OF AGRICULTURE, *Positioning Farmer Cooperatives for the Future :A Report to the Senate Agricultural Appropriations Subcommittee*. Washington, DC: USDA ACS, Oct. 1, 1987.

2

Principles

David Barton,
Kansas State University

INTRODUCTION

Cooperatives are unique in several respects when compared to other forms of business. Many people believe that uniqueness goes beyond the actual law and common practices of our own day and age and is based on a set of distinctive principles that are timeless and universally valid. They believe that only by understanding and adhering to these principles can members, patrons, and managers fully understand and operate cooperatives effectively in a capitalistic market economy.

This belief has a powerful appeal, but to what extent do such principles really exist? And if they exist, what are they? How were they determined? How have they evolved over time? How valid are they today? To what extent is cooperative practice in conformity with the stated principles? What cooperative principles, if followed, will optimally position cooperatives in the future?

Significant disagreement exists as to what constitutes a correct set of cooperative principles. We will present several viewpoints, most of which fit into four sets of so-called principles. Each set will be referred to as a class or school of principles. The classes are the Rochdale, traditional, proportional, and contemporary principles. When classifying cooperatives, those who advocate and adhere to a certain class of principles can be categorized using one of these names. Some disagreement exists as to what exact principles each class should contain. As much as possible, a consensus has been sought based on the literature of this century.

The first two classes have a strong historical basis. They represent the thinking of cooperative observers over a significant period of time. The Rochdale class was developed and widely practiced in the nineteenth century; the traditional class was developed and widely practiced in this century. The proportional class was proposed in this century but never became popular. The contemporary class represents a recent restatement that captures the essential elements of the definition of a cooperative and makes explicit implied concepts in traditional principles. Most cooperative leaders currently express adherence to the traditional principles. Some, however, believe there has been, will be, or should be a transition from traditional to proportional or contemporary principles.

Although good principles are essential for cooperative survival, there is a danger in attributing the success or failure of a particular cooperative, or cooperatives in general, just to the validity of principles. Business success is based on many factors. Assuming that the external environment provides adequate opportunity for success of similar businesses, including competing investor-oriented firms (IOFs), three factors are important: (1) validity of principles, (2) understanding and adherence to principles, and (3) effective management. Even if a cooperative adopts principles that are valid, it may fail the market test if either the principles are not understood or adhered to or if its management is ineffective. To paraphrase Milton, a good principle, not properly understood, may prove as damaging as a faulty principle (Roy, p. 249). On the other hand, good principles, if properly understood and practiced, should lead to superior performance of cooperatives.

Like most institutions, cooperatives were created because of strongly perceived economic needs; the extent of their growth and development to this time has been due to their ability to effectively meet the needs of their patrons in a capitalistic, competitive economy characterized by a changing economic environment. Their future will be determined by their ability to continue meeting those needs. Therefore, the questions naturally arise: "To what extent are principles timeless and universal?" and "To what extent are they changing due to changing conditions?"

BACKGROUND AND ISSUES

Definitions

Much of the controversy and misunderstandings surrounding principles stems from whether certain guidelines are considered mandatory *principles*, wise and strongly recommended *policies*, or common *practices*. A definition of each will facilitate the discussion of the issues. Dictionary definitions as well as common perception of these three terms overlap. Our definition will maintain distinctions among these terms.

Principle   A principle is a governing law of conduct, a general or fundamental truth, a comprehensive or fundamental law. Abiding by the definition of a cooperative and its principles should preserve the essential objectives and uniqueness of the cooperative form of business.

The term *law* in specific application implies that a ruling authority will prescribe and enforce a particular rule. In fact, many rules concerning the conduct of a cooperative's activities are prescribed by laws which differ from the rules governing other businesses. Many of those laws, especially state incorporation statutes, reflect principles (Baarda). In fact, one approach to specifying principles is to rely on the law of the land as the source. However, the laws concerning control, ownership, and distribution of net income vary from state to state and are subject to change. Furthermore, most laws are based on a legislative body's view of appropriate principles and practices rather than vice versa. The goal here is to identify principles that transcend human-made laws and are more on the order of canons that express a fundamental truth, at least for a cooperative private enterprise in a market economy in the present day. Most of the principles of the past do not meet this very demanding requirement in an unqualified manner.

Principles, if commonly accepted, may in turn lead to a change in the laws of the land where conflicts exist. Current laws are largely a result of an earlier generation's understanding of a cooperative and the principles by which it should operate.

Policy   A *policy* is a wise or expedient rule of conduct or management. It is not a universal, unchanging truth but a highly recommended course of action, given the situation. Many so-called principles could in fact be more accurately thought of as fundamental cooperative policies. Policies must change with conditions to allow and encourage cooperatives effectively to fulfill their essential purpose as previously described. Policies should reinforce or at least be compatible with the cooperative definition and principles.

Practice   A *practice* is a usual method, customary habit, action, or convention; a frequent or usual action. Substantial flexibility exists in designing and using practices to meet the objectives of the business while respecting the cooperative definition, principles, and policies. However, over time it is common for some traditional practices and policies to become obsolete, antiquated, or even detrimental because of changes in the social or economic environment. A change in environment sometimes requires a change in policies, which causes conflict with traditional practices. Many cooperative observers believe that some so-called principles are in fact obsolete, given current practices.

A review of cooperative development indicates that this is indeed the case. For example, cash trading only was called a principle of the first successful, permanent cooperative in England. This small, local, consumer cooperative, known as the Rochdale Society, was established in Rochdale, England in 1844. We could debate whether cash trading was in fact a principle for that time and place, especially since the pioneer founders of the Rochdale Society never issued a formal list of principles. Cash trading certainly was a practice and probably should be categorized as a policy for that situation. However, today cash trading only is not even a common practice, let alone a policy, and especially not a principle of the modern agricultural cooperative.

We may find that traditional practices of today are or soon will be obsolete, given the current or expected environment of a particular type of cooperative. However, cooperative principles still should be generally applicable to all consumer and producer cooperatives. Many of the policies should also apply. Practices, on the other hand, may vary widely. Differences in policies and practices between different types of cooperatives, or even between cooperatives of the same type, should not be surprising since situations may differ widely.

The criteria by which proposed principles, policies, and practices should be evaluated are (1) compatibility with the cooperative definition and (2) performance compared to the standard of doing what is best for the patron, especially for the member-patron.

History and Sources

It is not clear who first thought of identifying and proposing the concept and content of cooperative principles. However, it is generally agreed that current-day principles evolved from "rules of conduct and points of organization" put forth by the Rochdale Society, probably for the first time in its *Annual Almanac* of 1860 (Abrahamsen, p. 48). Catherine Webb listed nine rules or points in her study of cooperatives (Webb). Others have chosen slightly different lists, probably because the rules enumerated by the Rochdale Society continued to evolve from its founding in 1844 during the 16 years up to its 1860 publication and thereafter. The most prominent early American to write about cooperatives was Edwin G. Nourse. His article "The Economic Philosophy of Cooperation" is recognized as a key reference point among American observers. He listed three key cooperative principles which are still recognized as the most important of the traditional principles that most cooperatives advocate and follow today.

Cooperative principles, policies, and practices also have continued to evolve since that time, as cooperative development has spread to new times and places.

The primary source of principles has been successful experiences of cooperatives over the years, beginning with the experiences of the Rochdale Society. Since that beginning, others have proposed principles based primarily on their additional experiences.

Other important and complementary sources have been philosophies about how best to meet the economic needs of members when using group action or cooperation to achieve common economic goals. Included in philosophies are the beliefs and theories about the nature of humankind and of economic systems. Cooperatives, more than other types of business, have tried to combine the theories of social and economic systems in designing a business.

Theory, however, has been used much less than experience or belief systems to derive principles. Although these theory-based analyses of principles are few, we will note some of the research results available.[1]

Uses and Controversies

The primary use of principles is to serve as guidelines to cooperative directors, managers, and legislators. If the principles are fundamental truths and laws of conduct, they should be invaluable in helping decision makers make cooperatives effective in meeting the economic needs of member-patrons.

Another use of principles is to compare and classify cooperatives. Espousal and adherence to a certain class of principles clearly identifies a philosophy as well as a relationship to policies and practices.

A related method of cooperative classification is the use of the categories *true* and *quasi*. Those who use this method distinguish between *true* cooperatives, which adhere to traditional principles, and *quasi* cooperatives, which do not. This type of judgmental approach is often used to discourage wayward cooperatives and leaders from unacceptable or suspect behavior. Unfortunately, it is also based on the assumption that principles have an almost divine origin, leaving the impression that cooperatives are part of a religious-like movement and that adherence to specific cooperative principles is a prerequisite to being among the faithful.

This approach is intolerant and exclusionary, and it is probably detrimental to the development of cooperatives as effective businesses in a dynamic and changing environment. Thus, rather than use this method, we will rely only on the cooperative definition as the measure of whether

---

[1]W. P. Watkins recently made a nontraditional approach to deriving principles based on both belief system premises and general social and economic experience. He derived six principles which he calls association (or unity), economy, equity, democracy, liberty, and education. You will find his book *Cooperative Principles: Today and Tomorrow* (Watkins) a valuable source of information on the history and validity of cooperative principles from a British perspective.

a business is a cooperative or not. This definition permits substantial flexibility in viewpoints and choices regarding principles.

In the eyes of many there is a cooperative movement whose purpose is generally to improve our society and economy and specifically to improve the economic situation of patrons whose needs are not being met efficiently. Such a broad objective is laudable and has been supported by a broad spectrum of people. In some ways it is not unlike the rationale and support given to democracy, capitalism, and the market system. However, care must be taken to focus on the ends and then to evaluate the means in that light. Cooperatives and their principles are a means to an end. The end is effectively meeting the needs of patrons.

CLASSES OF PRINCIPLES

The fundamental method of classifying businesses that claim to be cooperatives is to measure them against the definition of a cooperative business. A more refined way, however, is to classify them according to the principles they support and follow. The four classes in Table 2.1 group together particular principles, in order to simplify both classification and discussion. However, you should be aware that variations from these basic classes are possible in particular situations, or cooperatives. A comparison of these classes will illustrate both the evolutionary change in principles that has occurred and the change that some people are expecting or proposing. It should be noted that the proportional principles are based on a narrow and exact definition of a cooperative. As such, they imply a definitive set of principles. They are not a part of the evolutionary continuation, except as they influence the creation of new principles.

Rochdale Principles

The Rochdale principles are a set of guidelines that grew out of the experience of the Rochdale Society. This cooperative operated for the first eight years under the Friendly Societies Act of British law. In 1852 they incorporated. The 28 founders of the Rochdale Society, often called the Rochdale pioneers, were from a variety of professions. The group consisted of nine weavers; two each of cloth manufacturers, woolsorters, shoemakers, and joiners; and 11 in other trades. They formed a consumer cooperative selling primarily consumer goods such as food and clothing because of dissatisfaction with the retail shopkeepers in their community. About half were believers in socialism and the other half supported capitalism. Many were followers of or sympathetic to Robert Owen and were interested in establishing a cooperative commonwealth as a long-term goal. They relied heavily on the experience of earlier cooperative efforts in putting together their own guidelines (Abrahamsen).

TABLE 2.1 Four classes of principles of cooperatives

Business decision Factor	Class of Cooperative Principles			
	Rochdale	Traditional	Proportional	Contemporary
Control	1. Voting is by members on democratic (one-member, one-vote) basis 2. Membership is open	1. Voting is by members on democratic (one-member, one-vote) basis[a] 2. Membership is open	1. Voting is by members in proportion to patronage	1. Voting is by member-users on a democratic or proportional basis
Ownership	3. Equity is provided by patrons 4. Equity ownership share of individual patrons is limited	3. Equity is provided by patrons 4. Ownership of voting stock is limited	2. Equity is provided by patrons in proportion to patronage	2. Equity is provided by patrons
Benefits	5. Net income is distributed to patrons as patronage refunds on a cost basis 6. Dividend on equity capital is limited 7. Exchange of goods and services at market prices	5. Net income is distributed to patrons as patronage refunds on a cost basis[a] 6. Dividend on equity capital is limited[a] 7. Business is done primarily with member-patrons	3. Net income is distributed to patrons as patronage refunds on a cost basis	3. Net income is distributed to patrons as patronage refunds on a cost basis
Other	8. Duty to educate 9. Cash trading only 10. No unusual risk assumption 11. Political and religious neutrality 12. Equality of the sexes in membership	8. Duty to educate		

[a]Traditional hard-core principles.

The Rochdale pioneers were more realists than idealists. They wanted a cooperative business that would meet their needs and survive as a business. Using the experience of others as well as their own experience, they formulated and refined rules of conduct and points of organization for guiding the business affairs of the society. The Rochdale pioneers did not propose a set of principles. However, many years later their rules were studied and called the Rochdale principles.

Because the rules evolved over time somewhat different sets of rules have been suggested by various authors. As we have noted, Catherine Webb suggested nine principles based on the publication of nine rules of conduct and points of organization in the Rochdale Society's Annual Almanac of 1860. Ewell Roy also suggested that 10 principles comprise the most popular set. Our list of Rochdale Principles includes those 10 plus two more principles included in the rules of conduct in the 1860 Annual Almanac. Table 2.1 contains these 12 principles, grouped according to the primary business decision factor it addresses, such as control, ownership, and benefits. We have also reworded them in some cases to conform to current language (including the terminology used in this book) and to clarify their meaning without significantly changing the original meaning.

Each of these Rochdale principles has been subjected to scrutiny by many cooperative observers. Their evaluations differ. Some revere them as canons of cooperative philosophy and action, seeing them all as true principles that are timeless and of universal validity. Others claim that the principles are based on the unique situation in which a pragmatic group of Rochdale pioneers found themselves. One thing is certain. The conditions prevailing in Rochdale, England in the 1840s are far different than those in North America in the 1980s and 1990s. So too are the cooperatives of today far different.

Consequently, most modern observers have concluded that only one of the Rochdale principles is timeless and universal, characterizing all types of cooperatives, at all times, in all places. This principle is *business at cost*, which requires net income to be distributed pro rata to patrons based on patronage. Two or sometimes three additional Rochdale principles have been added to a group which some believe comprise the traditional core principles of cooperatives. They are the first, *democratic control*, the sixth, *limited dividends*, and the third, *patron ownership*. Some reservations and qualifications are expressed about each as essential principles; they might better be considered wise policies, given our strict definition of principle. However, the current agricultural cooperative law in most states requires adherence to democratic control or limited dividends (Baarda). For this reason these two guidelines are considered principles even though they many not necessarily be timeless or universal.

Democratic control also can be considered an essential principle, especially if *democratic* is interpreted according to its broadest meaning. The narrow and common meaning is equality in voting power, meaning that each member receives one vote. The broadest interpretation, however, is that of government by the people or control by the members on some acceptable basis. One practice some might accept as democratic is granting voting power proportional to patronage. In fact, many state statutes allow this. However, we will use "democratic" in its narrowest meaning.

Three Rochdale principles, limited individual patron ownership (4), open membership (2), and duty to educate (8) are also popular and have strong support from several sources. Some state statutes limit the amount of common stock that any one member (except other cooperative associations) may own, usually to 5% of outstanding stock. Since common stock is usually voting stock, this regulation may be intended as more of a limitation on voting than on ownership in states where other than democratic control is allowed.

The contributions of the Rochdale Society to cooperative principles are substantial. As many as seven of the 12 principles are still considered highly relevant guidelines as either principles or policies. Principles 1, 5, and 6 are widely accepted today as the hard-core principles most characteristic of traditional cooperatives. Principles 5 and 6, concerning the distribution of net income benefits to patrons and owners, are expected to continue indefinitely into the future.

Traditional Principles

The principles recommended most by observers and supported by cooperatives during this century are called the traditional principles (Abrahamsen). They coincide with seven of the Rochdale principles, four of which many agree are hard-core principles (1, 3, 5 and 6, Table 2.1.). Traditional principles 2, 4, and 7 (open membership, limitation on voting stock, and doing business primarily with member-patrons) are not universally listed as traditional principles, although most cooperatives follow them. Other cooperatives, however, do not. For example, several cooperatives have closed or restricted membership policies. The seventh traditional principle, business is done primarily with member-patrons, was not listed as a Rochdals principle by previous writers. We can safely assume it was a common practice of the Rochdale pioneers.

Proportional Principles

Many people concerned about the future of cooperatives have suggested that a major factor in their success or failure will be the principles that cooperatives choose to follow. These people believe that significant departures from some of the traditional principles are needed.

The proportional principles are based on a narrower, more specific definition of a cooperative as a strictly proportional enterprise with respect to voting control, equity ownership investments, and profit distribution. The formal definition we have fashioned may be stated as follows:

A *cooperative* is a private business organized and joined by members to fulfill their mutual economic needs as patrons of the business, with the key control, ownership, and income distribution decisions based on patronage proportions; namely, member voting, equity capital investment by patrons, and distribution of net income to patrons are proportional to use of the cooperative.

(See Table 2.1 for the specific wording of the three key principles directly implied by this definition.) This definition extends the concept of patronage proportionality as a basis for determining net income allocations (business at cost), voting rights, and equity capital investment obligations of the patrons.

Contemporary Principles

Contemporary principles are simple, flexible, and few in number. They avoid including specific points some people may consider policies or practices. At the same time they encompass a latitude of practices such as open or closed membership and one vote per member or proportional voting. Abraham Lincoln said, "Important principles may and must be flexible" (Mencken, p. 974). In spite of their simple format, the contemporary principles and associated definition were only recently espoused and given prominence in a special report prepared by the Agricultural Cooperative Service (U.S. Department of Agriculture).

Methods of putting contemporary principles into action are briefly touched on here and covered in detail in the remainder of the book. Ownership of cooperatives is created by direct investment, retained patronage refunds, and per-unit capital retains. Members typically control cooperatives by one vote per member, but voting may be proportional to patronage or to equity investment. Benefits are reflected by returning net income to patrons in proportion of use, by favorable prices, and by gaining access to markets, supplies, and services.

FURTHER REVIEW

Another approach to principles, which some term new or futuristic, is subjective and controversial. It is based on the evolution of cooperative principles and a collection of contemporary ideas. It does not adopt in its entirety any one of the other classes of principles. It does, however, conform to the basic definition of a cooperative. Some of the most prominent contemporary contributors to the futuristic line of thinking are Fredrickson, Knutson, and Schwendiman. Their opinions are resisted by defenders of traditional and contemporary principles.

The chief proponents argue that in light of the changing environment in which cooperatives of the future must compete traditional principles in general are incomplete or incorrect guidelines. For example, there are no specific guidelines for ownership and pricing among the three hard-core, traditional principles. Creators of the futuristic oriented principles have attempted to address the ownership deficiency by adopting a proportionality approach. The democratic control principle is seen as incorrect. Voting proportional to patronage or equity has been proposed. Pricing policies need more attention. For example, prices need to be fair for both large- and small-volume patrons.

A commonly referenced source of principles is the International Cooperative Alliance (ICA). This body has attempted to gain an international consensus. In 1966 they listed six principles, the first five of which coincide with five of the first seven Rochdale principles. The six are (1) open membership, (2) democratic control, (3) business at cost, (4) limited dividends (return) on capital, (5) duty to educate, and (6) cooperation among cooperatives.

In 1976, Abrahamsen settled on five basic principles that coincide with five of the Rochdale principles, except that he relaxes the democratic control requirement to member control. Member control allows voting proportional to patronage or to ownership if the members so choose. His five basic principles are (1) member control, (2) limited dividends (return) on equity, (3) business at cost, (4) member ownership, and (5) duty to educate.

In 1981, Roy suggested five of the first six Rochdale principles were relevant principles to some extent. He supported the principle of service at cost without reservation but did not include or evaluate the third principle—equity capital provided by patrons—in his list of Rochdale principles. The remaining four of the first six he supported with some reservations, and the last six Rochdale principles, including the duty-to-educate principle supported by most observers, he rejected as nonprincipled.

Fischer recently completed an in-depth review of the hard-core principles and of three other principles: offering open membership, making ownership proportional to patronage, and doing business primarily with members. He concluded that the three other principles are not necessarily desirable from the members' perspectives and that even the desirable principles (the hard-core principles) are often difficult to implement in practice.

The idea of a cooperative as a proportional enterprise[2] in these key activities is not new, but it has received increased support in recent years. The idea is thought to have originated in Germany by Robert Liefmann, who probably influenced Ivan Emelianoff, an exiled Soviet economist. Emelianoff in turn influenced Frank Robotka of Iowa State University, and

---

[2]See Trifon for a critical review of the principles of proportionality.

Robotka influenced Richard Phillips of Kansas State University. Phillips's 1953 article "Economic Nature of the Cooperative Association" was one of the early prominent descriptions and advocates of the proportionality concept. Roy supports the concept further in his textbook *Cooperatives*, stating:

> The concept of proportionality, rather than equality, is a dominant one and is becoming more widely accepted as one of the most basic economic concepts of cooperation. This concept is of strategic importance in highly developed countries where competition with profit-type corporations is accelerating (Roy, p. 260).

> The net effect of this principle might be to greatly accelerate the joining of cooperatives by large farmers which could result in farmers' cooperatives gaining a much larger share of the farm supply, marketing, credit and service business (Roy, p. 262).

The proportionality concept has received stronger support for ownership and profit distribution than for voting control. Several, including Schaars, Cobia, and Abrahamsen, have supported the proportional ownership concept. Abrahamsen, however, does not advocate strict proportionality; he simply suggests that ownership should be in the hands of active users.

SUMMARY

One of the distinctive features of the cooperative form of business is the promotion and adherence to a set of principles. Other forms of business, such as IOF corporations, do not openly claim to abide by a set of principles. Principles play a central role in the cooperative culture and define to a great extent the nature and role of cooperatives. An understanding of the nature and role of cooperatives, including their definition, principles, policies, and practices, is of critical importance to you if you want to help make them effective businesses or if you want to compete with them in the most effective way.

Cooperative principles are embodied in the law and in common practice. Some claim that they are also timeless and universally valid. However, significant disagreement exists over what constitutes the correct set of principles. Much of the controversy is caused by what we really mean by the terms "principle," "policy" and "practice."

Cooperative principles have evolved over the last century and a half, beginning with the Rochdale principles. For purposes of discussion we have identified four distinctive classes of principles: (1) Rochdale, (2) traditional, (3) proportional, and (4) contemporary. Current practice is most closely aligned with traditional and contemporary principles. Many believe that further evolution will or should occur if cooperatives are to continue as effective economic institutions. The proportional principles are one possible set of guidelines for managing control, ownership and benefits.

DISCUSSION QUESTIONS

2-1. Why are cooperative principles important?

2-2. What are the differences between principles, policies, and practices? Give an example of each.

2-3. Why are the Rochdale principles of interest today? Which ones are still supported and practiced?

2-4. What are the major similarities and differences between the Rochdale and traditional principles?

2-5. Why have principles evolved over the last 100 years?

2-6. Why are some people suggesting a change in the traditional principles?

2-7. If you could design your own set of cooperative principles what would they be? How do they compare to the traditional, proportional and contemporary principles presented in this textbook?

2-8. To what extent do you believe that cooperative businesses must strictly follow a commonly accepted set of principles? What are the advantages and disadvantages of doing so?

REFERENCES

ABRAHAMSEN, MARTIN A., *Cooperative Business Enterprise*. New York: McGraw-Hill, 1976.

BAARDA, JAMES R., *Cooperative Principles and Statutes*. Washington, DC: USDA ACS RR 54, Mar. 1986.

COBIA, DAVID W., et al., *Equity Redemption: Issues and Alternatives for Farmer Cooperatives*. Washington, DC: USDA ACS RR 23, Oct. 1982.

FISCHER, MARTIN LEE, "Financing Agricultural Cooperatives: Economic Issues and Alternatives," Ph.D. thesis, Univ. of Minnesota, 1984.

FREDRICKSON, C. T., "Cooperatives—Hard Times and Hard Realities," *Coop. Accountant*, (Summer 1985):9.

INGALSBE, GENE, "Cooperative Principles and Practices Redefined for 5-Minute Explanation," *Farm. Coop.* (Jul. 1984):12.

KNUTSON, RONALD D., "Cooperative Principles and Practices: Future Needs," *Farmer Cooperatives for the Future*, ed. Lee F. Schrader and William D. Dobson, workshop, St. Louis, Mo.; Dept. of Agr. Econ., Purdue Univ. 1985.

MENCKEN, H. L., ed., *A New World Dictionary of Quotations on Historical Principles from Ancient to Modern Sources.* New York: Alfred A. Knopf, 1952.

NOURSE, EDWIN G., "The Economic Philosophy of Cooperation," *Am. Econ. Rev.* 12(1922):577.

PHILLIPS, RICHARD, "Economic Nature of the Cooperative Association," *J. Farm Econ.* 35(1953):74.

ROY, EWELL PAUL, *Cooperatives: Development, Principles, and Management,* 4th ed. Danville, IL: The Interstate Printers & Publishers, Inc. 1981.

SCHAARS, MARVIN A., *Cooperatives, Principles and Practices.* Univ. Center for Cooperatives A1457, Univ. of Wisconsin-Madison, 1971.

SCHWENDIMAN, GARY, "The New Cooperative," unpublished paper presented at the National Farm Credit Conference, Sept. 1984.

TRIFON, R., "The Economics of Cooperative Ventures—Further Comments," *J. Farm Econ.* 43(1961):43.

U.S. DEPARTMENT OF AGRICULTURE, *Positioning Farmer Cooperatives for the Future: A Report to the Senate Agricultural Appropiations Subcommittee,* Washington, DC: USDA ACS, Oct. 1, 1987.

WATKINS, W. P., *Cooperative Principles.* Manchester, England: Holyoake Books, 1986.

WEBB, CATHERINE, *Industrial Cooperatives—The Story of a Peaceful Revolution,* 5th ed. Manchester, England: Cooperative Union, 1912.

Part II.

Structure and Scope

3

Structure and Scope
of Agricultural Cooperatives

Robert Cropp,
University of Wisconsin-Platteville
Gene Ingalsbe,
Agricultural Cooperative Service, USDA

AGRICULTURAL COOPERATIVES AT A GLANCE

Cooperatives exist in nearly every kind of agricultural business activity and are organized in a variety of ways. Like other types of businesses, cooperatives range in size from a few farmers joining to auction their products only during the harvest season to massive and complex agribusiness organizations involving thousands of members.

Cooperatives can be classified into a few basic structural types in terms of membership and organization. But operationally, in terms of products handled and functions performed, their relationships with members, other cooperatives, and other businesses are rich in variety.

U.S. farmers are well known for setting standards of excellence in ingenuity and productivity in agriculture. Their public image as family farmers, however, belies their involvement and achievements in cooperative business enterprises beyond the farm fence. Consider the following: More than a dozen cooperatives that farmers own appear on *Fortune* magazine's list of the 500 largest industrial corporations in the country (Table 3.1). CF Industries, Long Grove, Illinois, is the largest fertilizer manufacturing firm in the nation. Sioux Honey Association,

Sioux City, Iowa, with 1,000 members scattered over 35 states, is the world's largest honey-marketing organization. Farmers have managed to unite in Land O'Lakes, Inc., Minneapolis, Minnesota, to market both butter and oleomargarine, traditional competitors, and then to become an industry innovator in marketing a blend of the two products. Innovative operating efficiencies have also been employed. Blue Diamond Growers burns its almond shells to produce energy for itself, two other large processing firms, and 10,000 homes. Finally, in addition to Land O'Lakes and other familiar cooperative brands, such as Sunkist, Welch, Ocean Spray, Sun-Maid, Sunsweet, and Tree Top, more than 350 brands of consumer products are owned by 106 farmer cooperatives.

TABLE 3.1 Sales of 50 largest U.S. agricultural
cooperatives and Fortune largest 500 rank, 1986

1986 Rank	Cooperative	Total sales (Millions)	Fortune 500 rank[a]
1	Agway Inc.	$3,511	112
2	Farmland Industries, Inc.	2,700	142
3	Associated Milk Producers	2,536	
4	Land O'Lakes, Inc.	2,215	167
5	Harvest States Cooperatives	1,837	
6	Gold Kist Inc.	1,433	239
7	Mid-America Dairymen, Inc.	1,420	237
8	CENEX	1,093	260
9	Dairymen, Inc	987	
10	Indiana Farm Bureau Co-op Assn.	892	
11	Countrymark, Inc.	834	
12	Sunkist Growers, Inc.	831	
13	Milk Marketing, Inc.	816	
14	GROWMARK, Inc.	781	
15	CF Industries, Inc.	766	365
16	National Cooperative Refinery Assn	716	378
17	Southern States Cooperatives, Inc.	646	
18	Tri/Valley Growers	644	
19	Ocean Spray Cranberries, Inc.	636	410
20	Ag Processing Inc.	625	408
21	Northwest Dairymen's Association	615	
22	Calcot, Ltd.	570	
23	Union Equity Cooperative Exchange	568	
24	Riceland Food, Inc.	532	454
25	California & Hawaiian Sugar Co.	526	459
26	Sun-Diamond Growers of California	496	470
27	Wisconsin Dairies Cooperative	462	482
28	Western Dairymen Cooperative, Inc.	447	
29	Michigan Milk Producers Assn.	432	494
30	CA Almond Growers Exchange	418	
31	MFA, Inc.	398	
32	Prairie Farm Dairy, Inc.	391	
33	Agri-Mark, Inc.	381	

TABLE 3.1   (cont'd)

1986 Rank	Cooperative	Total sales (Millions)	Fortune 500 Rank[a]
34	American Crystal Sugar Co.	354	
35	Dairlea Cooperative, Inc.	342	
36	Eastern Milk Producers Cooperative	309	
37	Citrus World, Inc.	285	
38	Maryland and Virginia Milk Producers Cooperative Assn.	280	
39	National Grape Co-op Assn.	269	
40	Tennessee Farmers Cooperative	240	
41	Norbest, Inc.	232	
42	Universal Cooperatives, Inc.	229	
43	Mississippi Chemical Corporation	226	
44	Swiss Valley Farms, Co.	217	
45	Upstate Milk Cooperatives, Inc.	216	
46	Tree Top, Inc.	192	
47	MFA Oil Company	186	
48	Knouse Foods Cooperative, Inc.	160	
49	MFC Services	160	
50	NORPAC Foods, Inc.	147	

Source: National Cooperative Business Association.
[a]Only cooperatives that met Fortune 500 criteria.

Total Activity

About 5,370 farmer marketing and supply cooperatives were operating in the United States in 1986. They varied widely in size, organizational structure, and function. The number of cooperatives peaked in 1930 (Figure 3.1) and has been declining since, primarily as a result of consolidations, changing industry structure, and declining farm numbers (Figure 3.2). However, 91% of all cooperatives still had annual sales of $15 million and 57% of them had assets over $1 million (Figure 3.3).

Over the past three decades and until 1981, farmers had gradually increased their use of cooperatives as a business extension of the farm firm. Cooperatives enjoyed uninterrupted growth and became a substantial sector in U.S. agriculture. However, a combination of macro economic factors brought on a severe recession in agriculture in the first half of the 1980s. The feed and food grains and related farm supply industries, with major cooperative involvement, were hardest hit. As a result, the growth in cooperatives' total business volume, net income, and shares of the farmer's marketing and purchasing business came to a halt (Figures 3.4 and 3.5). By the late 1980s, after considerable downsizing of the grains and farm supply sector and continued growth in some other sectors, signs began to point to a resumption of overall cooperative growth.

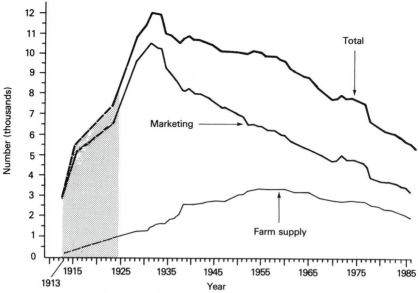

Figure 3.1 Total number of farmer cooperatives including primarily marketing and farm supply, U.S. *Source:* ACS USDA.

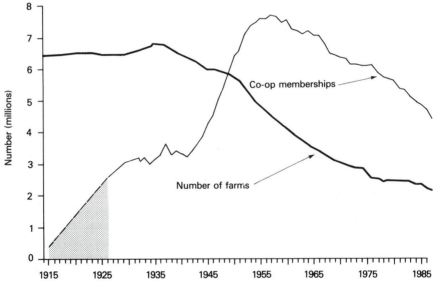

Figure 3.2 Number of farms and memberships in farmer cooperatives. *Source*: ACS USDA.

Although cooperative memberships have followed the decline in number of farms, market penetration by cooperatives has gradually increased. For example, farmers increased their purchases of major farm supplies through cooperatives from 19% in 1951 to 26% in 1985. Similarly,

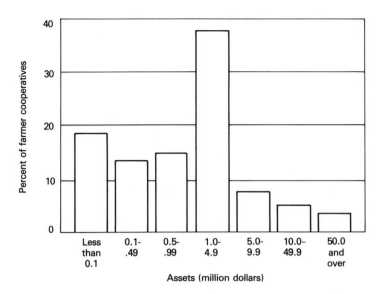

Figure 3.3 Distribution of farmer cooperatives by size of assets, 1984. *Source*: ASC USDA.

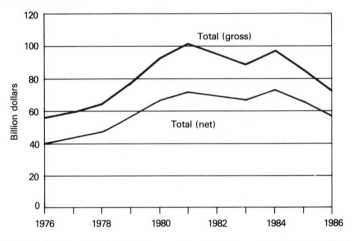

Figure 3.4 Business volume of farmer cooperatives. *Source* ACS USDA.

farmers increased the percentage of products marketed through cooperatives from 17% in 1951 to 28% in 1985. Although use has slightly declined since 1981, economic activity handled by cooperatives at the farm level is a substantial amount.

A majority of commercial farmers are involved in cooperatives, regardless of farm size. Also, data collected in 1986 indicate that the percentage of farmers holding membership or using cooperatives as nonmember patrons is about the same, regardless of farm size. The notable change during the period was that the

Structure and Scope of Agricultural Cooperatives 39

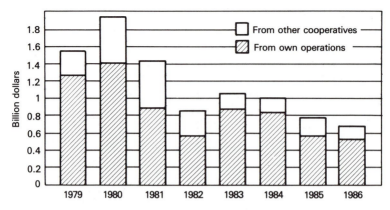

Figure 3.5 Net income of farmer cooperatives. Source: ACS USDA.

largest group of farmers (sales of $500,000 or more per year) increased their use of cooperatives from 65% in 1980 to 79% in 1986 (Kraenzle et al.). Dairy farmers are the most involved commodity sector in cooperatives, and that involvement increased from 83% in 1980 to 87% in 1986. Livestock producers are least involved, 49% in 1980 and 48% in 1986.

Not surprisingly, cooperatives are prominent in big agricultural states: the upper Midwest, California, Texas, and Florida. In Table 3.2, we have listed the top 10 states according to farm receipts in 1985. These states account for 50% of all marketing and purchasing cooperatives, and seven of these states in the upper Midwest account for 37% of the total. Cooperatives' share of marketing and purchasing activity tends to be higher than the national average in these states.

Table 3.2   Importance of cooperatives in the top 10 agricultural states, 1985

State	Number of co-ops	Cooperatives share[a] Purchased	Marketed
California	233	5	28
Texas	387	6	19
Iowa	373	27	43
Illinois	269	21	38
Nebraska	236	25	28
Minnesota	566	29	50
Kansas	233	24	26
Wisconsin	342	30	48
Indiana	82	24	29
Florida	66	6	28
U.S. total	5,625	20	30

[a] Latest data by state are for 1981.
Source: Biser and O'Day.

Supply cooperatives are most prominent in terms of market share in the northeast and least prominent in the west (Table 3.3). On the marketing side, cooperatives have a relatively greater share in the northeast and north central regions of the country and in Alaska and Hawaii. Cooperatives are a substantial extension of the farm business. In the mid-1980s, farmers had more than $28 billion invested in all kinds of cooperatives, averaging $12,000 per farm.

TABLE 3.3 **Market share held by cooperatives for farm supplies and products marketed in the five states with highest and lowest share, 1981**

	Supply			Marketing	
Rank	State	Share (%)	Rank	State	Share (%)
1	New Hampshire	51.6	1	Massachusetts	51.9
2	New York	51.6	2	Vermont	51.7
3	Massachusetts	50.5	3	Minnesota	49.9
4	Connecticut	47.9	4	Wisconsin	48.3
5	New Jersey	42.2	5	Alaska & Hawaii	45.5
45	California	4.7	46	South Carolina	9.7
46	Arizona	3.9	47	Maine	7.3
47–48	Alaska & Hawaii	3.2	48	Delaware	5.7
49	New Mexico	2.4	49	Wyoming	4.5
50	Nevada	0.4	50	New Mexico	4.1

Source: Biser and O'Day.

Total business volume is difficult to measure because of intercooperative business and because cooperative transactions vary from obtaining loans and purchasing insurance coverage to buying and selling supplies and products. One measure is that marketing and purchasing cooperative volume is nearly $60 billion annually, excluding business among cooperatives that would represent double counting (Figure 3.4). Another measurement of volume is based on loans outstanding from the cooperative Farm Credit System, totaling nearly $55 billion.

Benefits

In 1985, agricultural producers purchased 26% of their major supplies and marketed 28% of their products through cooperatives. Benefits from this patronage are both tangible and intangible. During the 1980-1986 period, difficult economically for agriculture generally, marketing and purchasing cooperatives generated an average of more than $1 billion per year in net income for their patrons (Figure 3.5). These patronage refunds fall far short of reflecting total benefits. They do not measure benefits from favorable prices nor the benefits from belonging to cooperatives using marketing pools. Additional economic benefits include bargaining power, access to or ability to create a needed product, extended control over functions affecting the farm business, a more dependable source of supply, lower costs, ability to provide or improve a service, better product quality, broader access to markets, higher returns for products, and accumulation of assets. Additional social and political benefits include a broader business education, leadership training, legislative influence, personal stature in the community, and a greater sense of achievement.

FINANCIAL STRUCTURE

Cooperatives are incorporated as capital stock or noncapital stock organizations. The type of capital structure is specified in the articles of incorporation.

Stock Cooperatives

In a stock cooperative, capital is divided into shares of common stock owned by members. Some stock cooperatives issue preferred stock and other forms of equity, such as membership certificates, which may or may not be owned by members. In contrast to IOFs (investor-oriented firms), which have general provisions for the ready transfer of stock ownership, stock cooperatives usually have restrictions on the transferability of stock by members. The purpose of the stock transfer restriction is to limit the ownership of the cooperative primarily to qualifying member-patrons. Most agricultural cooperatives are organized as stock cooperatives.

Nonstock Cooperatives

Nonstock cooperatives commonly use membership certificates and/or capital certificates to acquire equity. Members receive membership certificates upon payment of membership fees. Membership certificates are usually kept rather low—$50 or less—like the par value of common stock, and no interest is paid on membership fees. Thus membership certificates usually fall short of providing members all the necessary capital.

To supplement membership certificates, nonstock cooperatives use capital certificates in a manner similar to the way that stock cooperatives use preferred stock. Capital certificates may be sold directly to members or issued as part of patronage refunds. Capital certificates may also be sold to nonmembers, because the certificates usually carry no voting rights.

GEOGRAPHIC AREA SERVED

Cooperatives also differ in structure, depending on the size of the area they serve. We distinguish between these cooperatives by classifying them as local, regional, interregional, national, and international cooperatives.

Local Cooperatives

Local cooperatives are independent and operate in relatively small, geographic areas typically within a radius of 10 to 30 miles. Mergers and acquisitions have enlarged the operating size of many locals. Local cooperatives are separated out, however, to characterize the cooperatives serving smaller geographic areas. Local cooperatives may or may not be affiliated with other cooperatives.

Regional Cooperatives

Large federated and centralized cooperatives serving an area the size of one or several states are called regional cooperatives.

Interregional and National Cooperatives

Interregional and national cooperatives are organized by two or more regional cooperatives, which may serve a major portion or virtually all of the United States. The purpose of interregional and national cooperatives is to further reduce costs and improve the efficiency of providing certain farm inputs and services over what could be achieved by regionals operating independently. Although regional cooperatives are sometimes in competition with one another, they cooperate through these national associations to better serve their patrons. Certain products and services either would not be available or would not be price competitive without this level of cooperation.

Major interregional and national cooperatives include CF Industries and Universal Cooperatives. CF Industries, a major producer of fertilizer in the United States and the world, is owned by and supplies fertilizer to 21 regional cooperatives. This national cooperative, along with many regionals' own manufacturing plants, have enabled agricultural cooperatives to supply 44% of farmers' fertilizer needs, up from only 16% in 1951. Universal Cooperatives serves 37 farm supply and consumer cooperatives. Products supplied range from tires to detergents to food products. Universal owns the "Co-op Label" trademark.

International Cooperatives

Cooperatives operating on an international basis may have headquarters in the United States, or in other countries. Three examples are International Cooperative Petroleum Association (ICPA) in New York City, Cooperative Fertilizers International (CFI) in Chicago, and InTrade in New York City.

ICPA was organized in 1947 to provide lubricating oils and related services for national cooperatives in Belgium, Denmark, Egypt, France, Germany, the Netherlands, Scotland, Ceylon, Sweden, and Switzerland. Two U.S. regional cooperatives are members of ICPA. They supply lubricating oils to other cooperatives. CFI is a joint effort among U.S. regional cooperatives and cooperatives in India to produce fertilizer. In-Trade is a consortium of 11 German, French, Dutch, Canadian, and U.S. cooperatives. InTrade owns 51% of A.C. Toepfer, an international grain trading firm.

ORGANIZATIONAL STRUCTURE

We also can classify cooperatives according to their membership structure. Three types are (1) centralized cooperatives, (2) federated cooperatives, and (3) a combination of (1) and (2). Most cooperatives are centralized organizations whose farmers hold direct membership. In 1986, of the 5,369 farmer cooperatives counted, 5,197 were centralized. A federated cooperative is a cooperative whose members are other cooperatives. In 1986, 101 of these were identified. Some large cooperatives have both individual farmers and cooperatives as members. Seventy-one cooperatives have this mixed structure.

Centralized Cooperatives

Centralized cooperatives, in which farmers hold direct membership, include two types—local and regional. Most centralized cooperatives are local. They serve a relatively small area. Farmers generally vote directly, and board vacancies are voted on at large. They may only have one business location (Panel 1, Figure 3.6), or they may have several branch locations for supplies or services purchased, or products marketed (Panel 2, Figure 3.6). Regional centralized cooperatives invariably have several branch locations as depicted in Panel 2.

Local cooperatives have several advantages over larger cooperatives. First, because locals serve a small geographic area and have a relatively small business volume, members are more apt to know one another and to have close personal relationships. Communicating with members is also simpler. In addition, members have similar marketing and production problems and thus less chance of disagreements. They can more easily understand and conduct cooperative business; and they can operate on a more democratic basis, because every member has the opportunity to vote directly for the board of directors and on major business decisions. This leads to greater member support and loyalty as well. Further, members may have more confidence in the local manager whom they know personally and meet regularly. Finally, the local is free to shift its own patronage from or to any federated cooperative or other organization to capitalize on any short- or long-term advantage.

Despite these advantages, the number of local agricultural cooperatives is declining. Most disadvantages stem from a relatively small busi-

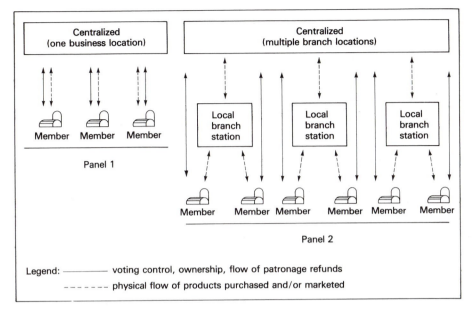

Figure 3.6 The structure of centralized cooperatives ranges from those with a single business location to those with several branch locations.

ness volume that limits both the cooperative's bargaining power and its ability to take advantage of processing economies and volume discounts and premiums.

A few centralized regional cooperatives operate over large geographic areas and have members in several states. Centralized cooperatives have no autonomous local association members. Instead, they have branches, retail outlets and in a few cases, franchise dealers (Panel 2, Figure 3.6). Operational control and authority are centralized in the headquarters of the cooperative. Individuals are direct members. The members elect, often through a delegate system, the board of directors, who in turn hire the chief executive officer or manager of the centralized cooperative. Operation of local units is vested in a manager hired by the management of the centralized cooperative.

The service area of the centralized cooperative is often divided into districts. Each district usually has a given number of delegates depending on the size of the cooperative membership. Members within each district elect the delegates, who in turn elect the board of directors of the centralized cooperative.

Regional centralized cooperatives can adapt more rapidly than federated cooperatives to changing economic conditions because of their centralized management. Adaptations may range from closing or creating local units to switching to bulk delivery or pickup services. Regional centralized cooperatives enjoy the advantage of being able to achieve considerable uniformity by operating all local units—carrying the same product lines and offering the same services as the centralized board of directors and manager wish. Consequently, their overhead and operating costs may be lower than that of a federated system due to better location and structure of local units.

Centralized regionals may obtain greater bargaining power, especially if they are marketing cooperatives, because they can centralize control of the handling and marketing of patrons' commodities. In addition, centralized cooperatives can handle some products and services more economically than locals can.

Disadvantages of centralized cooperatives are associated with membership problems. Members sometimes have difficulty identifying with a centralized cooperative if it has a relatively large number of members. Although members have a vote in the operation of the centralized cooperative through a delegate system or some other means, members often feel far removed from the cooperative. In contrast to a federated cooperative, the centralized cooperative has no local directors for members to elect or local annual meeting for them to attend. Thus member communication may be a greater challenge to large centralized cooperatives, especially since patronage refunds are often blended so that a more efficient local operation's benefits are shared throughout the system. However, members should be aware that greater efficiency at the headquarters or local levels may be made possible by membership in a centralized cooperative, thus increasing the earnings available for sharing.

Federated Cooperatives

A federated cooperative is a cooperative owned by other cooperatives. Control and flow of benefits between the federated cooperative and its member cooperatives corresponds to that between farmer members and their local. Farmers elect the board of their local cooperative and receive benefits in proportion to patronage. Local cooperatives elect (through their board or elected delegates) the board of the federated cooperative and receive benefits in proportion to their patronage with the federated cooperative (Figure 3.7).

Directors may represent geographic districts (as in Land O'Lakes), they may all be farmers (as in Harvest States), or they may represent a combination of farmers and managers (as in Farmland Industries). The number of voting delegates may be one per cooperative, or it may be based on membership size or business volume with the federation. Federated cooperatives are organized and controlled from the bottom up. However, a federated may supervise a local by a management contract that is approved by the local's board. On the other hand, if a local becomes finan-

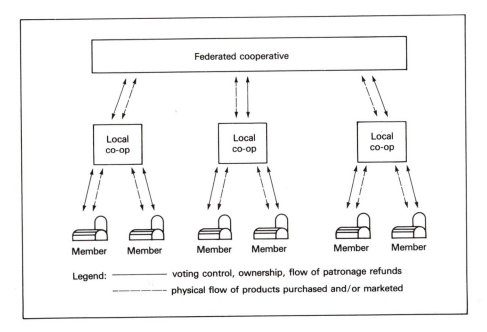

Figure 3.7 Structure of federated cooperatives.

cially encumbered to the federated, the federated may exercise some control over the local, as a bank would in a similar situation.

Cooperative federations maximize the benefits derived from the advantages of a republican form of government; that is, members of a local cooperative are represented at the federated level. Thus governance authority and control remain at the local level, more so than they do in centralized regional cooperatives. Because the federation is built and controlled in this manner (from the bottom up), the local members' interests and needs may be better expressed in federation-membership communications, and contact may be more readily maintained because of direct ties to a local. Members may also feel that they have a more direct involvement in the operation of the federation than in a centralized regional cooperative, because each member has a direct vote in the local unit. Of course, this depends to a large degree on the effectiveness of member communications.

In addition, by working through a federation rather than independently, local cooperatives are better able to obtain volume price discounts and marketing premiums, as well as to provide products and services at competitive prices. Federated cooperatives are also more easily financed than large centralized cooperatives, because managements of the locals in the federation better recognize the need for capital than individual farmers.

Furthermore, farmer members in a federated structure can realize added income from superior management of their local cooperative, independent of thé performance of the regional. In contrast, overall management of a centralized regional determines the income returning to the local level. Creating and expanding a federated structure from existing cooperatives is often easier than merging or consolidating several smaller cooperatives into a larger centralized organization.

Federated structure limits the risk of its member cooperatives. Financial failure or catastrophe destroys an entire centralized organization. On the other hand, when a federated regional fails, member cooperatives lose their investment in the regional but they still may survive. Energy Cooperative Inc. (ECI) and Farmers Export Company are examples of interregional federations that failed. ECI's failure was a contributing factor in bringing about Midland's merger with Land O'Lakes.

The main disadvantages of the federation are (1) less ability to control and coordinate the flow of product between regionals and locals, (2) the member local's freedom to defect totally or transact business with other firms at the expense of the regional, and (3) divided opinions of farmers about the regional and the local in terms of function, growth, and direction of growth.

Combination of Centralized and Federated Cooperatives

Some large regional cooperatives have features of both federated and centralized structure. In these cooperatives, both individuals and autonomous cooperatives are direct members. Harvest States, for example, has a centralized line elevator division and a federated affiliated member division. Both individual growers of Sunkist and local Sunkist packing houses are direct members. Southern States and Agway also have both individual and cooperative members, but their two types of members are not separated into divisions.

The major problem with the mixed structure is the difficulty in determining voting rights. One solution has been to allow one vote to each member cooperative and one to each individual, plus additional votes based on the volume of business each local does with the regional.

OTHER STRUCTURAL ARRANGEMENTS

Cooperatives can use several other types of structural arrangements to take advantage of economic opportunities. These arrangements include subsidiaries, marketing agencies-in-common, joint ventures, holding companies and contract agents, and franchises. The objectives of such arrangements are to gain efficiencies in operation, to enter into other activities, to enhance financial strength, to gain market entry, to increase market power, and to reduce competition among cooperatives.

A *subsidiary* corporation is a corporation organized, owned, and controlled either directly or through trustees by a parent cooperative. The purpose of the subsidiary is to assume certain duties and functions of the parent cooperative.

A *marketing agency-in-common* is organized by two or more marketing cooperatives to market the output of member cooperatives. It is, in fact, like a federated cooperative whose sole responsibility is to serve as a marketing agent of its members. It does not physically handle products, and it generally does not take title to them. Its sole responsibility is to arrange for the sale of its members' products.

A *joint venture* is an association of two or more participants, persons, partnerships, corporations, or cooperatives to carry on a specific economic operation, enterprise, or venture. The identities of these participants, however, remain separate from their ownership or participation in the venture. Use of joint ventures among cooperatives involves a partnership arrangement between two or more cooperatives. This type of activity has been commonplace among regional cooperatives marketing cotton, grain, dairy products, and fruits and vegetables. Regional purchasing (supply) cooperatives have formed joint ventures to manufacture feed and fertilizer or to refine petroleum products. More recently, cooperatives have become increasingly involved in joint ventures with IOFs. One such venture, often considered a success, is Agway's majority- owned subsidiary arrangement with Curtice Burns and H.P. Hood.

A *holding company* is a corporate entity with a controlling ownership in one or more operating companies. This degree of ownership can vary widely, as long as the holding company can exercise control through the operating company's board of directors. Normally, the holding company generates no revenues from operations. Income is limited to returns from investments in the operating companies. Northwest Dairymen's Association, Seattle, Washington, for example, describes itself as the holding company for Darigold, Inc., which is its operating company subsidiary. Agway Inc., Syracuse, New York, has the characteristics of a holding company, with substantial operations itself in addition to control of several operating subsidiary companies.

On a similar scale, cooperatives use *contract agents* and *private dealers*. For example, a county or community cooperative may organize, owning nothing but contracts and paying only the money necessary to hire an agent to handle goods and keep patronage records. The cooperative then pays patronage refunds on the basis of the records turned over by the agent. Therefore, the patronage refunds encourage individuals to trade more with the agent or franchise.

Another alternative is for the cooperative to set up a private dealer as a franchise. In this case, the dealer keeps patronage records. If the franchiser cooperative makes money and pays patronage refunds, these go to the dealer's patrons, and the dealer is paid a commission on sales.

A third alternative is to hire an agent, who buys from the cooperative at the regular price of the cooperative. The agent then sells the goods at whatever price customers will pay, keeps no patronage records, and gets no patronage refunds. This is considered strictly nonmember business for the cooperative.

Common Names

Because of common usage, the word "cooperative" may be dropped when referring to a particular type of cooperative. For example, a local cooperative, a federated cooperative, or a regional cooperative is simply referred to as a local, federated, or regional. This book will continue with the tradition to maintain brevity.

FUNCTIONS PERFORMED

Other criteria for classifying cooperatives are the functions they perform or the commodities they handle. Three general classifications are market, supply, and service. These groups may be further designated according to function. For example, marketing cooperatives may bargain, process or manufacture, and sell products. A supply cooperative (also called a purchasing cooperative) may purchase in volume, manufacture, process or formulate, and distribute. Service cooperatives provide services such as artificial insemination and transportation.

A cooperative's major activity is used to determine its classification by type, even though it may engage in multiple functions. For example, Mid-America Dairymen, Inc., is a dairy marketing cooperative; Farmland Industries, Inc., is a farm supply cooperative; CF Industries, Inc., is a fertilizer manufacturing cooperative; Michigan Agricultural Marketing Association (MACMA) is a bargaining cooperative; and Servi-Tech, Inc., provides agronomic and animal health services.

Production Cooperatives

Cooperative farming is found in some countries but is rare in the United States. A few farmer production cooperatives were organized in North Dakota from 1932 to 1981 when that state's Anti-Corporation Farming Law was in effect. That law allowed individuals to organize their farms as cooperative corporations. Equity in the farm cooperative was linked to individuals; thus they were able to maintain private property, a characteristic often not associated with social living arrangements such as that of the

Hutterites. A few farm cooperatives have been created in California and other states as well.

Marketing Cooperatives

Statistics on agricultural cooperatives lump all bargaining and processing cooperatives, as well as marketing cooperatives, into a single category of marketing. Classifying them separately would be artificial because very few are strictly one or the other. In Figure 3.8 we have given a capsule statistical picture of cooperatives' share of total marketing activity by selected commodities.

The primary function of marketing cooperatives is the marketing of farm products for members. Of the 5,369 agricultural cooperatives, 3,260 are classified as marketing cooperatives (1986 data). They account for about 45% of the membership. The volume of these cooperatives plus the marketing volume of other cooperatives account for 71% of the business volume done by agricultural cooperatives. They vary greatly as to functions performed. Some receive, grade, process, package, label, brand, store, distribute, and merchandise products. Others perform only one or two marketing functions.

Ranging in size from small local cooperatives to large centralized or federated cooperatives, marketing cooperatives are becoming more vertically integrated by increasing their ownership and control of facilities beyond the first buyer level, and in some instances, all the way to the retail

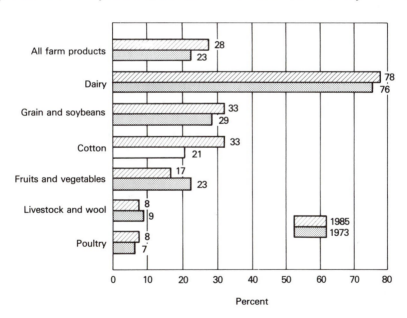

Figure 3.8 Cooperative's share of marketing activity. *Source:* ACS USDA.

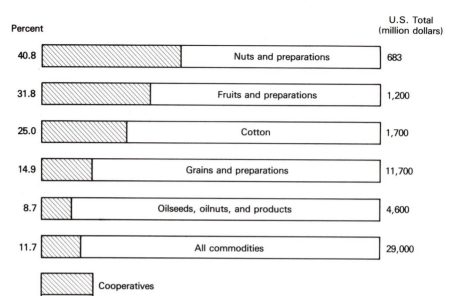

Figure 3.9 Cooperative's share of U.S. agricultural exports, 1985. *Source:* ACS USDA.

level. Regional marketing cooperatives especially have established well-recognized brand names such as those mentioned earlier.

Marketing cooperatives are major suppliers to the hotel, restaurant, and institutional trade. It is common to find identifiable cooperative products, such as butter and juices in restaurants and fruit and nut snack products on airplanes. However, cooperatives are not normally in food retailing. An exception once was Capitol Milk Producers Cooperative, Inc., headquartered in Maryland in the Washington, DC, metropolitan area. Capitol Milk was owned by dairy farmers primarily in Maryland and Virginia and operated several hundred retail convenience stores along the east coast. However, in 1986, members sold the profitable operation to a competing IOF. On a smaller scale, Blue Diamond Growers operates several retail stores handling nut products and other specialty food items; and Eastern Milk Producers operates six retail dairy stores, some of which offer full-service deli and take-out lunches, along with dairy products.

Cooperatives are involved in exporting many of the products they handle, both fresh and processed. Cooperatives are relatively more important in the export of nuts, fruits, and vegetables, but grain and soybeans account for the greatest absolute value and volume (Figure 3.9).

Bargaining Bargaining cooperatives (sometimes called bargaining associations) bargain or negotiate with processors and other first handlers for better terms of trade for members. Two or more operating cooperatives that are involved in the physical handling and/or processing of products may form a cooperative bargaining federation (often referred to as a marketing agency-in-common) to perform the bargaining function. Many dairy cooperatives combine bargaining with other operating and marketing functions. A pure bargaining cooperative does not physically handle or take title to the product involved but merely bargains for price and other terms of trade. In 1985, about 125 pure bargaining cooperatives, handling primarily fruits and vegetables, bargained for crops valued at $767 million.

Processing or Manufacturing The main emphasis of processing cooperatives is in the processing of raw farm products rather than on bargaining or marketing. For example, several dairy cooperatives that manufacture cheese, butter, and powdered milk concentrate leave marketing to brokers or to a regional cooperative such as Land O'Lakes, Inc. Other examples include sugar and vegetable processing cooperatives.

Scope of Marketing Cooperatives in Major Commodities

Five commodities account for 90% of the marketing volume handled by cooperatives. Two-thirds of the volume is accounted for by grain and dairy products. The other three commodities are fruits, vegetables, and nuts as a group; livestock; and cotton. Profiles of cooperative activity in these commodities follow.

Grain and Oilseeds Corn accounts for 45% of all grain volume handled by cooperatives. The remainder, in order by volume, are wheat, soybeans, sorghum, barley, sunflower, oats, and other grains such as rye and flaxseed. About 2,000 cooperatives handle grain, valued at the farm level at $12 billion (Richardson). Most are local-area cooperatives. Their farm-level purchases of grain account for 33% of U.S. farm sales.

The marketing cooperative's role is most extensive in the assembly function of marketing and diminishes as grain moves into further processing and export markets. Cooperatives handling grain account for 28% of U.S. elevator facilities and 38% of total U.S. elevator storage capacity. They operate 21% of the soybean processing industry's crushing capacity. Regional cooperatives account for about 25% of farm-level purchases of grain, and they handle about 60% of the grain sales by local cooperatives. Regional grain cooperatives also move about 40% of U.S. grain to export positions, but they directly export only 8%. Further processing operations by grain cooperatives include operating soy oil refining plants, rice mills, flaxseed and sunflower seed crushing plants, a durum flour mill, and a corn wet-milling plant producing syrup and starch.

Cooperatives handling grain have experienced a decade of significant structural and operational change. Unit train rates and rail branchline abandonment have had an especially dramatic impact. Continuing factors prompting adjustment include the changing structure of production agriculture; governmental fiscal, monetary, regulatory, and agricultural policies; and world supply-demand fluctuations. A relatively uniform pattern of assembling grain through local cooperative elevators has been transformed by bankruptcies, mergers, and withdrawals from grain marketing. Assets sometimes have been sold to strategically located and more successful grain cooperatives or other firms. At the same time, innovative restructuring has created new processing and marketing organizations either jointly owned by cooperatives or by cooperatives and IOFs. These dramatic changes have created a structure characterized by an increasing number of large super-locals or mini-regionals with one to six trainloading facilities that receive grain from farmers, other cooperatives, and IOF elevators.

Dairy Cooperatives are more important in the dairy industry than in any other major agricultural commodity. Cooperatives handle 78% of the country's milk as it leaves the farm. They also account for sizable shares in the manufacturing of dairy products.

Dairy farmers were among the earliest to use cooperatives, forming them to address different marketing problems. Evolving dairy cooperatives can be generally categorized into four types: manufacturing cooperatives, federated sales cooperatives, raw milk sales cooperatives, and bottling cooperatives (Roof). Manufacturing cooperatives started with the purpose of converting milk into products such as butter and cheese. Federated sales cooperatives developed out of the recognized need by small milk manufacturing cooperatives to coordinate their grading, packing, and marketing efforts. Many raw milk sales cooperatives began as pure bargaining associations to market raw whole milk. Later, most of them acquired facilities to manufacture reserve milk supplies into other products, becoming known as bargaining-operating cooperatives. Bottling cooperatives began because producers believed they could get a higher net price for their milk. Consolidations have produced multifunctional and highly integrated dairy cooperatives, with large amounts of capital invested in manufacturing and processing facilities, research, and merchandising activities.

Some 350 dairy cooperatives handle products valued at $15 billion. The largest 20 account for about two-thirds of the total marketing activity. Cooperatives account for four-fifths of the dry milk products, two-thirds of the butter, and half of the cheese marketed in the United States. Their product lines also include ice cream, ice milk, bulk condensed-milk products, condensed whey, dry whey, and frozen product mix.

Dairy cooperatives are heavily involved in brand merchandising, accounting for more than half of all the cooperatives that market products under their own brands. However, they sell nearly all of their branded products to wholesale outlets. Only about a dozen cooperatives are involved in retailing.

Dairy cooperatives have derived some market stability from federal dairy price support programs and federal marketing orders. However, perishability of the raw product and chronic overproduction continue to motivate dairy farmers to be in the forefront of engaging in group action. Consequently, dairy cooperatives have led the way in organizational consolidation, in further processing, in product diversification, and in product promotion.

Fruits, vegetables, and nuts Consumers are probably most familiar with the brands of fruit, vegetable, and nut cooperatives such as Blue Diamond, Ocean Spray, Sunkist, Sun-Maid, Sunsweet, Tree Top, and Welch's. One-fourth of the cooperatives marketing branded products are in this category. Most of these cooperatives are organized along commodity lines. Although some are primarily single-commodity associations, many market a multicommodity line and perform two main functions: marketing products in fresh or processed form and bargaining for terms of trade.

Major commodities that cooperatives handle are oranges, grapes, apples, almonds, and potatoes, but cooperatives also market nearly every other fruit, vegetable, and nut grown in the United States. Growers of these commodities own 400 cooperatives that market products valued at $6 billion. These sales account for 17% of the U.S. total marketing activity. Two dozen cooperatives account for 41% of U.S. exports in nuts and 32% of fruits.

Probably no group of commodities handled by cooperatives have been more affected by changes in food marketing than this one. The food retailing industry demands large volumes of uniformly sized, graded, and packed produce. At the same time, competition motivates a changing and increasing variety of product uses and packaging. Federal and state marketing orders have helped stabilize the market for various fruits at the raw product level, contributing to improved uniformity and quality. But cooperatives are on their own in product innovation. Examples of such innovations include soft-drink combinations, concentrates, and preserves in squeezable plastic containers. Similarly, the cranberry has wound up in drink combinations in aseptic packages with attached straws. At least 500 ways have been found to cook with almonds, along with packaging them as snack foods separately or with a wide variety of dried fruits.

Livestock and wool Livestock cooperative functions can be grouped into marketing, integrated operations, and other services.

Marketing activities include selling on commission (terminal marketing, country selling, private treaty direct selling, auction marketing, and tel-o-auction and computerized marketing); buying on commission (order-buying feeder livestock for producers and slaughter livestock for packers); and dealer operations (country buying stations with a central sales desk buying feeder livestock for members, and buying slaughter and feeder livestock from members and others). Many of the livestock and wool cooperatives are largely shipping associations and local wool pools that have little or no net worth.

Integrated operations include production of feeder animals; contract hog production; and slaughtering, processing, and meat distribution. Other services include providing sources for high-quality breeding stock, providing artificial insemination, processing meat by-products, financing, futures trading, and offering market information.

About 530 cooperatives handle livestock, wool, and animal products valued at $3.0 billion. Although this dollar volume is the fourth largest in cooperative marketing activity, most of it is accounted for by live animal marketing and thus accounts for only 8% of total U.S. livestock marketing volume at the first handler level.

A few regional cooperatives established a substantial position in slaughtering and processing red meat in the 1970s but sold these facilities in the early 1980s. Farmland Industries continues to be highly integrated in pork marketing, participating in activities ranging from conducting swine research to processing branded consumer products.

Wool marketing and related services include assembling, grading, pooling, storing, and selling either on consignment or taking title and reselling to manufacturers. Cooperatives also provide information on financing and production, marketing, and management.

Cotton Farmer cooperatives play a major role in performing and coordinating most of the services involved in the marketing of U.S. cotton and cottonseed products. General functions they perform are ginning, warehousing, lint merchandising, and cottonseed processing.

Slightly more than 430 cooperatives handle cotton products valued annually at $1.4 billion. This volume amounts to 33% of total U.S. cotton marketing activities both domestic and export. Cooperatives own 20% of cotton compress and warehouse capacity and nearly 40% of the cottonseed crushing capacity.

In 1971, four regional cooperatives took a major step in strengthening their marketing programs by forming AMCOT, a joint marketing association whose primary functions were to provide regionals with market information, to obtain greater global coverage for their different cotton varieties, and to arrange either domestic or export transactions. However, each regional still made and continues to make its own sales decisions. About 30% of U.S. cotton production today is sold through this marketing system. Two other innovative activities are an electronic marketing system called TELCOT and forward integration by producers into textile manufacturing through a denim fabric plant at Littlefield, Texas, owned by Plains Cotton Cooperative of Lubbock.

Other marketing activities In addition to processing or marketing the five major products just discussed, cooperatives are involved in marketing nearly every agricultural product. The annual value of sugar, poultry, and rice products handled is about $1 billion for each. Other commodities include sweeteners, such as honey, sorghum, and maple syrup; dry beans, peas, and lentils; tobacco; seed; forest products; hay; safflower; hops; cut flowers, bulbs, and nursery stock; fresh ginger; and coffee. Some of these products are handled by cooperatives whose primary business overshadows the small volume marketed of the special crops.

Supply Cooperatives

Supply cooperatives, often referred to as purchasing cooperatives, include those handling all types of farm production supplies and equipment, such as feed, seed, fertilizer, petroleum products, farm equipment, hardware, and building supplies. Some also handle such items as heating oil, lawn and garden equipment, household appliances, and grocery items.

One of the prime benefits of supply cooperatives is to provide farmers with a steady, dependable supply of farm inputs at competitive prices. Even in periods of shortages of inputs such as petroleum or fertilizer, purchasing cooperatives have still been able to meet member-patron needs.

Supply cooperatives have become more vertically integrated by manufacturing more of the supplies they distribute through the use of federated and centralized cooperatives. Extensive vertical integration especially has occurred in manufacturing feed, fertilizer, and petroleum products.

In 1986, total supply volume was valued at $15 billion. Purchasing has gradually accounted for an increasing portion of total cooperative activity. Purchasing, relative to total activity, rose from 20% in 1950 to 26% in 1986 (Figure 3.10).

In 1986, about 4,150 farmer cooperatives handled production supplies, but only for 2,000 of them did supplies account for more than 50% of their volume. Clearly, many cooperatives handling supplies performed other functions (such as marketing) as well. Profiles of activities involving the most important farm supplies—petroleum, feed, fertilizer, and farm chemicals—follow.

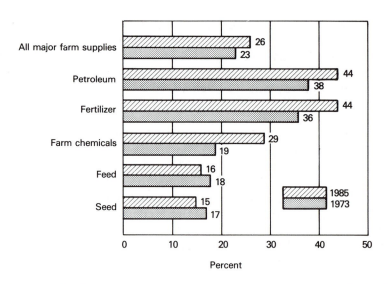

Figure 3.10 Cooperatives' shares of purchasing activity. *Source:* ACS USDA

Petroleum Cooperatives play a vital role in providing petroleum products to farmers and others in rural communities. Their range of activities includes exploring for crude oil and natural gas, refining and manufacturing, wholesale and retail distribution, and related operations such as research and product testing. By value, petroleum products are the largest component of cooperative purchasing activity. In 1986, more than 2,500 cooperatives handled petroleum products valued at $5.0 billion. This amounted to 44% of farmers' fuel purchases.

A dozen regionals solely or jointly own six refineries that provide about 85% of the requirements of retailing cooperatives. However, less than 12% of oil input for these refineries comes from sources owned or leased by cooperatives. Facilities at the retail level range from bulk delivery service only to full-service car, truck, and tractor care centers.

Bulk deliveries to farm storage tanks typically are transported by tank trucks carrying as much as 2,500 gallons. Direct deliveries from pipeline terminals and other wholesale facilities in 7,000– to 8,000–gallon tanker trucks are made to large farms with sufficient storage. One local supply cooperative had 11 such patrons. Full-service centers provide on-farm maintenance service, self-serve stations, and stock parts; repair work; and sell tires, batteries, and other automotive accessories.

Feed Feeds were some of the earlier and larger volume supplies farmers bought through cooperatives. Activity began more than 100 years ago with pooled car lot purchases of sacked feed. Cooperative feed services have evolved to a highly scientific business involving a wide range of activities, including research, ingredient formulation, and manufacturing. Most large regionals conduct their own research, demonstration, and on-farm testing. Cooperative Research Farms carries out these activities for the 19 regionals that created it.

Nearly 3,000 cooperatives handled feed in 1986 valued at $2.9 billion. Most were local-area cooperatives, but about two dozen regionals also retailed feed. This volume amounted to 16% of farmers' total feed purchases.

Cooperatives manufacture feed for practically every type of livestock, including horses and household pets. Related services include bulk delivery, custom grinding and mixing, contract feeding, and animal health products and services. Cooperatives are important providers of computerized livestock management systems and of the latest information on feeds and feeding practices.

Fertilizer Cooperative activity in supplying fertilizer is more than a century old, beginning with a farmers' buying club whose agent contracted to have Peruvian guano shipped to Riverhead, New York. Today, the largest fertilizer manufacturing company in the United States is the interregional farmer cooperative CF Industries, headquartered at Long Grove, Illinois.

Farmers have used cooperatives to integrate backward in the fertilizer industry just as they have in petroleum and feed industries, to secure fertilizer sources of raw materials. These cooperatives mine potash and phosphate rock. They also own numerous plants to manufacture, formulate, and mix dry, liquid, and gaseous fertilizers. They have also developed comprehensive transportation, storage, delivery, and application systems, including a barge line, railcars, truck transports, and pressurized storage tanks. In 1986 more than 3,100 cooperatives handled fertilizer valued at $2.9 billion. This represented 38% of their total fertilizer requirements.

A few regionals have been formed primarily to offer agronomic services. These organizations test soil, make fertilizer recommendations, and give other crop management advice. Multifunctional regionals offer similar services as a part of their overall fertilizer programs. Most have fertilizer test plots at their demonstration and research farms. In addition to supplying products, local-area cooperatives typically offer application equipment for lease or rental, for custom application, and for bulk delivery. Occasionally, they set up demonstration plots. A few cooperatives own airplanes and helicopters for aerial application.

Farm chemicals Initial cooperative efforts to supply farm chemicals began in the early 1900s and were focused primarily on obtaining citrus fumigants. Other early cooperatives provided dairy farmers with fly sprays, cleansing powders, and disinfectants.

Supplying farm chemicals has been the fastest growing of any activity in recent years. Cooperatives' increasing involvement has paralleled farmers' increasing use of farm chemicals generally. Farmers quadrupled their purchases of farm chemicals during the period 1970-1986. And from 1973-1985, farmers increased their percentage of purchases through cooperatives from 19% to 29%.

Chemical supplies include insecticides, fungicides, herbicides, rodenticides, repellents, fumigants, defoliants, seed inoculants, soil treatments, and wood preservatives. More than 3,100 cooperatives handled these farm chemicals valued in 1986 at $1.4 billion.

Interregionals and regionals manufacture; formulate; conduct research; and develop specialized storage, transportation, and application equipment. Local-area cooperatives provide packaged and bulk supplies, custom application, and other related services.

Usually, regionals and interregionals will handle farm chemicals as a division of their total operations. However, some specialized regionals have been organized to provide integrated pest management and agronomic services. A trend is to handle more and more pesticides in bulk. This allows cooperatives to custom apply pesticides with liquid fertilizers.

Other supplies Although the four major supplies discussed account for 81% of total purchasing activity, cooperatives also provide significant amounts of other farm production inputs. These include seed, building materials, farm machinery and equipment, meats and groceries, and containers.

In 1986 more than 3,000 cooperatives handled seed valued at $513 million, or 17% of farmers' total seed purchases. Seed research is conducted primarily at the regional and interregional level, but both regionals and locals set up demonstration plots. One regional, NC+ Hybrids, Lincoln, Nebraska, is among the 15 largest seed businesses in the United States. The principal interregional research organization is FFR Cooperative, Inc., Lafayette, Indiana. Owned by six regional cooperatives, FFR has breeding programs to develop varieties of hybrid field corn, soybeans, alfalfa, red clover, orchard grass, tall fescue, timothy, sorghum-sudan, and sudangrass.

Regional cooperatives are involved in the manufacture and assembly of wood and steel components for buildings and grain bins, feeders and waterers, paint, various containers and packaging material, milking equipment, and fencing. Regionals offer turnkey facility services that include determining feasibility, designing, providing materials, and contracting construction projects. Many local cooperatives are virtually one-stop shopping centers, handling the full range of farm and home supplies, including hardware, appliances, lumber, clothing, and even meats and groceries.

Cooperatives have not penetrated the farm machinery market, even though this market accounts for 20% of all farm cash expenditures. Very few local cooperatives have maintained a successful dealership over an extended period. Only during the initial post-World War II period was there a significant penetration of this market.

Service Cooperatives

Farmers use cooperatives to obtain a wide variety of specialized services related to farm purchasing and marketing. These can be grouped into four categories: credit, utilities, insurance, and specialty services. In some cases, these services may be provided as a division or subsidiary of a cooperative whose primary function is either marketing or purchasing. USDA's Agricultural Cooperative Service reports a category of Related Services. This category documents income that would not be listed as part of marketing or purchasing income. Related services are defined as those affecting the form, quality, or location of farm products and supplies handled by cooperatives, such as artificial insemination, ginning, trucking, storage, grinding, and drying. In 1986, more than 3,600 cooperatives reported related service income of $1.7 billion. Most of these were marketing and purchasing cooperatives, but 138 were cooperatives whose major income came from services.

Nationwide cooperative systems have developed to provide a specific service. Discussion of the farm credit system, rural utilities, and farm-related specialty services are covered in this chapter. Selected utility, credit, and insurance systems, serving both rural and urban members, are covered in Chapter 4. Cooperative trade organizations providing services on a state or national basis in the areas of legislation, public relations, education, administrative support, and international activities are discussed in Chapter 6.

Farm credit system Two distinct types of credit cooperatives are (1) those in which the participants pool funds and borrow from each other and (2) those in which the participants join to procure funds collectively from investors. The first type is referred to as a credit union and is discussed chap. 4. The second type is illustrated by the farmer-owned cooperative Farm Credit System (FCS). In agriculture, FCS is by far the most extensive type. FCS provides about one-third of all agricultural credit. Loans outstanding in 1986 totaled nearly $55 billion.

At the beginning of 1988, FCS was organized into 12 farm credit districts (Figure 3.11). Each district has three types of banks: a Federal Land Bank (FLB), a Federal Intermediate Credit Bank (FICB), and a Bank for Cooperatives (BC).

The farm financial crisis of the early 1980s brought substantial change to FCS structure, management, and supervision. Increases in nonperforming loans and reductions in the volume of lending resulted in combined system losses of $2.6 billion in 1985 and $1.9 billion in 1986.

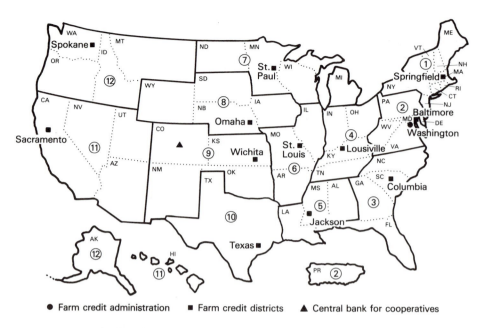

Figure 3.11 The farm credit system districts.

In response to the crisis, the trend to joint management of district banks accelerated, consolidations reduced the number of local-area banking associations to nearly half their original numbers, the supervising government agency restructured, proposals under consideration would merge one more type of district bank, and an infusion of government capital had been requested from Congress and was approved in January 1988. Reports in the public press, and in current periodicals such as those issued by FCS banks and *Farmer Cooperatives*, are sources for subsequent change.

The Federal Farm Loan Act of 1916 created the *Federal Land Bank*. Farmers borrow for a variety of purposes, including the purchase of farmland, machinery and equipment; livestock; paying off other debt and refinancing mortgages; making land improvements; and for purchasing, constructing, and repairing the farm home and other buildings. Loans have terms of from 5 to 40 years at amounts up to 85% of the market value of real estate security, or to 97% if government guaranteed. A statutory condition to obtaining a loan requires the borrower to purchase stock in the Federal Land Bank Association (FLBA) equal to at least 5% of the amount borrowed. The FLBA buys an equal amount of capital stock in the district Federal Land Bank. These requirements place the ownership and control of the banks in the hands of the farmer-borrowers.

The Agricultural Credit Act of 1923 established 12 *Federal Intermediate Credit Banks* (FICBs) as wholesalers of short- and intermediate-term loans. After the

Farm Credit Act of 1933, the primary retailers of FICB loans were *Production Credit Associations* (PCAs). Although the PCAs get most of their funds from the FICB for their district, they may participate with the FICB and commercial banks in making loans. Farmers borrow from PCAs for production and operating purposes. Most loans mature in a year, but farm and rural home loans may have terms of up to 10 years and aquatic product loans of up to 15 years. Stock purchase requirements are the same as for FLB loans.

Banks for Cooperatives (BCs) were established also by the Farm Credit Act of 1933. BCs were chartered and organized in each of the 12 districts. Additionally, a Central Bank for Cooperatives was established to participate in loans exceeding the lending capacity of individual banks and to handle loans connected with international transactions of borrowing cooperatives. In July 1988, eight of the district banks and the Central Bank for Cooperatives voted to merger, thus creating a National Bank for Cooperatives.

BCs provide cooperatives with four types of loans: (1) short-term or seasonal operating loans, (2) intermediate-term loans for working capital needs, (3) long-term loans for capital construction and improvements, and (4) export assistance loans. To borrow from a BC, a cooperative must (1) have 80% of its voting control in the hands of agricultural or aquatic producers or federations of cooperatives—voting control may be as low as 60% for certain farm supply cooperatives; (2) do a minimum of 50% of its business with members, except that business done with the government and services and supplies furnished by the cooperative as a public utility are excluded; and (3) use one-member, one-vote governance or limit the capital stock dividend rate to the lesser of the minimum allowed under state law or 8%.

Farm Credit Administration (FCA), an independent government agency, supervises the Farm Credit System. The Farm Credit Amendments Act of 1985 restructured FCA into a more arm's-length regulator of the system with cease-and-desist powers. FCA had been governed by a 13-member part-time board. Combined boards of banks in each of the 12 Farm Credit Districts nominated a director for the President to consider in appointing, and the Senate in confirming, a Federal Farm Credit Board. The thirteenth member was appointed by the Secretary of Agriculture. The Act replaced the 13-member board with a three-member full-time administrative board appointed by the President. The board chairman serves as chief executive officer.

Rural utility systems In manner of speaking, rural America was kept in the dark for 50 years after the invention of the electric light bulb.[1] The principal reason was that companies could not provide electricity to rural areas because farmers lived too far apart and were unable to pay the rates necessary to make service profitable. By the mid-1930s, only one in 10 farms

[1]See Pence for the fiftieth anniversary pictorial account of rural electrification.

had electric power, and rates were far higher than those paid by people living in towns (Angevine).

The potential of extending electric power to the countryside was widely recognized by government leaders and particularly by farmers, even though IOF power companies expressed little interest in expanding service. The Rural Electrification Act of 1936 set up the Rural Electrification Administration in the U.S. Department of Agriculture as a lending agency to finance the extension of electric power to rural areas. Nonprofit organizations were given first preference on funds, but existing power companies meeting REA loan provisions could also receive funds. Farmers moved quickly through cooperatives to take advantage of the new program. As a result, a formidable argument could be advanced that rural electric cooperatives are responsible for bringing about one of the more profound changes in U.S. agriculture.

In 1969, the National Rural Utilities Cooperative Finance Corporation (CFC) was established as a private source of funds to supplement and extend the REA financing programs. A Rural Telephone Bank was established by Congress in 1971 to be administered by REA.

In 1986 in rural America, 98.8% of the 2.4 million U.S. farms enjoyed central station electric service. REA finances 2,140 rural electric and telephone systems—1,244 of them cooperatives—providing electric and telephone service to more than 40 million people in rural areas of 47 states and other territorial possessions.

On October 28, 1949, the REA was authorized to loan money necessary to improve and extend telephone service in rural areas. The Rural Telephone Bank (RTB) was established within the USDA in 1971 as a source for financing rural telephone companies. Established as a government agency with government and user ownership, the RTB would become private whenever 51% of the government stock has been fully redeemed and retired. About 250 of the country's rural telephone systems are cooperatives. With 10 million members and $300 million in annual revenue, they provide telephone service to 5.5 million subscribers in rural areas of 46 states, Puerto Rico, and several Pacific Island territories.

Specialty service Although cooperatives providing specialty services are not as extensive, they may signal farmers' changing needs, technical or biological developments, or the evolving market environment. Some examples best illustrate the types of specialty services farmers are obtaining through cooperatives.

To track and improve the performance of dairy cows, farmers have formed dairy herd improvement associations. Computer technology has greatly expedited the data collection and analysis. In 1985, a little more than 1,000 associations had nearly 38,000 memberships and 3.2 million cows were tested.

Cooperatives have been the predominant organizational approach to improving beef and dairy cattle through artificial insemination. After experimentation, largely at land-grant universities, in the late 1930s proved that the practice held great potential, many breeding associations were formed. These associations reached a record number in 1950 of 97, with 83 of them formed as cooperatives (Herman). In that year, cooperatives inseminated 80% of the 2.6 million dairy cows. All firms, including cooperatives, have followed a consolidation trend similar to other businesses supporting agriculture. By 1988, only 19 artificial insemination cooperatives could be identified.

Integrated pest and agronomic management originated as a cooperative activity in the mid-1970s. Servi-Tech, Inc., of Dodge City, Kansas, began as a federated regional. CENEX, a farm supply cooperative, fostered the formation of a federated CENTROL cooperative in 1979 as a joint venture with Control Data Corporation. Grove-care cooperatives serve citrus growers. Typical activities of these cooperatives include pest and disease detection and control, soil testing and fertilizer recommendations, application or delivery services, irrigation and soil conservation advice, and crop management analysis.

Farmers' irrigation cooperatives have played an important part in the development of agriculture in several western states. Through grazing associations, beginning as early as 1900, ranchers have teamed up with the federal government to use, and improve while using, certain public and private land.

These few examples demonstrate the adaptability of cooperative structure to most any situation. Additionally, farmers have extended the scope of cooperatives through organizational combinations.

SUMMARY

You will discover that agricultural cooperatives occupy a prominent position in the nation's food industry—if you know where to look. *Fortune's* annual list of the 500 largest industrial corporations include more than a dozen cooperatives. At your local supermarket, you could literally fill your shopping cart with brands offered by cooperatives—such as Sunkist, Welch's, Ocean Spray, Land O'Lakes, Blue Diamond, Sunsweet, and many more of the 350 brands offered by more than 100 cooperatives.

Cooperative's numbers and memberships have followed the decline in numbers of farms and farmers. But as farms have become fewer and larger, so have the cooperatives that support them. A majority of farmers use cooperatives, and they have gradually increased their use. By the mid-1980s, they were purchasing 26% of their major supplies and marketing 28% of their products through cooperatives.

In terms of how they are owned and controlled, cooperatives can be structurally classified as centralized, federated, or a combination of the two. Federated cooperatives are owned by other cooperatives, while farmers hold direct membership in centralized cooperatives. Operationally, cooperatives' organization and internal management control is as varied as other businesses in terms of sales outlets, subsidiaries, and joint arrangements with other cooperatives or IOFs.

Cooperatives may also be classified by function (marketing, supply, processing, bargaining, and service). Marketing cooperatives number 3,400 and account for 60% of all agricultural cooperatives, 46% of the membership, and 72% of the business volume done by agricultural cooperatives. Bargaining cooperatives bargain or negotiate with processors and other first handlers for better terms of trade for members. Pure bargaining cooperatives have no physical handling, ownership, or control over the actual commodity. Processing cooperatives emphasize the processing of raw farm products rather than on bargaining or marketing.

Dairy and grain account for the biggest dollar volume of products marketed through cooperatives. Petroleum, fertilizer, and feed are the biggest dollar volume of supplies purchased.

Supply cooperatives handle all types of farm production supplies and equipment. Some also handle various consumer items. Agricultural service cooperatives provide farmers with a wide variety of services, including credit, utilities, insurance, artificial insemination, irrigation, and others.

DISCUSSION QUESTIONS

3-1. Discuss trends in cooperatives in terms of numbers, memberships, business volume, and farmer use.

3-2. Which type of cooperative organizational structure may best serve the needs of modern-day agriculture? Would this vary depending on whether the organization is a marketing, purchasing, or service cooperative?

3-3. Cooperatives' market share varies considerably by commodity in both marketing and purchasing. What are some of the major differences, and why the difference?

3-4. Explain the major geographic classifications of cooperatives.

3-5. Why are cooperatives relatively more important in some parts of the country than in others?

3-6. Most agricultural supply cooperatives are organized as multicommodity cooperatives in that they handle several types of farm inputs. In contrast, most agricultural marketing cooperatives are single-commodity cooperatives. What might be the major reason(s) for this contrast?

3-7. How is the Farm Credit System organized to meet the credit needs of cooperatives? Based on information you have gleaned from current periodicals, explain how that organization and function has changed from what is described in this book.

3-8. What are some major service areas related to agriculture that have significant activity?

3-9. Contrast the advantages of a federated structure with that of a centralized structure.

REFERENCES

ANGEVINE, ERMA, et al., *People—Their Power: The Rural Electric Fact Book, Washington, DC: National Rural Electric Cooperative Association (1980).*

BISER, LLOYD, AND LYDEN O'DAY, *Growth and Trends in Cooperative Operations, 1951-81.* Washington, DC: USDA ACS RR 37, 1984.

HERMAN, HARRY A., *Improving Cattle by the Millions, NAAB and the Development and Worldwide Application of Artificial Insemination.* Colombia, MO: University of Missouri Press (1981).

KRAENZLE, CHARLES A., J. WARREN MATHER, AND KATHERINE C. DEVILLE, *Cooperative Historical Statistics.* Washington, DC: USDA ACS CIR 1, Sec. 26, 1987.

PENCE, RICHARD A., *The Next Greatest Thing.* Washington, DC: The National Rural Electric Cooperative Association, 1984.

RICHARDSON, RALPH M., et al., *Farmer Cooperative Statistics, 1986.* Washington, DC: USDA ACS SR 19, 1987.

ROOF, JAMES B. and GEORGE C. TUCKER, *Dairy Cooperatives.* Washington, DC: USDA ACS CIR 1, Section 16, 1986.

4

Cooperatives In Other Industries

Ann Hoyt,
University of Wisconsin–Madison

INTRODUCTION

The cooperative form of business described in Chapter 1 is flexible and can be applied to nearly every conceivable business activity. In this chapter we describe applications of the cooperative form to businesses owned by consumers, by employees, by businesses other than those owned by farmers, and by government and nonprofit organizations. Stakeholders in these cooperative businesses are not necessarily rural or primarily agriculturally oriented. A summary of data on selected types of cooperatives is given in Table 4.1.

CONSUMER COOPERATIVES

A consumer cooperative is formed by the ultimate users of a good or service who share a need that is not being filled in the marketplace. They may want affordable housing, credit, a source of high-quality or natural foods, dignified burial, or day care for their children–all at reasonable prices. People can pool their talents, resources, and energies and form a cooperative to provide for themselves services they all need.

TABLE 4.1 Selected cooperative numbers and membership[a]

Type	Number	Membership (Thousands)
Agriculture (marketing-supply)	5,369	4,402
Commercial[b]	258	77
Communications	110	45
Consumer goods	6,897	1,700
Day care/nursery	540	42
Employee-owned[c]	1,500	1,500
Finance		
Credit unions	18,318	52,000
Farm credit system	484	840[d]
Other	103	1,502[e]
Fishery	102	10
Handcraft	200	25
Housing	2,000	550
Health care	13	862
Insurance	2,034	18,600
Memorial societies	160	350
Rural electric systems	1,084	10,077
Rural telephone systems	255	1,575
Student housing	109	20
Student retail, credit, other	63	80
Worker	300	unknown
Total	39,899	94,257

Source: Ingalsbe.
[a]Latest data and estimates available, but primarily for 1985.
[b]Describes various types of businesses that have united to form cooperatives, such as food wholesalers, retail hardware stores, florists, pharmacies, petroleum retailers, taxicabs, and miscellaneous other services.
[c]Employee-owned businesses through Employee Stock Ownership Plan (ESOP), in which employees own at least 51% of the stock.
[d]Includes 3,000 cooperative borrowers from 13 Banks for Co-operatives.
[e]Includes 1,411 Cooperative borrowers.

Retail Business

The first retail cooperative in the United States was started in 1864 by the Union Cooperative Association No. 1 in Philadelphia. By the 1920s cooperative general stores could be found throughout the country. Today there are nearly 7,000 consumer-owned retail businesses in the United States. They sell food, books, optical supplies, recreational equipment, yard goods, clothing, furniture, pharmaceuticals, heating oil, and many other products.

Just as product lines vary, the size and type of cooperative retail businesses also vary. For example, the largest retail cooperative in the country is Recreational Equipment Inc. (REI), headquartered in Seattle, Washington. REI sells recreational equipment primarily through mail-order catalogs. It also owns 15 retail outlets. In 1986, REI was owned by 1.8 million members and had sales in excess of $132 million. REIs bylaws require the cooperative to return 85% of net income to the members. This translates to consistent refunds of 10% or more on their member purchases. Over $18 million in patronage refunds were returned to members in 1986.

At the other end of the spectrum, there are hundreds of preorder food cooperatives in the country. These businesses operate as groups of consumers who buy food in case lots directly from wholesalers. Purchases are delivered to a central location (e.g., a member's garage or a church basement). By eliminating the overhead of a retail operation and performing order distribution and bookkeeping tasks themselves, members are able to save substantially on food costs. Cooperatives have been responsible for many innovations in retailing as well. Among these are in-store education, unit pricing, nutritional labeling, and bin sales of produce and grains.

Consumer Services

The cooperative business form also has worked well to help consumers get needed services. These include insurance, health care, and banking; 160 memorial societies, which provide funeral and burial services; 540 preschools and child care centers; and prepaid legal services.

In 1752, Benjamin Franklin started the country's first and oldest consumer cooperative, the Philadelphia Conservatorship, a mutual fire insurance company. In 1986 over 18 million Americans were served by 2000 mutual insurance companies owned directly by consumers, by other cooperative businesses, or by firms that are closely affiliated with cooperatives. Mutual insurance companies involve people who pool savings and spread the risks they face over a lifetime. In most cases, these organizations serve both rural and urban members, businesses, and individuals.

Roy has noted that cooperatives have introduced several innovations in the insurance industry. For example, they pioneered multiple-line protection, family life insurance plans, credit life insurance for credit union members, and group and pension plan coverage to cooperative employees. The largest of these cooperatives is the family of Nationwide Insurance Companies, originally organized by the Ohio Farm Bureau Federation.

Application of the cooperative form to health care delivery has fundamentally altered that industry. While working in a rural Oklahoma region that had no hospital, Michael Shadid pioneered the concept of user-owned, prepaid health care in the 1920s. Shadid's work provided the foundation for prepaid health maintenance organizations (HMOs), which dominate the delivery of health care services in much of the country today. The Oklahoma cooperative was an adaptation of a cooperative benevolent society started in 1849 by a group of French gold miners in San Francisco. They formed a prepaid health care provider which has become the French Hospital in San Francisco, a cooperative that continues to be controlled by its member-users.

Although many of today's HMOs are not user-owned, there are 20 consumer-owned and consumer-controlled HMOs in the United States serving nearly 1 million members. The largest of these is Group Health of Puget Sound, which serves 350,000 consumers in Washington state.

The concept of prepaid health care has provided an effective means to reduce the escalating costs of medical services. Because the consumer pays for health care before it is needed (as one would pay for insurance), and because the customer pays a set amount regardless of the extent of care needed, there is a built-in business incentive to keep the HMO members healthy, emphasize wellness, and decrease hospital stays. Prepayment also encourages the user to seek early medical treatment, which helps to reduce long-run costs and allows consumers to plan for and budget medical costs on a monthly basis.

Except for Massachusetts, which specifically provides for user-owned, full- service banks, all states provide for consumer ownership and control over banking-related activities through credit unions. A credit union, by law, must be an association of people who have a common bond. They may all live in one community, work for the same employer, or belong to the same association. Most credit unions are restricted to providing consumer financial services but still provide most of the services consumers can get from commercial and savings banks. Like other consumer cooperatives, credit unions are owned and controlled by member-users.

Credit union members pool their funds and borrow from each other. In this way they have been able to provide credit at reasonable rates to consumers who may not have access to the commercial banking system. Net income from the operation is sometimes distributed to members as dividends after adequate reserve requirements have been filled. Credit unions are the largest and most influential of the country's cooperatives in terms of members (52 million), total assets (over $100 billion), number of cooperatives (over 18,000), and industry penetration (15% of the nation's consumer loans are made through credit unions). The volume of business and number of members in credit unions accounts for about half of the total cooperative activity in the country. Most credit unions are in urban areas. Rural credit unions numbered 944 in 1984 and had 2.67 million members with share savings of $4.3 billion. Rural credit unions are usually rather small and obtain funds primarily from members' deposits or savings but may rediscount loans with the Federal Intermediate Credit Banks. Often they operate in conjunction with farmers' purchasing or marketing cooperatives.

Utilities

Rural electric and telephone cooperatives first modernized the countryside and connected it to the rest of the world. These cooperatives are discussed in Chapter 3.

Consumers have succeeded in buying cable television franchises which are owned and operated by the viewers. Member-viewers capitalize the cable cooperative with their purchase of membership shares. After purchasing a membership or a share of stock in the cooperative, subscriber-members pay a monthly rate for cable TV. Net savings are returned in proportion to subscription charges, and policies are established by a member-viewer board of directors. Cable cooperatives have had greater success in rural areas than in urban areas. Although strong attempts have been made to secure a cooperatively owned cable franchise in many urban areas, they have been successful in only a few cases.

During the summer of 1986, the National Rural Electric Cooperative Association and the National Rural Utilities Cooperative Finance Corporation formed a new cooperative to provide satellite communication and television services for rural consumers. After its first year of operation the National Rural Telecommunications Cooperative had 200 members, 187 rural electric systems, and 13 rural telephone systems.

Housing

Sometimes housing units (e.g., apartment buildings, duplexes, single family dwellings, rooming houses) are also operated as a cooperative corporation. Residents buy shares in the cooperative. The cooperative holds title and signs mortgages on the property or properties it owns. Ownership of a share in the cooperative gives each member the right to occupy a specific dwelling unit. A member share, then, is similar to a down payment for a private real estate transaction.

Members are responsible for monthly charges, which include proportional payments on the cooperative's mortgage, property taxes, insurance, assessments, ground rent, operating expenses, required reserves, legal and audit expenses, and any other expenses that may arise. Members must also sign various agreements pertaining to use and maintenance of each individual unit.

Control of the cooperative is similar to other types of cooperatives. Each member has one vote. Members elect a board of directors which sets basic policies regarding monthly charges, member responsibilities, expansion, and member services. The board may hire a professional management company to sell vacant units and perform bookkeeping, maintenance, and other functions necessary to keep cooperatives on a sound financial basis.

The major advantages of living in cooperative housing include the following: no profit is paid to a landlord—any accumulated funds at the end of the year are returned to the member or reinvested in the cooperative; when a member moves, the cooperative usually accepts responsibility for resale of the member's share in the cooperative; members are not subject to eviction at the end of a lease; members are able to deduct their proportion of mortgage interest and property tax payments from their personal income taxes; individuals are not personally liable for the mortgage; the cooperative handles major maintenance, insurance, and repairs; and the members exercise control over their housing.

Over 600,000 families now live in over 2,000 housing cooperatives in the United States. The largest U.S. housing cooperative is Co-op City in the Bronx, New York, owning over 16,000 units and holding the largest single mortgage in the world. Housing cooperatives are scattered throughout most of the 50 states, with a concentration in New York, Michigan, and California.

Although there are some very prominent exceptions (particularly in New York City), housing cooperatives are used primarily to provide low-cost, good-quality housing for low- and moderate-income families and for people with special housing needs (e.g., the elderly or the handicapped). Because construction or renovation of this type of multifamily housing is quite complex, most new housing cooperatives are sponsored by a local development or charitable organization. Sponsors have included churches, other cooperatives, veteran's groups, community development agencies, local governments, trade unions, and private foundations.

About 20,000 college students belong to 200 student cooperatives which provide housing, meals, groceries, records, and used books. Major student housing cooperatives include those at the University of Michigan at Ann Arbor, the University of California at Berkeley, the University of Texas at Austin, and Oberlin College in Ohio. Most student housing cooperatives began during the depression, when students needed low-cost services to get through school. The keys to their success have been the cooperative organization and the philosophy of shared work. Students reduce cleaning, maintenance, and cooking costs by performing the work themselves. This can reduce monthly rent 5 to 20% below the market prices.

Observations on Consumer Cooperation

Consumer cooperatives have never been as significant in the United States as they have been in Europe. They have gone through several periods of widespread enthusiasm and success, followed by disinterest and failure. One such wave was during the Depression of the 1930s. Americans were struggling for institutions that would allow them to overcome unemployment, poverty, and possible exploitation by monopolist elements in the economy. Although some of the cooperatives that were strong during the Depression have survived, most disappeared with increasing affluence. Other waves of activity have also occurred: cooperatives flourished in the housing sector in New York following World War II, limited resource cooperatives were popular during the civil rights movement, and food purchasing cooperatives sprang up among the middle class during the supermarket strikes of the 1960s. During the 1980s growth areas have been in day care for preschool children, in prepaid health care, in housing (particularly for low-income and elderly citizens), and in credit unions.

There are many explanations for the absence of widespread, ongoing interest in consumer cooperation. Private enterprise has developed a remarkably effective distribution system which provides consumers with a wide variety of relatively high quality goods at relatively low prices. With the exception of cooperative housing programs and the National Cooperative Bank, consumer cooperative development has been random, with little or no government support. Most consumer cooperative development is experimental and unorganized, with little evidence of a coherent local, regional, or national plan. Local cooperatives have had poor management, inadequate capital investment by members, and inadequate business functions. Until very recently, there were few businesses engaged in cooperative development. Only two states have cooperative development departments, and only one university, the University of Wisconsin, has a faculty member specifically assigned to consumer cooperative research and education.

In 1964, a National Commission on Cooperative Development listed four major factors in successful consumer-owned cooperatives. The factors continue to be as necessary today as they were in the 1960s. They are (1) members must express and act upon a need for the cooperative; (2) both the manager and the board of directors must be competent; (3) the business must be adequately capitalized; and (4) the members must be supportive, informed, and educated.

WORKER-OWNED COOPERATIVES

During the 1980s many people stressed the need to create employees' ownership interest in the firms they work for in order to preserve jobs, to improve productivity, to encourage employee investment in the firm, to spread ownership of capital resources more broadly, and to establish democratic work places. Employee-owned cooperatives may be created to take advantage of favorable tax laws. Owners of IOFs have also voluntarily sold their companies to employees upon retirement. The forms of employee ownership and participation are as varied as the motivations that lie behind them. They range from increased employee participation in management, profit sharing, and Employee Stock Ownership Plans (ESOPs), to total employee ownership and control of a business.

Within this context, the cooperative business form has been adapted to provide for ownership and operation of a business by the people who work for the enterprise. These businesses are called worker, industrial, or production cooperatives as well as employee-owned cooperatives or collectives. Employee cooperatives are structured much like consumer and agricultural cooperatives. Each member has one vote, elects a board of directors from among the members, and helps to decide company policy at general meetings. The board is responsible for hiring and evaluating management. Management may or may not include members of the cooperative. These cooperatives pay limited interest on invested capital and return net income to the employee-owners in proportion to the labor they contribute to the enterprise. Labor may be measured in terms of hours worked or relative pay. Although employee cooperatives can (and do) hire nonmember employees, only employees can be member-owners. Employee Stock Option Plan (ESOP) companies may or may not be structured as a cooperative.

Because there are more employee cooperatives in the United States than ever before, it is tempting to think that this is a new form of cooperative business. In truth, employee-owned cooperatives are the country's oldest form of employee ownership. Just as the first consumer cooperative was started in Philadelphia, so was the first employee cooperative. That business was organized by carpenters following a strike during the 1770s.

Just as interest in consumer cooperatives has fluctuated, interest in employee cooperatives has waxed and waned since 1770. Jackall and Levin (1985) divide the history of U.S. employee cooperatives into three cycles. The first (1790–1900) was primarily an outgrowth of the labor movement with needs focused on employment security and changes in the nature of work due to industrialization. The second cycle (1900–1940) peaked during the Great Depression, when employee cooperatives were organized to combat widespread unemployment. The most successful employee cooperatives organized during this period operated in the plywood industry in the Pacific Northwest. Twenty-four plywood cooperatives are still in operation. The third cycle (since 1940) saw a decline in employee cooperatives after World War II, and fairly little interest until the late 1960s. Hundreds of small employee cooperatives grew out of the social unrest of the period, the reevaluation of traditional notions of work careers.

Although it is too soon to tell, a new cycle may have started in the 1980s as employee cooperative leaders learn to adapt favorable parts of the U.S. tax code to cooperatives (particularly the employee stock ownership plans discussed below). Today, employee ownership is seen as a feasible alternative to save jobs and to retain local ownership and control over business enterprises. It has also become an attractive tool for community economic development.

Although employee-owned cooperatives appear to flourish in relation to social and economic conditions, as individual firms they are fairly short lived. A survey of small employee cooperatives cited by Jackall and Levin found that almost all were less than 10 years old, and the median age was about 6 years. Historically, the same pattern has been evident in every periodic cycle in which democratic enterprises have been established.

Although there are businesses owned by the employees on a cooperative basis in a wide variety of industries, use of the cooperative business form has tended to be concentrated in a few industries. These include plywood manufacturing, taxicab, and a variety of alternative retail businesses (e.g., natural food stores, bookstores, and natural foods restaurants). Recently, cooperatives have been organized to buy out supermarkets in the Philadelphia area. They have also been used to provide jobs for low-income people in an effort to encourage economic development in depressed areas. Employee Owned Sewing Company, a cooperative of 50 low-income women in North Carolina, has provided a successful model for this type of cooperative.

When a cooperative business is financially successful, as the plywood cooperatives have been, it encounters a difficult problem in attracting new members because the value of the voting/membership share appreciates considerably. Sherman Kreiner of the Philadelphia Association for Cooperative Enterprise (PACE) describes the organizational structure that has been developed to overcome the problem.

> As the value of the assets increases, the value of the shares increases, until it becomes financially prohibitive for new workers to buy in. To rectify this problem, an internal capital account system was developed. Under this system, a fixed membership fee is set by the by-laws, and an internal capital account is started for each member. Profits that accrue during the period the person is a member are allocated to the account. When the person leaves the cooperative, he/she receives his/her membership fee. The balance in his/her internal capital account becomes a debt of the entire cooperative to the member who has left and is paid out to him/her with interest over a period of time (e.g., five years) set in the by-laws. A new worker coming in is simply required to pay the fixed membership fee in order to become a member (Kreiner, p. 1).

During the early 1970s the U.S. Congress enacted changes in the federal tax code which provided for employee participation in ownership of businesses through a special type of pension plan benefit called an employee stock ownership plan (ESOP). A detailed description of ESOPs, beyond the scope of this chapter, can be found in the writings of Corey Rosen. For our purposes, it is sufficient to note that significant tax advantages are available to business owners who sell stock in their businesses to their employees through an ESOP. The tax code does not require that ESOPs be democratically structured. However, they may be. They may also be combined with a cooperative business structure and thus provide significant capital investment for an employee-owned firm. According to Rosen et al. (p. 271), "ESOPs are a mechanism for capital expansion of a firm through the transfer of stock to the employees of the firm with favorable tax benefits for both the firm and the employees." Because of the relatively high costs associated with establishing an ESOP, this form of pension plan has been used primarily by firms with 40 or more employees.

Increasingly, national concern about retaining jobs in local areas and employees having an equity stake in the net income (and losses) of the firms that employ them has led to spectacular growth in all forms of employee ownership, among them the employee-owned cooperative. In light of current inquiries and negotiations, employee-owned cooperatives will be one of the fastest-growing group cooperatives, in both terms of number of members and total assets, for the remainder of the century. It is also likely that cooperatives owned by farmers and consumers alike will increasingly develop various hybrid forms of cooperative ownership in order to give employees an equity stake in the business successes and failures of their cooperative employer.

NONAGRICULTURAL BUSINESS COOPERATIVES

Nonagricultural business cooperatives have been particularly successful in providing a wide range of wholesale distribution services to locally owned retailer-members. For example, independent pharmacies have formed a cooperative that provides bulk purchasing, computerized distribution and payment systems, as well as a common identity and joint advertising for over 400 pharmacies in the upper Midwest. In addition, cooperatives of independent franchisees provide shared management information systems and accounting services, improved access to financing, and reduced group insurance premiums. Some well-known franchises that have organized purchasing cooperatives include Kentucky Fried Chicken and Dunkin Donuts. Well-known transportation and delivery service cooperatives include the Railway Express Agency and Florists Telegraph Delivery Service. Over 20,000 independent food retailers own food wholesaling cooperatives. Retailer-owned wholesaling hardware cooperatives are among the largest and most successful wholesalers in that industry.

The Associated Press (AP), which distributes the news to its investor-oriented firm (IOF) member newspapers, is one of the nonagricultural business cooperatives in the country. Each member newspaper has one vote in the affairs of the cooperative, and net income from operations of AP is returned to the members in proportion to use. The Mutual Broadcasting

Company is another example of the use of cooperative business in the communications industry. The cooperative's members are independent radio stations that use the cooperative to produce programs jointly.

In the nonprofit sector, some of the more interesting users of the cooperative business are hospitals. The Rural Wisconsin Hospital Cooperative provides a number of services for hospital-members, including pathology consultation, physical therapy and respiratory therapy services, an infection control program, management development, reimbursement, and legal consultation. At the Houston Medical Center in Texas, hospitals jointly own a large power plant which delivers heating and air-conditioning to the member hospitals. Several of the same hospitals also have formed a laundry cooperative.

Other nonprofit organizations have used cooperatives to provide a shared telecommunication system, production facilities for film and video, and data processing services. CSPAN, which broadcasts national news and information from Washington, DC, is a nonprofit cooperative committed to providing access and a higher standard of public information.

Government entities use the cooperative form to their advantage as well. In Pennsylvania, a group of cities formed a cooperative to share transportation equipment. Many municipalities and school districts have formed cooperatives for bulk purchasing of supplies and equipment. The purchasing departments of several large universities have formed a cooperative to acquire supplies and equipment.

THE NATIONAL COOPERATIVE BANK

The most important recent development for nonagricultural cooperatives of all kinds was the establishment of the National Cooperative Bank (NCB), which opened in 1980. Modeled after the Banks for Cooperatives, the NCB was created by the U.S. Congress in 1978 to provide financing to cooperatives not eligible to borrow from the Banks for Cooperatives or the REA. Originally structured as a mixed-ownership corporation, the Bank was privatized at the end of 1981 and is now totally owned by its borrowers. With assets of over $300 million, it provides a variety of financing services nationwide. One of its affiliates, the NCB Development Corporation, provides risk capital to new cooperatives.

Some examples of cooperatives that have been financed by NCB and its affiliates include health maintenance organizations, employee cooperatives, telecommunication cooperatives, franchisee-owned purchasing cooperatives, a sporting goods cooperative, retailer-owned wholesale food warehouses, housing cooperatives, shopping centers owned by the businesses in the center, retail food cooperatives, and a housing cooperative for the hearing-impaired.

Even with the help of the NCB, the difficulties of starting new businesses make it clear that new cooperative development on a large scale will be dependent on initial sponsorship by community organizations, churches, business associations, labor unions, credit unions, other cooperatives, or nonprofit foundations. Although some government support has come from local and state governments, such support is not nearly as strong in the United States as it is in many other countries. The nature and scope of cooperatives throughout the world and government participation in the development of cooperative businesses is the topic of the next chapter.

SUMMARY

Although the primary focus of this book is on cooperatives in agriculture, the business form has been used in a wide variety of industries throughout the United States. These businesses have been organized by consumers, employees, and businesses to provide themselves with consumer goods and services, utilities, business services, and in the case of employees' cooperatives, an ownership stake in the firm for which they work. Governmental units and nonprofit organizations have also organized cooperatives. Growth in both number and size of all these cooperatives has been fostered by the National Cooperative Bank.

DISCUSSION QUESTIONS

4-1.   Identify and discuss the various forms of consumer cooperative businesses, noting their similarities and differences.

4-2.   What are advantages of cooperative housing?

4-3.   Discuss the U.S. consumer cooperative movement by identifying the periods of popular support. Has the focus of business activity changed?

4-4.   What factors have contributed to the lack of interest in consumer cooperatives and the sustainability of interest?

4-5.   Identify and discuss the four major factors of a successful consumer cooperative.

4-6.   Discuss the motivation for employee ownership and the various forms of employee ownership.

4-7.   What are the characteristics of an employee cooperative? How is an employee cooperative similar to a consumer cooperative?

4.8.   Identify and discuss the cycles of employee cooperatives in the United States.

4-9. What are some of the functions performed by nonagricultural business cooperatives?

REFERENCES

ELLERMAN, DAVID, *What Is a Workers' Cooperative?* Somerville, MA: Industrial Cooperative Association, undated.

HALKETT, JAN E., WILLIAM R. SEYMORE, and GERALD E. ELY, *The Cooperative Approach to Crafts.* Washington, DC: USDA ACS CIR 33, 1985.

HARDY, KENNETH G., and ALAN J. MAGRATH, "Buying Groups: Clout for Small Businesses," *Harvard Bus. Rev.* 5(Sept.–Oct. 1987):16.

INGALSBE, GENE, *Cooperative Facts,* Washington, DC: USDA ACS CIR 2, 1987.

JACKALL, ROBERT, and HENRY LEVIN, eds., *Worker Cooperatives in America,* Berkeley, CA: Univ. of California Press, 1985.

KREINER, SHERMAN, *Forms of Employee Ownership,* Philadelphia, PA: Philadelphia Association for Cooperative Enterprise, 1982.

NATIONAL COOPERATIVE BUSINESS ASSOCIATION, *Cooperative Business in the United States,* Washington, DC: NCBA, 1985.

ROSEN, COREY M., KATHERINE J. KLEIN, and KAREN M. YOUNG, *Employee Ownership in America: The Equity Solution,* Lexington, MA: Lexington Books, 1986.

ROY, EWELL P., *Cooperatives: Development, Principles, and Management,* 4th ed. Danville, IL: The Interstate Printers & Publishers, Inc., 1981.

5

Cooperatives in Other Countries

Ann Hoyt,
University of Wisconsin-Madison

INTRODUCTION

The modern cooperative was first developed in Europe. During the late nineteenth century, cooperation spread to other industrializing countries as a self-help method to attack the extreme conditions of poverty, which often accompanied industrialization. From humble beginnings among isolated industrial workers and small farmers organizing for self-help, cooperation has grown to be a worldwide movement. The International Labour Office has estimated there are cooperatives in 140 of the world's 171 countries.

In this chapter we present the size and scope of international cooperative activities outside the United States. It explores the role of agricultural cooperatives in market, planned, and developing economies. It also presents a continuum of public policy toward cooperatives with examples from around the world. There is a short discussion of the role of cooperatives in centrally planned economies and the potential difficulties created by top-down government organization of cooperatives. An introduction to international development activities of cooperative organizations completes this chapter.

It is beyond the scope of this chapter to give a comprehensive view of the wide variety of cooperatives throughout the world. For example, we do not discuss the *moshavim* (cooperative small-holder villages) or the *kibbutzim* (communal settlements) that have been crucial to the development of Israel (Klayman; Stettner, 1977; and Konopnicki), the Mondragon system of industrial employee-owned cooperatives in the Basque region of Spain (Ellerman), or consumer cooperation throughout Europe. Several references are available on consumer, employee, housing, and credit cooperatives in other countries (Lundberg; Roy; Sayin; Ralph Nader Task Force). In any case, insight into the rich international cooperative experience can be gained by looking at the characteristics of agricultural cooperatives in other countries.

AN INTERNATIONAL QUESTION: WHAT IS A COOPERATIVE?

The issue of what a cooperative is and the relationship between cooperatives and the government in planned and developing economies has created controversy. Chapter 1 gives a foundation for this issue by developing a definition of the term cooperative and distinguishing cooperatives from investor-oriented firms (IOFs). Chapter 2 contains a discussion of the cooperative principles and the six principles adopted by the International Cooperative Alliance (ICA) in 1966. Those principles are open membership, democratic control, service at cost, limited return on capital, duty to educate, and cooperation among cooperatives.

The controversy arises because not all countries require businesses to conform to the ICA principles in order to be considered cooperatives. In particular, the principles that assure distribution of net income on the basis of use and democratic control through the one-member, one-vote concept may be violated. For example, according to Watkins (p. 69), the Polish government had so controlled the cooperatives that ICA was unable to accept Poland's request for membership. It was not until after the cooperatives were allowed some autonomy within the national economic plan that membership in ICA was granted.

Cooperatives in some countries are controlled by the state or must conform to some aspects of a state plan. Ownership may not be traceable to individuals but instead is vested in some aspect of the state. Management may be selected by the state. This control by state authorities or the ruling party is exercised through state representatives on the boards of cooperatives. There are changes in some planned economies that are moving these organizations more in the direction of direct member control.

Organizations in many countries do not meet our definition of a cooperative but are called cooperatives in their respective countries; therefore, we should be aware of them and how they are organized. These organizations possess vestiges or elements of local control and sharing benefits which are characteristic of cooperatives in the United States.

QUANTIFYING INTERNATIONAL COOPERATION

Measuring the role and impact of cooperatives on a global basis is difficult even though several international organizations collect information on cooperatives worldwide. The ICA, a membership organization of cooperatives from 72 countries, collects data only from its members. Many of the large agricultural cooperatives are not ICA members. The World Council of Credit Unions (WOCCU) collects data only on the credit union sector. The Committee for the Promotion of Aid to Cooperatives (COPAC) collects statistics that refer only to developing countries. The International Labor Organization (ILO) and Food and Agricultural Organization (FAO) among others also publish cooperative statistical data. Data in some countries are quite accurate but in many others are accumulated infrequently, so the time of reporting varies. There are some countries for which no data on cooperatives are listed. COPAC (p. 2) states that "no systematic method exists for the collection of information at the international level."

Additional difficulties with international cooperative statistics exist. No attempt at reconciling available data bases has been published. Double counting of individual cooperative members is frequent. For example, one farmer may belong to six or seven separate cooperatives. Cooperatives are defined in different ways; therefore, we cannot be sure what the numbers represent. An assessment of the significance of cooperatives in a country should take into account its political system, organizational characteristics, and its share of population and the share of economic activity accounted for by cooperatives.

Keeping all of these cautions in mind, we are able to make some generally accepted estimates of the size of the international cooperative movement. In 1985, the 72 country-members of the ICA represented 740,000 cooperatives on five continents. Although ICA member cooperatives represent many economic sectors (agriculture, consumer, credit, fishery, housing, employee, and others), the worldwide cooperative movement is dominated by the agricultural, credit, and consumer sectors. Slightly more than one-third (35%) of ICA's 1985 members were agricultural cooperatives; 28% were financial thrift and credit cooperatives. Although only 9% of ICA members were consumer cooperatives, 26% (129 million) of the individuals represented by ICA cooperatives belonged to consumer cooperatives.

Individual member totals by country show that cooperatives are present in both centrally planned and market economy countries. According to ICA member statistics, the People's Republic of China has the largest number of cooperative members (132 million). They are followed by India (68 million), the USSR (60 million), the United States (58 million), Japan (18 million), Romania (15 million), France (14 million), Poland (12 million), Canada (11 million), and the United Kingdom (10 million).

COPAC's summary of cooperatives in developing countries is presented in Table 5.1. Recall that the COPAC data do not include the more developed economies, capitalist or socialist, nor are they constrained by the requirement that a cooperative meet the ICA principles in order to be counted.

Agriculture and finance cooperatives are dominant in developing countries. In terms of the number of cooperatives and the number of members, Asia and the Pacific are by far the most active regions. As urbanization and its attendant problems become more acute, the role of urban cooperatives is growing in developing countries, particularly in Latin America. Thus consumer, thrift, credit, housing, insurance, and industrial cooperatives are providing significant services and are receiving increases in external support.

COPAC reports 359,775 agricultural cooperatives with 176 million members in the developing countries. ICA reported that there were over 256,000 agriculture cooperatives in the 72 member cooperatives in 1985. Agricultural cooperatives represented 66.6 million farmer-members.

Industrialized Market Economies

In most industrialized countries, cooperatives have taken their place as a significant or, in some cases, major component of the market economy (Tables 5.2 and 5.3). They handle 25 to 50% of the farm supplies and marketing in western Europe and North America. For example, the dairy industry in Denmark, the grain industry in Canada, and rice marketing in Japan are all dominated by cooperatives. Cooperatives are also strong in dairy marketing in all the countries selected. The dairy industry is dominated by cooperatives (more than 97% of marketing activity) in Ireland, Norway, Sweden, and Finland (Armstrong). The cooperative banks of Europe and Japan are among the largest financial institutions in the world. Consumer cooperatives are dominant in food retailing in many European countries.

Because of their important position in agriculture, cooperatives can play a significant role in farm policy. Cooperatives may negotiate with the government in establishing what the general farm policy should be and have responsibility for implementing aspects of it (Foxall; Sargent).

TABLE 5.1 Number of Cooperatives and Their Members in Developing Countries

Industry	Africa		Asia/Pacific		Latin America		N. Africa/Near E.		Total[a]	
	Co-ops	Members (Thousands)	Co-ops	Members (Thousands)	Co-ops	Members (Thousands)	Co-ops	Members (Thousands)	Co-ops	Members (Thousands)
Agriculture	44,451	12,583	275,073	154,229	15,427	2,876	24,824	6,719	359,775	176,407
Consumer	17,128	2,637	49,575	16,581	4,013	3,014	6,604	4,022	77,320	26,254
Credit	12,597	1,538	161,323	88,045	6,667	7,572	389	488	180,976	97,643
Fisheries	942	31	12,249	1,546	1,167	79	152	53	14,510	1,710
Housing	250	23	35,004	2,198	2,789	505	2,180	775	40,230	3,500
Industrial/ worker	3,692	135	59,112	4,668	3,052	370	1,067	61	66,923	5,235
Other	3,193	196	106,947	14,501	6,445	5,562	5,662	124	122,247	20,384
Total[a]	83,634	16,981	702,706	283,253	38,072	19,568	32,809	19,568	858,221	331,954

Source: COPAC.
[a]Reconciliation of discrepancies in the totals are not clear in the source.

Cooperatives in Other Countries   85

TABLE 5.2 Cooperatives' shares (percent) of agricultural marketing
activity, selected countries and commodities

Country	Dairy	Grain	Fruits & veg.	Slaugh-tering	Wine	Poultry
Belgium	70	—	65	15	—	—
Denmark	87	15	—	90	—	55
Finland	97	73	—	94	—	60
France	44	52	28	37	68	—
Ireland	100	—	34	20–35	—	64
Netherlands	87	—	80	25	—	17
Norway	100	—	40	74	—	73
Sweden	99	80	—	80	—	70
United States	78	41	17	—	—	8
West Germany	79	55	46	30	35	—

Source: Armstrong.

TABLE 5.3 Cooperatives' shares (percent) of
agricultural purchasing activity, selected
countries and commodities

Country	Feed	Fertilizer	Seed	All Supplies
Belgium	20	15	16	—
Denmark	50	43	53	45
Finland	—	—	—	70
France	19	—	—	—
Ireland	53	50	55	—
Netherlands	53	—	—	55
Norway	—	—	—	—
Sweden	—	—	—	60
United States	18	38	17	27
West Germany	60	60	—	50

Source: Armstrong.

U.S. agricultural cooperatives are much larger as a group, in terms of
business volume, than anywhere in the world. But the largest cooperative
organization is Japan's all-purpose agricultural cooperative, Zen-noh, a
federation of more than 4,200 smaller cooperatives with more than 5 million members. It has more than seven times the dollar volume of the largest
cooperative in the United States.

In some respects European cooperatives have evolved a more sophisticated institutional framework. Periodically, they have been blueprints for
study by students of cooperation. Randall E. Torgerson, administrator of

ACS, USDA, in an unpublished talk, reported that European cooperatives, compared to those in the United States, had the following characteristics:

- National organizations of cooperatives are common.
- Vertical integration is greater in farm commodity marketing, but less so in production supplies.
- In some countries, such as Denmark, cooperatives are unincorporated and therefore have unlimited liability.
- Pooling and membership contracts are more prevalent in marketing farm products.
- Domestic agricultural policy results in cooperatives operating in a more stable political and economic environment.
- Cooperatives generally handle only one product, therefore, a single farm may be served by several (six to seven) cooperatives.
- Membership is often tied to the farm rather than the farmer.
- Education of members is taken more seriously; for example, the understanding of principles and operations are advanced by study circles.
- Very little competition is permitted between cooperatives.

Centrally Planned Economies

Cooperatives are important in all phases of the food production cycle in centrally planned economies. They are more dominant in agricultural production and retailing than their market economy peers. Cooperative business is quite important in the socialist world in terms of membership and volume of activity. According to Thordarson, cooperatives in Bulgaria account for 33% of retail goods sales and 70% of agricultural land under cultivation. In Hungary, cooperatives contribute about 20% of the national income. Polish cooperatives account for up to 60% of retail trade and nearly 65% of urban housing construction. In the Soviet Union, consumer cooperatives have 59 million members and control 30% of retail trade.

A description of the role of cooperation in a planned economy food system is found in the following adaptation of Waszak's (p. 31) description of the cooperative movement in Poland. Different types of cooperatives developed in different sections of Poland. In the Russian sector, the prevailing forms of cooperation were consumer cooperatives; in the Prussian sector, craftsmen's and peasants' banks or commercial agricultural cooperatives; and in the Austrian sector, agricultural associations and credit cooperatives. Today, cooperatives are strong in agriculture, food products, trade, and housing. In these fields the cooperative market share is 70 to 80%. In towns, consumer, housing, employee, and craft cooperatives are prominent. In rural areas, there are supply and marketing cooperatives, collective farms, and mutual loan societies.

In a state-controlled economy, cooperatives conduct business according to the needs of the members and the assumptions of the central and regional social and economic plans. Cooperatives are obliged to join one of the central cooperative unions, which perform control functions over local cooperatives. A cooperative can belong to more than one cooperative union or association if they have common social and economic aims.

Developing Countries

In developing countries, agricultural cooperatives focus more on food marketing than on production. Attempts to organize farmers into cooperative production units in the developing countries have not been entirely successful. Although cooperatives have the potential to supply farm inputs and market produce (both important to agricultural development), there have been relatively few successes. Food-processing cooperatives, for example, have been successful in Argentina and Brazil. In India, cooperatives are especially strong in processing and marketing milk, sugar, and oil seeds. Credit cooperatives, however, have been crucial to the success of agricultural cooperatives in many developing countries.

Cooperatives, as a business form, came into being in response to the excesses and injustices of the industrial revolution. More often than not, the business form was used to solve the problems of poverty in the midst of capitalist growth and expansion. The concentration of income and wealth in the hands of a few and the extreme gap between the rich and the poor created conditions for violent political revolution. Cooperation, in its early days, came to be seen as a peaceful method of social betterment.

The impulse toward fostering cooperatives in today's less-developed countries stems from the same impulse in the development context. Successful cooperatives provide many positive benefits for the country as well as the individual members. To paraphrase Thordarson, cooperatives mobilize resources internally, provide economies of size and services to the populace not otherwise available, and build permanence through creating an institutional framework. Most important cooperatives potentially build human resources.

GOVERNMENT AND COOPERATIVES

Underlying the attempt to identify what factors contribute to success (and failure) in cooperative development is an ongoing debate over the proper relationship between each country's government and cooperatives. Experience throughout the world has shown that government policies can impede or enhance independent cooperative development. The debate centers on the need to preserve autonomy and democratic control of the cooperative by its members, while recognizing the cooperatives' need, in some countries, to receive management and financial support from the government and to operate in a favorable legislative environment.

A Continuum of Public Policy

Government policy toward cooperatives can be viewed as a continuum from overt hostility to total control. The continuum in Figure 5.1 describes public policies toward cooperatives. A brief discussion of each policy level follows.

Destructive policy Countries at level 1 are hostile toward cooperatives. They are suspicious of them and attempt to restrict, suppress, or outlaw them as the fascists did in Italy and as the government of Chile has more recently attempted to do. The Uganda government allowed cooperatives to operate by law, but set agricultural prices in such a way that cooperatives could not compete. Roy cites the case of Indonesia where cooperatives were deprived of all legal rights, many were forcibly disbanded, and state assistance to cooperative, was terminated during the 1960s. (Since 1966 the government of Indonesia has encouraged cooperatives.) Governments that actively oppose cooperatives are usually authoritarian regimes, which are unable to tolerate freely operating democratic institutions. By their nature, cooperatives are a threat to the power of these governments.

Neutral policy At level 2 the government does not actively attempt to destroy cooperatives, nor does it give them special treatment. In effect, cooperative businesses operate in the same climate as all other businesses. This limited involvement by government has been typical of industrialized countries.

In western Europe, cooperatives came into being through the action of popular movements. The cooperative movement in France, Sweden, the United Kingdom, and other western European countries has a strong tradition of independence, voluntarism, and self-help.

Type of policy:	Destructive	Neutral	Supportive	Participating	Controlling
Level:	1	2	3	4	5
Description:	Antagonism, hostility, violent destruction	No public policy, positive or negative	Creation of a favorable legal/business environment for cooperatives	Active provision of support services for cooperatives; may include management	Total control over cooperative management & decision making

Figure 5.1 Continuum of public policy toward cooperative business.

Supportive policy   At level 3, governments demonstrate a positive attitude toward cooperatives as a tool that citizens can use to improve their economic well-being and participate in economic democracy.   Artificial barriers to cooperative operations are removed. For example, special legislation is passed to make it easier to organize and operate them. Education, research, and technical assistance programs are initiated to help cooperatives be successful. The aim of government is to encourage the development of cooperatives; however, responsibility for initiating and carrying through this development rests with members. Because the members benefit, cooperatives should succeed or fail according to their own performance. Although the government may provide services and incentives, which make the cooperative an attractive form by which to conduct business, the government is not actively involved in the day-to-day affairs of the cooperative, does not participate in cooperative management, and does not have representatives on the board of directors.

Although their early actions were laissez-faire, many western European countries have created special legislation, under which cooperatives operate, and have provided information and technical assistance. Cooperative operations may be monitored to guarantee that they continue to function as cooperatives.

An example of government support which falls between levels 3 and 4 (short of actual participation in management) is the cooperative law enacted by Egypt in 1980 (Krasheninnikov). That law regulates the establishment of cooperatives, regional federations or unions of local cooperatives, and national cooperative federations. It also establishes a cooperative bank. The law guarantees cooperatives' autonomy as well as support by the state which is to actively promote cooperative development and stimulate the activity of cooperative enterprises. Egyptian law also provides cooperatives with significant business advantages. For example, Egyptian cooperatives have the right to duty-free import of equipment, a 5% discount on goods purchased from state-run companies, and a 10% discount on electric power. Egyptian cooperatives are exempt from liquid asset, profit, and local taxes. Further, these cooperatives receive priority in using public lands and concluding contracts with state-run companies (Thordason). These government provisions give a substantial competitive advantage to cooperative businesses over IOFs that do not share the same advantages.

Participating policy   At level 4 government is directly involved in organizing cooperatives and providing capital and management. This is common in developing countries, where the lack of capital and education among farmers must be overcome. Many third-world cooperative organizations are undercapitalized. To the extent that they depend on external financing, including government financing, they may be subject to manipulation. To the extent that development projects, including those carried out by cooperatives, attempt to meet credit needs rather than create debt capacity, they may contribute to the failure of those projects. Further, when cooperatives are carrying out the development programs of government, risks should be shared between government and cooperatives. The risk of providing services to the poor should not be assumed entirely by cooperatives.

Although many cooperatives, especially in developing countries, need direct government assistance in operations, there is a danger that the cooperative will not grow to be able to operate independently. For cooperatives to succeed as self-help businesses in the long run, governments must pursue strategies that allow them to withdraw their direct support and shift responsibility to the member-owners. As members become educated about the cooperative and their role, they can see the benefits and begin to take an active part in its control. Then as the cooperative begins to take hold, the government can withdraw. Roy (p. 583) points out that "unless a government does withdraw and shift responsibility to the people, then any so-called co-op development is but another name for state control and political domination."

Governments create problems when they attempt to impose their timing or priorities on cooperative development (U.S. Overseas Cooperative Development Committee). One reason that cooperatives fail is because they were set up under external pressure and the members are not convinced about what they are doing. In many instances villagers are expected to organize around an activity which is not necessarily their main concern and which may not take into account family and tribal customs and mores that may be in direct conflict with the loyalty expected of cooperative members.

An attempt by the government of Ecuador to establish an agricultural producers' cooperative in the Guyas Basin characterizes the problems encountered:

There had been, for some years, a strong peasant movement demanding title to the land. As part of it[s] agrarian reform program, the government encouraged the setting up of rural cooperatives. But ten years later, their "state of health" was disastrous. There was a major crisis of participation, of leadership and of economic stability and the national credit institutions were threatening to confiscate the land. What had happened?

The cooperative form of organization, alien to the sociocultural reality in the countryside and imposed entirely from outside and above, had been accepted by the peasants only as a means of obtaining land, credit and technical assistance.

The land, and later the credit, was distributed among cooperative members in a very unfair way. Some peasants received 20 to 30 hectares and other[s] only two or three. In fact there was never any genuinely cooperative production.... [The cooperative was plagued with] bad management by unqualified officials, distribution of cash without clear economic criteria, lack of any economic rationality among the peasants themselves (most of whom were neither literate nor accustomed to handling large amounts of money), [and] persistence of a market structure unsuited to their production and speculation . . .

Cooperatives as established in the rural areas of Ecuador have tended to be a mechanical transfer of the North American model of agrarian cooperativism, introduced and developed in completely different structural conditions.... The idea that the member represents his or her family nucleus, although perfectly logical in an industrial society, is irrelevant here, since the families are extensive and such standards of representation are not applicable (Vozza, pp. 38-40).

Controlling Policy   At the last level, governments use cooperatives as a tool to achieve their own agenda (e.g., to provide employment) and take control of the cooperative. Government hires management and dictates policy. In a number of countries, the government has the power to appoint and dismiss cooperative board officials and, in doing so, has undermined their autonomous and democratic character. Cooperative members have in these situations not only lost control over the management of the organization but are often required to submit to government controls on the production, pricing, and marketing of their products. Cooperatives become an arm of the state. For example, the Indonesian government sets the price of fertilizer and rice, for agricultural cooperatives, and the board of directors is obliged to follow government decisions. In these situations, there is often no difference between the rights and obligations of members and nonmembers.

Changes in Policy over Time

Radical shifts in policy toward cooperatives often occur in countries that have frequent and sometimes violent changes in government. Chile provides an example of shifting public policy toward cooperatives. In the 1950s and early 1960s the government had a neutral or laissez-faire attitude toward cooperatives (level 2). With the advent of the Alliance for Progress in the 1960s, cooperatives became an instrument of government policy along with land reform. Collective production cooperatives were organized. They were subsidized and controlled by the government with the intent that the government would withdraw (level 4). Cooperatives would then become independent of government support. Allende took control of these cooperatives (level 5) when he came to power in 1970. He viewed independent cooperatives with fear and as being antisocialistic. In

1973, Pinochet took power after a military coup. He also viewed cooperatives with fear, but for the opposite reason. He considered them to be socialistic and as bleeding the government. As a result, most Chilean cooperatives have been destroyed. Chile, in recent years, would be classified as a "1" in Figure 5.1.

It is also important to note that some countries, such as Argentina, Costa Rica, and Brazil, have had relatively stable policies toward cooperatives. In other countries, policies have been erratic.

Centrally Planned Economies

Several times we have used examples from socialist countries as we have discussed the continuum of public policy toward cooperatives. A major distinguishing feature between planned and market economies is the coordination of economic activities into a single national economic plan that establishes the common goals created for the entire society. The government is responsible for implementing the plan, and until recently, there was fairly little room for entrepreneurial business activities conducted by individuals for their own profit. A major distinguishing feature of cooperatives in planned economies is that they are expected to coordinate their economic activities with the national economic plan.

Watkins describes the impact this difference has had on cooperatives in planned economies. Socialist or communist governments have a bias toward cooperatives that should be stressed since these governments interpret the meaning of a cooperative from their own ideological standpoint. Within the planned economy, cooperatives offer "an alternative to private and capitalist enterprise" on the economic side (Watkins, p. 69). On the social side they offer a means of weaning the common people from individualism and acquisitiveness and of conditioning them to a regime under which service to the community takes precedence over private gain (Watkins, p. 69). Many cooperative leaders in capitalistic countries take exception to these expressions because they view cooperatives as at the apex of capitalism (private property, voluntary ownership, and distribution of benefits according to participation).

Watkins describes the importance of the national plan to cooperatives as follows:

> The scope and objectives of cooperation depend . . . upon the national plan, its immediate and ultimate aims. At one time targets would be set for the cooperatives by the central planning authorities, but in recent years the tendency has been apparent for the cooperative organizations to be allowed to draft their own plans for approval of the authorities which, of course, retain and exercise their over-riding powers to amend or reject. In connection with planning, control is exercised by central authorities over finance. Organizations which cannot finance their development from their members' savings and their own accumulated reserves must have recourse to state banks (Watkins, p. 69).

The difficulties for cooperative operations under socialism have been expressed by Krasheninnikov, a representative to ICA from the USSR consumer cooperative Centrosoyus.

> The state often, without any plausible reason, interferes with the management of coops, thereby infringing upon their interest and hampering their development. For this reason in relations between the state and coops, it is very important to create conditions favorable for the development of coops as autonomous and economically viable organizations having the right to make independent decisions (Krasheninnikov, p. 23).

Cooperatives in socialist countries often carry out many activities that developed capitalistic countries generally leave to various civil divisions. According to Thordarson, this is a result of a different allocation of responsibilities between the state and cooperatives than is found in the West. Socialist cooperatives may take an active role in education, cultural activities, athletics, and child care.

INTERNATIONAL COOPERATIVE DEVELOPMENT ASSISTANCE

Since the earliest successes of the European cooperative movement in the last part of the nineteenth century, visionary cooperative leaders were convinced that cooperative organization could not and should not be limited by national boundaries. The primary focus of cooperative assistance efforts has been to develop and strengthen the International Cooperative Alliance (ICA). Founded more that 80 years ago, ICA is the primary international organization dedicated solely to the promotion of cooperation. ICA is the apex organization of the world cooperative movement. It provides technical assistance, financial aid, and educational opportunities for its member cooperatives.

In addition to ICA activities, development projects are often undertaken by international development banks (e.g., World Bank, African Development Bank, and Asian Development Bank) and are co-supported by the governments of several countries. The Inter-American Development Bank has a special fund for small projects, approximately 40% of which has gone to cooperatives.

Many governments of the developed countries provide assistance for cooperative development in the third world. They rely on close cooperation between the sponsoring country's government, its cooperatives, and the developing country's government. The developed country's governmental role is primarily a financial one, which assures retention of control over project approval. For the most part, actual project management is conducted by cooperative national federations or apex organizations. For example, the Swedish Cooperative Centre (SCC) receives financial aid from the Swedish International Development Authority (SIDA) to provide development and technical assistance to cooperatives in Kenya, the United Republic of Tanzania, and Zambia.

Canadian cooperative organizations receive financial support from the Canadian International Development Agency (CIDA) for cooperative projects in developing countries. Cooperatives are also included in a few of the large-scale projects operated by CIDA itself.

In the United States, the U.S. Agency for International Development (USAID) has a development strategy oriented to providing benefits for the majority of the population in developing countries. Assistance is generally on a grant basis. Cooperatives are identified as a priority under the U.S. Foreign Assistance Act. The United States is unusual in that federal funds flow through several cooperative organizations rather than through one national apex organization. Funds flow through Agricultural Cooperative Development International, the National Cooperative Business Association, the Credit Union National Association, the Cooperative Housing Foundation, the National Rural Electric Cooperative Association, and the Volunteer Development Corps. In 1980, USAID initiated a policy that provides for the continued use of U.S. cooperatives in bilateral assistance, but that also encourages the cooperatives to relate directly to the developing country's cooperative counterparts outside the context of government-to-government bilateral programs. The United Nations has come to call this activity movement-to-movement assistance.

SUMMARY

Cooperatives exist in a wide variety of industries in almost every country in the world. In fact, U.S. cooperatives have much to learn from their often more sophisticated international counterparts. This is particularly true in the areas of greater cooperation between agricultural cooperatives, the potential for development of employee-owned cooperative businesses, and the exploration of international cooperative development efforts through government agencies.

Internationally, cooperatives conduct business under a continuum of political conditions, ranging from destructive through supportive to total control. These conditions vary from country to country and can change through time as new political regimes gain power.

Cooperatives have played a special role in planned economies, where they are very strong. This role has created a controversy in the international cooperative community as to what constitutes a cooperative. In some countries cooperatives must conform to state plans, ownership cannot be traced to individuals, and benefits are not distributed on the basis of patronage—all of which violate fundamental cooperative principles.

Cooperatives have been a tool in efforts to augment development in third-world countries. This, too, has created a debate over the proper role of government in cooperative development efforts. Most development assistance flows from the developed countries' governments and their cooperative organizations to assist in developing of cooperatives in the third world.

DISCUSSION QUESTIONS

5-1. Compare and contrast the role that agricultural cooperatives play in market, centrally planned, and developing economies. How does the role of government differ between economies?

5-2. Identify the characteristics of government policy toward cooperatives at each of the five positions on the continuum of public policy scale.

5-3. Discuss the relationship between credit cooperatives and agricultural cooperatives in developing countries.

5-4. What do you think is the best relationship between a country's cooperatives and its government? Justify your answer.

5-5. Discuss the pitfalls of a top-down policy of cooperative development.

5-6. Discuss the role of cooperatives in planned economies.

5-7. How does national economic planning affect cooperatives in centrally planned economies?

5-8. Identify and discuss the effect of activities of cooperative development organizations from developed countries on cooperatives in developing countries.

5-9. What are the problems or obstacles facing cooperative organizations in developing countries?

5-10. Discuss the advantages that cooperatives offer to developing countries.

REFERENCES

ARMSTRONG, JACK H., "Cooperatives Today—US and Worldwide," paper presented at N.I.C.E., Univ. of Tennessee, Knoxville, TN, Aug. 1986.
COMMITTEE FOR THE PROMOTION OF AID TO COOPERATIVES (COPAC), *The Cooperative Network in Developing Countries: A Statistical Picture.* Rome, Italy, Feb. 1987.
ELLERMAN, DAVID P., "Entrepreneurship in the Mondragon Cooperatives," *Rev. Soc. Econ.* 42(Dec. 1984):272.
FOXALL, GORDON, *Cooperative Marketing in European Agriculture.* Hampshire, England: Grower, 1984.
KLAYMAN, MAXWELL I., *The Moshav in Israel: A Case Study of Institution- Building for Agricultural Development.* New York: Praeger, 1970.

KONOPNICKI, MAURICE, "The Public and Co-operative Sectors in Israel," *Ann. Public Coop. Econ.* 42(1971):47.

KRASHENINNIKOV, ALEXANDER, "The State and Cooperatives in Developing Countries," *Centrosoyus Rev.* 2(1986):20.

LAIDLAW, A. F., "Cooperatives in the Year 2000," paper presented at the 27th Congress of the International Cooperative Alliance, Cooperative Union of Canada, Ottawa, 1980.

LUNDBERG, W. T., *Consumer Owned: Sweden's Cooperative Democracy.* Palo Alto, CA: Consumer Coop. Pub. Assn., 1978.

RALPH NADER TASK FORCE ON EUROPEAN COOPERATIVES, *Making Change? Learning from Europe's Consumer Cooperatives.* Washington, DC: Center for Study of Responsive Law, 1985.

ROY, EWELL PAUL, *Cooperatives: Development, Principles, and Management,* 4th ed. Danville, IL: The Interstate Printers & Publishers, Inc., 1981.

SARGENT, MALCOLM, *Agricultural Co-operation.* Hampshire, England: Grower, 1982.

SAYIN, EROL, "Kent Ko-op's Batikent Project (Turkey)," *Rev. of International Co-operation.* 80(1987):51.

STETTNER, LEONORA, "The Role of Cooperatives in Israel." *Int. Rev. Coop.* 70(2)(1977):87.

STETTNER, LEONORA, *Chinese Cooperatives: Their Role in a Mixed Economy.* Oxford, England: Plunkett Foundation for Cooperative Studies, 1984.

THORDARSON, BRUCE, "Global Review of the Role of Cooperatives in Economic and Social Development," in *The Cooperative Network in Developing Countries: A Statistical Picture.* Rome: Committee for the Promotion of Aid to Cooperatives, 1987.

UNITED NATIONS GENERAL ASSEMBLY, *National Experience in Promoting the Cooperative Movement: Report of the Secretary General.* Economic and Social Council, A/42/56, E/1987/7, Dec. 11, 1986.

U.S. OVERSEAS COOPERATIVE DEVELOPMENT COMMITTEE, *Why Cooperatives Succeed . . . and Fail,* synopsis of workshop, Washington, DC, Oct. 10–11, 1985.

VOZZA, GUISEPPE W., "Ecuador: The Crisis of Rural Cooperatives and the Quest for Alternatives," *Cult. Surv. Q.* 11(1)(1978):38.

WASZAK, ZBIGNIEW, *Cooperation in Poland,* The Society of Cooperative Studies, Bul. 53, Apr. 1985, p. 31.

WATKINS, W. P., *The International Cooperative Movement: Its Growth, Structure and Future Possibilities,* Manchester, England: Cooperative Union, undated.

6

Supporting Organizations

Robert Cropp,
University of Wisconsin-Platteville

INTRODUCTION

Cooperatives have organized educational, research, technical assistance, public relations, and trade-lobbying organizations to solve common problems and to provide other services that have broad application. Some of these organizations are regional or national in scope; others are international. Some engage in many activities, and others restrict their operations to a few specific functions. Nearly two dozen national organizations headquartered in Washington, DC, either represent specific cooperative interests regarding commodities, business functions, and business structure or else address across-the-board cooperative concerns.

TRADE ASSOCIATIONS

National Council of Farmer Cooperatives

The National Council of Farmer Cooperatives (NCFC) is the political voice for farmer cooperatives in Washington, DC. Established in 1929,

NCFC has been a guardian of agricultural cooperative foundations and laws for nearly 60 years. NCFC represents about 90% of the nearly 6,000 agricultural cooperatives in the nation. NCFC's membership includes the major farm marketing and supply cooperatives, all the banks in the cooperative Farm Credit System, and most of the nation's state cooperative councils. The NCFC is interested principally in federal legislative matters that might affect agricultural cooperatives. NCFC exists to promote a political and economic climate in which cooperatives can thrive and serve the needs of America's farmers.

This organization has worked hard to preserve the Capper-Volstead Act in an effort to protect farmers' rights to collectively market products and purchase supplies. It has been instrumental in preserving the single-tax principle as applied to patronage refunds. The organization also is working to develop new markets at home and abroad, to ensure adequate credit for farmers and their cooperatives, and to influence public policies affecting farmers and their cooperatives.

American Institute of Cooperation

The American Institute of Cooperation (AIC) was incorporated on January 22, 1925, and chartered as a university, under the laws of the District of Columbia. AIC is a national educational organization supported by agricultural cooperatives and endorsed by cooperative leaders, educators, and research workers.

AIC promotes educational programs by cooperatives and other organizations, stimulates research through an awards program and other means, and distributes educational materials and information. It also provides services to help member cooperatives develop special educational programs for youth, for young farm couples, and for women's groups.

Each year AIC holds an institute (National Institute on Cooperative Education, or NICE) during which lectures, panel discussions, symposia and reports on cooperatives are given. These papers are published in a proceedings called *American Cooperation*. These volumes have become the best compendium of information on agricultural cooperatives to be found in the United States. AIC is known as the university without a campus because NICE is typically held at campuses of land-grant universities.

AIC also publishes the *Journal of Agricultural Cooperation*. Its purpose is to encourage research on issues of importance to U.S. farmer cooperatives and to provide a forum for the review and exchange of research results among individuals in universities, cooperatives, and government.

AIC does not engage in legislative activity on either the federal or state level, nor does it adopt resolutions and policies for future action. Agricultural cooperatives support AIC voluntarily.

The Farm Credit Council

The Farm Credit Council (FCC) is a federated trade association organized to serve cooperative farm lenders by protecting and promoting the interests of cooperative farm lending institutions and by improving the business conditions for such institutions. The FCC's membership is comprised of district farm credit councils from each of 12 Farm Credit districts and from the Central Bank for Cooperatives.

Agricultural Cooperative Development International

The Agricultural Cooperative Development International (ACDI) was organized in 1963 to respond to the needs of agricultural cooperatives and farm credit systems in developing countries. Its membership consists of regional, national, and international cooperatives. ACDI's projects in client countries are under contract with the Agency for International Development and other international funding agencies.

NCFC Affiliated Organizations

Between 1983 and 1985, three national cooperative associations affiliated with the National Council of Farmer Cooperatives. These were the American Institute of Cooperation, the Farm Credit Council, and Agricultural Cooperative Development International. Under the plan, each organization remains a separate entity for legal and tax purposes. Each also maintains its own membership, board of directors, and management. The presidents of the three affiliated groups also serve as NCFC vice-presidents. This affiliation has resulted in greater organizational efficiency. The participants share a common newsletter and enjoy improved planning, better coordination of programs, and a united cooperative voice in Washington.

The National Cooperative Business Association

The National Cooperative Business Association (NCBA) is the oldest of the national cooperative organizations. Prior to 1985, it existed as The Cooperative League of the USA. Organized in 1916 to promote the interests of consumer cooperatives in the United States, it is a national federation of all types of cooperatives: agricultural cooperatives, credit unions, consumer societies, and organizations providing electricity, telephone, health, housing, and insurance services.

The NCBA has six major functions: (1) to advance public knowledge of cooperatives, (2) to improve the skills of cooperative directors and employees, (3) to encourage wise cooperative financing and operative policies, (4) to help cooperatives strengthen their member relations, (5) to seek federal laws and administrative decisions consistent with cooperative aims and purposes, and (6) to promote development in the world's less developed areas, both at home and abroad, through cooperatives. It is the only U.S. representative in the International Cooperative Alliance.

The NCBA has had an active program for providing assistance to cooperatives in developing countries since the early 1950s. Partially funded by the U.S. Agency for International Development, this program assists cooperatives in several developing nations. For example, NCBA has trained Peace Corps volunteers who have organized cooperatives in developing nations.

The National Rural Electric Cooperative Association

The National Rural Electric Cooperative Association (NRECA) is a service or trade association of rural electric cooperatives headquartered in Washington, DC. It was organized in 1942 to serve its member associations with information, insurance programs, research, and legislative representation. It has provided technicians and engineers to foreign countries interested in developing cooperative electric service as well. The organization is maintained through the voluntary dues paid by its members.

Other Trade Associations

Various other national commodity and service trade organizations provide educational, legislative, and informational services for affiliated members. Examples include the National Telephone Cooperative Association and the National Milk Producers Federation—each with headquarters in Washington, DC; the National Livestock Producers' Association located at Chicago, Illinois; and the National Pork Producers Association, headquartered in Des Moines, Iowa.

State Cooperative Councils

State cooperative councils, sometimes referred to as federations or associations, are statewide associations for cooperatives in 40 states. Some are organized to serve only agricultural cooperatives—others serve both agricultural and consumer cooperatives.

Most state councils are also members of the National Council of Farmer Cooperatives and the American Institute of Cooperation. They are organized to solve mutual problems of members; to monitor state legislative matters important to cooperatives; to assist cooperatives in improving their business operation through such means as member, director, manager, and employee educational programs; to be a public spokesperson for cooperatives; and to provide educational materials and information to educational institutions and the public in general.

State cooperative councils differ, however, in methods of financial support, in staff size, and in their general activity on behalf of and for members. They also differ in choosing the types of cooperatives they serve and the programs they sponsor.

National Farm Organizations

In this chapter we will not discuss national farm organizations such as the American Farm Bureau Federation, National Farmers Union, National Grange, or National Farmers Organization, because they are general farm lobbying and educational organizations in which farmers hold direct memberships. You should be aware, however, that for the most part these organizations support cooperatives.

COOPERATIVE PROFESSIONAL ASSOCIATIONS

Several national associations have been organized by employees of cooperatives to enhance their performance in their particular specialty or profession. The formality of the organizations ranges from full-fledged corporate entities to committee-directed activities. More informal organizations include the Advertising Council for Cooperatives International, economists and planners who meet annually, and members of the legal profession. The more formal organizations are discussed briefly below.

The National Society of Accountants for Cooperatives (NSAC) plays a critical role in disseminating technical information on accounting, tax, and legal issues. A paucity of these topics exists in standard tax and legal publications and in university-level textbooks. The organization, created in 1936, holds annual meetings and publishes a quarterly professional journal, *The Cooperative Accountant*. Most of its over 2,000 members are accountants and attorneys. It is organized into 13 districts, which also hold training and professional meetings.

Cooperatives Communicators Association was organized in 1952 to improve the effectiveness of cooperative communications, to advance professional standards, and to enable an interchange of facts, research techniques, and opinions among cooperative communicators.

Association of Cooperative Educators is composed of professionals in cooperative education and training. The focus of ACE is on advancing cooperative education in light of current cooperative operations and practices; promoting more effective use of techniques, materials, and methods; and improving professional capabilities and knowledge.

GOVERNMENT ORGANIZATIONS

Functions and actions of many government agencies affect cooperatives in the same way as they affect other businesses. One independent agency regulatory in nature is the Farm Credit Administration. On the other hand, the U.S. Department of Agriculture has a unique role in support of cooperatives. Several USDA agencies share this role, but three are particularly important: the Agricultural Cooperative Service, the Cooperative Extension Service, and the Rural Electrification Administration.

Agricultural Cooperative Service

The Cooperative Marketing Act of 1926 directed the establishment in the USDA of a division authorized to promote the knowledge of cooperative principles and practices and to cooperate with educational and marketing agencies, cooperatives, and others in promoting such knowledge. The division carrying out these functions is the Agricultural Cooperative Service (ACS), formerly the Farmer Cooperative Service. ACS provides research, management, and educational assistance to existing cooperatives and helps farmers and other rural people organize new cooperatives. This agency is the only organization (other than the Bureau of the Census) that collects nationwide data on the business activities of agricultural cooperatives. ACS then publishes its research and educational materials in several series of bulletins. Summaries of some of these materials and items of current interest are published in its monthly magazine entitled *Farmer Cooperatives.*

Cooperative Extension Service

Land-grant colleges were established through the passage of the Morrill Act of 1862. The Hatch Act of 1887 authorized federal support to each state that would establish an agricultural experiment station in conjunction with its land-grant college. These experiment stations developed useful new technology; however, it soon became obvious that this new technology was not getting to people in rural communities. In response to this need, the Cooperative Extension Service was established in 1914 by the Smith-Lever Act. This act provided an organizational structure enabling the Cooperative Extension Service to be funded cooperatively by the federal, state, and local governments and administered by the land-grant universities in each state. From early in their establishment, research projects and educational programs on agricultural cooperatives were undertaken by the land-grant colleges. The Cooperative Extension Service has played and continues to play a key role in carrying out educational programs that benefit cooperatives and in disseminating land-grant college research findings to cooperatives, to their member-patrons, and to the public.

International Cooperative Alliance

International Cooperative Alliance (ICA) headquarters in Geneva, Switzerland, is the apex organization of the world cooperative movement. After more than eight decades of activity, ICA remains the chief international organization dedicated solely to the promotion of cooperation. It provides technical assistance for ICA member organizations through its three regional offices: in New Delhi for southeast Asia; in Moshi, Tanzania, for east, central, and southern Africa; and in Abidjan, Ivory Coast of West Africa. ICA organizes seminars, workshops, exchange programs, studies, and consultancies, as well as research and project identification. At the head office, the Cooperative Development Fund and the Bonow Fund provides financial aid to small projects and scholarships. The ICA also channels applications to UNESCO for travel grants for studies in the field of cooperative education.

Committee for the Promotion of Aid to Cooperatives

The Committee for the Promotion of Aid to Cooperatives (COPAC) is a liaison body of UN agencies and international nongovernmental organizations set up to promote and coordinate assistance to cooperatives in developing countries. It was established in 1971. Membership includes the Food and Agriculture Organization of the United Nations (FAO), International Cooperatives Alliance (ICA), International Federation of Agricultural Producers (IFAP), International Federation of Plantation, Agricultural and Allied Workers (IFPAAW), International Labour Office (ILO), United National Secretariat (UN), and World Council of Credit Unions (WCOCCU).

Headquarters for COPAC is Rome, Italy. The Committee, consisting of a representative of each member organization, meets twice a year. The chairmanship is assumed by the representative of each member organization in rotation.

Originally concerned specifically with agricultural cooperatives, COPAC's activity now embraces all types of cooperative enterprises. COPAC sets up a forum for action-oriented consultation between agencies and organizations. The aim of such consultations is to exchange full information on programs and plans for cooperative development, to avoid duplication, and further, to ensure the greatest degree of complementary between programs. COPAC itself does not provide capital or technical assistance, although most of its individual member organizations do so.

SUMMARY

Cooperatives have seen the need to organize educational, research, technical assistance, public relation, and trade-lobbying organization to solve common problems and to provide other services.

The National Council of Farmer Cooperatives is the political voice for farmer cooperatives in Washington, DC. The American Institute of

Cooperation promotes educational programs on agricultural cooperatives and encourages research on cooperative issues. The National Cooperative Business Association is a federation of all types of cooperatives, agricultural and consumer cooperatives, and offers educational, training, and legislative services as well as promoting the development of cooperatives both at home and abroad.

Several states have state cooperative councils that perform similar educational, legislative, and cooperative promotional activities at the state level as national trade associations do at the national level. Several national associations have been organized by employees of cooperatives to enhance their performance in their particular specialty or profession.

The U.S. Department of Agriculture has a unique role in support of cooperatives. Two USDA agencies are particularly important to cooperatives: the Agricultural Cooperative Service and the Cooperative Extension Service.

DISCUSSION QUESTIONS

6-1.   Contrast the purpose and operation method of the AIC, the NCF, and the NCBA.

6-2.   What role has the Cooperative Extension Service played regarding cooperatives?

6-3.   What are some of the activities of the Agricultural Cooperative Service?

6-4.   Why have cooperatives in some states seen the need to organize state cooperative councils?

6-5.   What are the advantages of being affiliated with the National Council of Cooperatives?

6-6.   Do state cooperative councils have any influence at the national level?

7

Historical Development

Gene Ingalsbe,
Agricultural Cooperative Service, USDA

Frank Groves,
University of Wisconsin-Madison

WORLDWIDE DEVELOPMENT

Organized cooperation for recognized mutual benefit is so instinctive that discovering its beginning is impossible. However, we can estimate that cooperation as a business concept with a distinct set of operating principles is a relative infant of roughly a century and a half.

Cooperation as an idea is as old as civilization. No doubt, when human beings first started to live in groups, to domesticate animals, and to cultivate the land, they learned many tasks could be better accomplished through group action than by individual effort. Some of the earliest forms of cooperation probably occurred in group animal hunts, in united efforts to achieve mutual protection, and in community farming.

Ancient records and archeological discoveries in fact provide evidence that cooperative activity was common in early civilizations. For example, many years before the birth of Christ, the Chinese developed sophisticated savings and loan associations not too different from those we have today. In addition, Babylonians developed a way for farmers to cooperate and farm together, and craft and burial societies were common among ancient Egyptians, Greeks, and Romans.

During the dark ages, the idea of cooperation was kept alive primarily only in monasteries, but as Europe moved into the Renaissance, cooperation as an organizational concept emerged again. Expansion of trade and world exploration, coupled with the explosion of technology, marked the beginning of the industrial revolution and had a profound impact on business organization, operation, and ownership. Sums of money beyond the individual means of even the wealthy were needed to carry out economic endeavors. So the joint stock company was born, signaling the beginning of a new kind of cooperation—cooperation among business people.

For all its benefits, the industrial revolution also brought great social upheaval drastically altering the way people worked and even lived. The factory was born. People who were accustomed to working in small groups and frequently in their homes were now forced to make a living in a factory environment. Child labor was common. Working conditions were unbearable. Hours were long and wages were low—people often worked six days a week, 12 hours a day, for a few pennies. Many people advocated different economic systems as solutions for these problems. The economic, social, and political structures that evolved from these solutions are the source of some confusion and misconceptions about cooperatives in the United States.

The characteristics of cooperatives that operate today in the United States and in many parts of the world were formulated and identified at the same time that basic ideas of socialism and communal colonies were developed. Hence, in the minds of many people, cooperatives are in bad company. Adding to the confusion, cooperatives may vary greatly under different political systems. Some organizations called cooperatives in other countries are indeed communal or socialist in structure; however, they fail to meet the criteria of cooperatives in most capitalistic countries. U.S. cooperatives are an integral part of the nation's capitalistic private-enterprise system. In the United States, as in most capitalistic countries, ownership can be traced to individuals (private property), membership is voluntary, control is in the hands of members rather than the government, and benefits are distributed to member-patrons on a patronage and, to a lesser extent, an ownership basis.

The concepts evolving in response to the industrial revolution took off in vastly different directions. Some of these led to socialism and the destruction of capitalism, as advocated by Karl Marx and Frederick Engels. The Marx-Engels ideas eventually led to the communist brand of socialism practiced in the Soviet Union, eastern Europe, and China. Related concepts were practiced in communal colonies, as envisioned by Robert Owen, Charles Fourier, Louis Blanc, and others.

Some historians have called Owen the father of cooperation because several of the characteristics in his communal new-society villages have survived and continue to be espoused as principles or practices of cooperatives (Abrahamsen). Owen, a successful British industrialist and philanthropist, established a number of colonies on the theory that a society could be built around the principles of cooperation.

Owen and his colonies are recognized for aiding in the formative development of cooperatives because of their policies and operational characteristics of (1) voluntary membership, (2) democratic control, and (3) service to members, rather than operation for profit. However, Owen believed in joint ownership of property and relied heavily on other philanthropists' contributions for capital. He advocated that the colonies should be self-contained and not trade with each other and that they should be led by a guiding hand (his) until the colonies could operate in the manner he desired. None of the Owenite colonies lasted very long.

The most notable Owenite experiment in the United States was the New Harmony community in Indiana, 1825-1828. Owen purchased a 20,000-acre site that included a community developed by the Rappists, a religious sect. It included the facilities for a complete community, such as houses, community centers, a church, and factories or mills to produce goods.

Members were given work assignments by a committee. They received credits for their labor and debits to their account for goods provided from the public store. A balance was determined at the end of the year for each member, but no cash was paid without the consent of the committee. Members could give a week's notice to leave the community and withdraw their balance.

New Harmony was envisioned as a community of equality in every respect. The basic lesson learned was that people are not equal nor do all of them desire to be. Knapp observes that under Owen's scheme, there was "no penalty for idleness and no reward for industry and thus no outlet for ambition" (Knapp, 1969, p. 17).

In France, social philosopher Charles Fourier sought a system that would better balance the economic relationships between labor, capital, and talent. Contrasted to Owen's desire to eliminate competition and individual profit making, Fourier was more interested in producer's rights (Carr-Saunders et al., p. 27).

Fourier proposed planned communities or phalanxes, but they provided for individualized rewards to labor and payment of interest for the use of capital. About three dozen phalanxes were organized in the United States, of which two were most notable—Brook Farm in Massachusetts, 1842-1846, and Wisconsin Phalanx in Fond-du-Lac County, 1845-1850.

Brook Farm involved a group of intellectuals, including George Ripley, Margaret Fuller, Ralph Waldo Emerson, and Nathaniel Hawthorne. Their experiment grew out of a social consciousness of the ills of competition. Brook Farm emphasized voluntary participation, but the community began to unravel when it became necessary to assign tasks to keep it going. Reorganizing it as a phalanx extended its life only two years. Knapp (1969 p. 19) cites Brook Farm's significance in the realm of ideas. He notes that "by highlighting the communal weaknesses of utopianistic colonies, while forcing attention on the economic and social problems of the period, it gave encouragement to less grandiose and more practical schemes of cooperation."

Wisconsin Phalanx exhibited stronger similarities that can be identified with today's cooperatives. Physical structure was similar to New Harmony in that the community featured a school, employees' housing, and occupational facilities such as a gristmill, sawmill, and blacksmith shop. Its uniqueness involved an annual appraisal of real and personal property in which excess over cost was credited one-fourth to capital stock and three-fourths to labor. Failure of the phalanx was attributed in part to the subsequent conclusion that assigning any dividend to capital produced an irreconcilable conflict between recognizing the need to reward members according to their contributions and the utopian cooperation in which everyone was treated the same.

Independent of community experiments, several small self-help economic enterprises sprang up in the United States and in several countries of Europe during the latter part of the eighteenth century and the first part of the nineteenth century. Among these groups were penny capitalists in Scotland, U.S. colonists in protective unions, consumer societies in England, credit unions in Germany, and employees' cooperatives in France. The list of earliest known cooperatives in Table 7.1 is evidence of many origins of cooperative principles and their application to an economic endeavor. However, the success of one particular group of workers representing a variety of trades had perhaps the greatest singular impact on determining agricultural cooperatives' unique operating characteristics. In 1844, this group—the Rochdale Society of Equitable Pioneers, Ltd.—formed a consumer cooperative in Rochdale, England, and formulated a set of basic operating rules from a two-year study of cooperatives, including some that had failed. The unusual success of the Rochdale Pioneers attracted the attention of various writers, who studied its rules of operation to find out what made it so successful (Erdman). George Jacob Holyoake wrote three books (the first in 1858) about this innovative society. The books were eventually translated into many languages. A U.S. edition of the first book was published in 1859 by Horace Greeley, editor of the *New York Tribune*. Holyoake's practical guidebook on cooperative organization presented fourteen primary attributes of the Rochdale System. Subsequent historians have refined them into fewer principles and practices, which are discussed in Chapter 2.

TABLE 7.1 Earliest recorded cooperatives in selected countries

Year	Country	Type
1696	Great Britain	Fire insurance
1752	United States	Fire insurance
1816	Poland	Agriculture
1842	Spain	Consumer
1848	Belgium	Bakery
1849	Germany	Credit
1850	Sweden	Consumer
1851	Norway	Consumer
1853	Italy	Cattle insurance
1863	Bulgaria	Credit
1866	Denmark	Consumer
1876	Netherlands	Consumer

As the Rochdale system was proving its worth, other experiments of economic cooperation were occurring in Germany. These would have significant influence on the development of cooperative enterprise in the United States. Herman Schulze-Delitzsch, beginning in 1852, and Friedrich Wilhelm Raiffeisen, in 1864, were developing credit cooperatives, the Landschaften Cooperatives (forerunners of our Federal Land Banks), and credit unions. Raiffeisen is generally given credit for developing the rules for the operation of modern-day credit unions. Credit cooperatives, including both the cooperative farm credit system and credit unions, make up half of all types of U.S. cooperatives in both association numbers and membership.

AGRICULTURAL COOPERATIVE DEVELOPMENT IN THE UNITED STATES

Cooperatives account for a larger share of total economic activity in Europe than they do in the United States. However, they are nonetheless a fundamental part of American tradition—both past and present—especially in agriculture.

Pilgrims coming to the new world on the *Mayflower* in 1620 signed the Mayflower Compact, whose rules and regulations described the operations of an organization (or constitution) with cooperative characteristics. They immediately applied informal cooperative activity to clear land, to build homes and communities, to start farming, and to protect themselves.

In 1752, nearly a quarter-century before the birth of the country, Benjamin Franklin, one of the signers of the Declaration of Independence, helped start what is considered the first formal cooperative business in the

TABLE 7.2   Selected early cooperatives in the United States

Year	Cooperative
1752	Philadelphia Contributionship for the Insurance of Houses from Loss by Fire, PA
1810	Dairy at Goshen, CN, and cheese at South Trenton, NJ
1820	Hog marketing, slaughtering, packing, Granville, OH
1853	Irrigation, Tulare County, CA
1857	Grain elevator, Madison, WI
1862	Tobacco marketing, CT
1863	Purchasing, Riverhead, NY
1867	Fruit marketing, Hammonton, NJ
1874	Poultry marketing, IL
1877	Cattle rustling protection, TX
1885	Citrus (now Sunkist Growers, Inc.), CA
1887	Cotton gin, Wagner, TX

United States (Ingalsbe). Franklin already had formed the Union Fire-Fighting Company in 1736, which became the model for volunteer firefighting companies. To further mitigate losses from fire, he and other members of the firefighting association then formed in 1752 the Philadelphia Contributionship for the Insurance of Houses from Loss by Fire, which continues to operate today.

Group activity specifically among farmers began with the idea of improving agricultural techniques. The first known organization in the United States was the Philadelphia Society for Promoting Agriculture, created in 1785. However, the first formal farmer cooperatives were formed in 1810—a dairy cooperative started in Goshen, Connecticut, and a cheese manufacturing cooperative in South Trenton, New Jersey. On the heels of these earliest organizations, others involving different commodities were formed in many parts of the country in a sporadic and independent fashion (Table 7.2).

Growth Factors

Many factors have been cited as major influences in the development of agricultural cooperatives, but these factors logically can be grouped into three types, all interrelated: (1) economic conditions, whether produced by war, depression, technology, national economic policy, or the marketplace; (2) farmer organizations, including their leadership, their motivational capability, their enthusiasm for promoting cooperatives, their effectiveness in influencing public policy, and their longevity; and (3) public policy, as determined by presidential interest, by legislative initiative (at both state and federal level), and by judicial interpretation.

Not until after the Civil War were cooperatives recognized as a feasible, self-help business alternative by agricultural producers. An early developmental factor was their promotion by the Order of the Patrons of Husbandry (commonly known as the Grange), the first of several general farm organizations. Founded in 1867 by a U.S. Department of Agriculture employee named Oliver Hudson Kelley, the Grange was conceived as a fraternal order to restore good feeling between people of the north and south.

Due to economic conditions of the time, however, the Grange's social function was quickly subordinated to an emphasis on cooperatives for economic improvement. Difficulties with early attempts at forming cooperatives led the Grange to adopt the Rochdale system in 1875 and to distribute its rules to Grange stores. A burst of new cooperatives and reorganization of earlier joint-stock stores was short-lived as the Grange began to decline.

The Grange continues to exist today, occupying a philosophical position between the National Farmers Union and the American Farm Bureau Federation. Although it is smaller now than it was originally, historian Joseph G. Knapp (1969) credits the Grange with several significant contributions. The organization, claims Knapp, has (1) popularized cooperatives as a business concept throughout the nation; (2) proved the power of cooperative action; (3) demonstrated the value of the Rochdale principles, showing they could be applied to marketing as well as purchasing; and (4) identified the necessity of sound business management and operations.

As the Grange declined in influence, other farm organizations took their turn at fostering the development of cooperatives. Major ones, though short-lived, were the Farmers Alliance, formed in 1875, and the American Society of Equity, formed in 1902.

Another major farm organization, formed in 1902 and existing today, was the Farmers Educational and Cooperative Union of America, which evolved into the National Farmers Union.

Today's largest farm organization, the American Farm Bureau Federation, began in 1919. The American Farm Bureau and the National Farmers Union, by their long-term survival and organizational strength, have contributed the most among the farm organizations to the development of farmer cooperatives. Their support has been both in developing economic organizations and in exerting influence in the political arena.

Several of the largest cooperatives today have a farm organization heritage. Among these are Union Equity Cooperative Exchange, Inc., of Enid, Oklahoma; CENEX (Farmers Union Central Exchange, Inc.), St. Paul, Minnesota; and Harvest States Cooperatives (formerly Farmers Union Grain Terminal Association), St. Paul, Minnesota. Farm Bureau fostered Growmark, Bloomington, Illinois; Indiana Farm Bureau Cooperative Association, Indianapolis; and Nationwide Insurance Companies, Columbus, Ohio. Consolidations of organizations with Grange and Farm Bureau beginnings resulted in Agway Inc., Syracuse, New York. One statewide organization that sponsored many cooperatives is the Missouri Farmers Association, which operated for many years as a farm organization with an affiliated federation of cooperatives.

Many cooperatives, some of the largest operating today, developed largely independent of farm organization backing. In a few cases, neutrality with respect to farm organizations was intentional, viewed as an important practice synonymous with political neutrality. Such a strategy allowed the crossing of farm organizational lines to build cooperative organizations large enough to manufacture farm supplies as well as to process farm products. One such cooperative that has followed this strategy is Farmland Industries, Inc., Kansas City, Missouri.

In other situations, the nature of the commodity, such as its perishability or its compressed production region, was the greater influence. Examples are large cooperatives such as Associated Milk Producers, Inc., San Antonio, Texas; Ocean Spray Cranberries, Plymouth, Massachusetts; and Blue Diamond Growers, Sacramento.

Federal Government Support

As discussed in Chapter 5, public policy, as exercised by governments, has significantly influenced how cooperatives have evolved. Four of the five attitudes (Figure 5.1) have been present at either the state or national level at some time throughout the history of U.S. cooperatives. Before 1890 the U.S.'s policy toward cooperatives was level 2 (neutral or laissez-faire). Until that time there were not many cooperatives and what few there were functioned without the benefit of any government attention. Then beginning in 1890 with the passage of the Sherman Antitrust Act, attempts were made to in effect declare farmer cooperatives illegal through court action. Some of these attempts were inadvertently placing the United States at level 1 (antagonism or destructive). Beginning in 1914 the pendulum swung to level 4. Not only were laws passed to create a favorable environment, but the government became involved in the creation and funding of cooperatives in agricultural credit.

In two books covering the history of U.S. cooperatives from 1620 to 1945, Joseph G. Knapp documents the impact of government policy and its interrelation with other cooperative growth factors. He (1973) observes that, beginning in 1920, cooperative marketing activity doubled by 1925. Postwar depression was so severe that the agricultural situation captured the attention of three successive presidents—Warren Harding, Calvin Coolidge, and Herbert Hoover. As part of that attention all three presidents made strong endorsement for agricultural cooperatives.

The stage had been set by earlier presidential commissions. In 1908, Theodore Roosevelt had established the Country Life Commission, which called attention to the success of credit cooperatives in Europe. In 1913, Woodrow Wilson had sent a commission to Europe to study cooperative development. As a result, the Federal Farm Loan Act was passed, setting in motion events leading to the establishment of the cooperative Farm Credit System.

Major activity took place in the land-grant university system developing from the passage of the Morrill Land-Grant College Act in 1862. Also in 1913, the U.S. Department of Agriculture established an Office of Markets that broadened research and advisory programs to cooperatives, thus allowing them to conduct the first national statistical survey. In 1914, the Smith-Lever Act formalized cooperative agricultural extension work in the U.S. Department of Agriculture. Many cooperatives today owe their existence to the extension system, which continues to provide major educational, research, and technical assistance to agriculture and to cooperatives.

Government encouragement of cooperatives approached its most active stage directly during the period 1920–1932 and indirectly for an additional five years afterward. Most states established cooperative statutes during this period. Nine federal actions significantly promoted and supported cooperative growth.

The Capper-Volstead Act of 1922 especially facilitated cooperative growth by clarifying antitrust law treatment of farmers' agricultural cooperatives. The act permitted farmers to "act together in associations" to process and market their products collectively, and to form marketing agencies-in-common. Previously, farmers had been prosecuted under antitrust laws for collective action, particularly concerning agreements on pricing and terms of trade. However, antitrust provisions concerning business conduct continued to apply to cooperative operations in the same manner as for other businesses. In subsequent court decisions, cooperatives found they lost even their organizational exemption when they combined operations with other business types.

Congress was concerned about the implications of the broad mandate for growth that it was giving farmers. Realizing farmers could create a monopoly, Congress included a section 2 in the act, providing judicial redress if farmers' organizations used a monopoly position to "unduly enhance price." No other type of business was or is subject to such a consumer protection law.

For the most part, however, congressional action during this time promoted cooperative business organizations in agriculture. The Cooperative Marketing Act of 1926 especially broadened and formalized the U.S. Department of Agriculture's support and encouragement of farmer cooperatives. This act continues to be the legislative authority for USDA's Agricultural Cooperative Service. It directed USDA to create a division of cooperative marketing for the purposes of (1) collecting and analyzing historical and statistical data; (2) conducting studies of cooperation, particularly regarding economic, legal, financial, and social aspects; (3) surveying and analyzing business practices; (4) helping farmers organize cooperatives; (5) collecting and analyzing data on agricultural production and marketing; (6) promoting the knowledge of cooperative principles and practices; and (7) conducting other studies in the United States and in foreign countries that could prove useful to cooperative development.

The Agricultural Marketing Act of 1929 also included several provisions important to cooperatives. It provided for advisory commodity committees from cooperative associations. It established the Federal Farm Board, with a major policy of expanding and strengthening the cooperative movement. And it provided a $500 million revolving fund to make loans available to cooperatives for merchandising, acquiring marketing facilities, forming clearing house associations, extending membership, and extending credit to enhance farm prices.

The parent organization of today's Agricultural Cooperative Service was transferred from USDA to the Federal Farm Board by the 1929 act and to the Farm Credit Administration in 1933. Then it was returned to the USDA in 1939 when the FCA became part of the USDA. Thus cooperatives were poised to take advantage of the opportunities that would be presented by the New Deal administration of President Franklin D. Roosevelt. Knapp observes:

> The New Deal was a response to a crisis so intense that it called for deep-seated measures and long-applied effort. The concept of cooperation was endemic, for the New Deal represented a national effort to work together for national goals. Government agencies brought forth by the New Deal had a profound effect on cooperative enterprise. The Agricultural Adjustment Administration [1933] relieved marketing cooperatives of the major responsibility for production control and reemphasized the importance of marketing efficiency. The Farm Credit Administration [1933] broadened and strengthened credit services for agricultural cooperatives and gave encouragement to the use of cooperative business organizations. The Tennessee Valley Authority [1933] demonstrated the feasibility of cooperative rural electrification, thus opening the way for a national system of rural electric cooperatives financed through the Rural Electrification Administration [1935]. Moreover, TVA helped build the foundations for the development of regional and national fertilizer cooperatives. The Resettlement Administration (1935) and its successor agency, the Farm Security Administration (1937), made use of cooperatives for rural rehabilitation and gave stimulation to the cooperative idea (Knapp, 1973, p. 225).

State support of cooperatives has varied, also, usually following the lead of federal policy. Some states encouraged the formation of cooperatives through state extension services and state departments of agriculture.

Other states have taken a more neutral attitude. Differences in state attitudes toward cooperatives continue today. All together, states have 85 different statutes relating to the operation of cooperatives.

Cooperative Growth in Response

During the decades of the 1920s and 1930s when government encouragement and support were highest, cooperatives took steps to form organizations that would serve them well when government moved toward a more neutral position. In addition to forming regional and interregional economic organizations, farmers formed national associations to provide commodity information, to offer educational and other services, and to influence legislation.

Some examples of these associations are the Cooperative League of the USA (now the National Cooperative Business Association) and the National Milk Producers Federation, formed in 1916; the American Institute of Cooperation, formed in 1925; the National Council of Farmer Cooperatives, formed in 1925; and the Credit Union National Association, formed in 1934. Electric cooperatives followed the trend of organizing national trade associations a few years later by establishing the National Rural Electric Cooperative Association in 1942.

During this period of cooperative growth, two major schools of thought concerning the organizational approach and the role of cooperatives had developed. These were the Sapiro monopoly approach and the Nourse yardstick approach. A third idea, the cooperative sector approach, was present in Canada and in several European countries.

Aaron Sapiro was a California attorney who worked with cooperatives in the early part of this century and through the 1920s. As a lawyer, he emphasized a legalistic approach to cooperative businesses, advocating the formation of legal monopolies by agricultural producers on a commodity basis. He was a forceful and dynamic speaker who was able to sway large numbers of farmers toward his way of thinking. As a result, during the 1920s many cooperatives were formed based on the ideas of Sapiro. The major points he advocated were as follows:

1. Cooperatives should be organized around a single commodity such as wheat or tobacco.

2. Membership should be restricted to agricultural producers.

3. A cooperative should maintain long-term contracts with its members.

4. A high proportion of the farmers producing a particular commodity should sign the contracts before the contracts take effect.

5. Products should be pooled on grade, and the members should receive an annual average price based on the grade of their particular product.

6. Cooperatives should be organized on a nonstock basis.

7. Cooperatives should adopt sound merchandising principles and techniques.

8. Marketing should be orderly and spread over the course of the year, not concentrated at harvest time.

9. Cooperatives should be organized on a centralized (rather than federated) basis.

A somewhat different approach was advocated by Edwin G. Nourse, an Iowa State University professor and agricultural economist. He felt cooperatives should operate in the capitalistic system as a competitive yardstick, that is, as efficient businesses that would keep other businesses in line and make the whole economic system operate more efficiently. The following list outlines Nourse's major guidelines for establishing such cooperatives:

1. Establish cooperatives with a bottom-up democratic base.

2. Develop cooperatives as an integral segment of the existing capitalistic system.

3. Build business efficiency of the total economic system.

4. Control a modest share of a commodity, supply, or service market.

5. Keep other business forms competitive with cooperatives, serving as a balance wheel or yardstick and thereby helping correct many of the excesses associated with capitalism.

6. Preserve individual producer freedom of decision making.

7. Limit government assistance to cooperative development primarily to enabling legislation, or, in other words, to establishing the legal framework for cooperatives to organize and cooperate.

The cooperative sector approach was advanced mainly by Moses Coady and Alex Laidlaw in Canada and by Jerry Voorhis in the United States. This idea is also supported in the Scandinavian countries, in Israel, and in Japan. Laidlaw described the cooperative sector model as follows: "A cooperative sector concept merely says that cooperatives will strive to do only those things they can do best, and they will leave the rest to others, either public enterprise or private business. In many situations, cooperatives will be content to carry on in a state of peaceful coexistence with the other two. This is the whole idea of the mixed economy."

Another approach was a *commonwealth* or *cooperative democracy* in which cooperatives would become the predominant form of economic activity in all sectors of the economy. This approach has few supporters today, however.

Cooperative Business Concept Matures

World War II ushered in the nuclear age, and another explosion of technology led by television and computers hastened the acceptance of

TABLE 7.3 Total farmer marketing, purchasing, and related service cooperatives

Year	Number of co-ops	Membership	Business volume (millions)
1900[a]	1,233	[b]	$ [b]
1915[c]	5,424	651,186	1
1921	7,374	[b]	1,256
1929–30	12,000[d]	3,100,000	2,500
1935–36	10,500	3,660,000	1,840
1940–41	10,600	3,400,000	2,280
1945–46	10,150	5,010,000	6,070
1950–51	10,064	7,091,120	8,147
1955–56	9,894	7,731,735	9,756
1960–61	9,163	7,202,895	12,409
1965–66	8,329	6,826,275	15,608
1970–71	7,995	6,157,740	20,556
1975–76	7,535	5,906,379	40,051
1980	6,282	5,378,888	66,254
1985	5,625	4,781,216	65,601
1986	5,369	4,401,757	58,395

[a]Earliest known figure published as part of USDA report.
[b]No data.
[c]First nationwide co-op survey by USDA.
[d]Largest number of co-ops according to USDA.

cooperatives as a mature business concept. Boom decades of the 1950s, 1960s, and 1970s brought enormous growth in their business activity (Table 7.3).

Almost continuous record sales volumes were achieved despite a continuing numerical decline in agricultural cooperatives (from a peak of 12,000 in 1930) and in members (from a peak of 7,731,735 in 1956) (Kraenzle and DeVille). Although farms, farmers, and cooperatives became fewer in number, remaining farmers gradually increased their use of cooperatives. Individual cooperatives began to show up on lists of the largest corporations in the country, with more than two dozen among the largest 1,000 by the mid-1980s.

The characteristics of cooperative development over the past 40 years can be summarized as follows:

1. Government's approach to cooperatives has been to maintain an interest in cooperatives with minimal support in the form of advice, education, and broad research.

2. More of an arms-length relationship, although for the most part friendly and complementary, has developed between general farm organizations and cooperatives. Only one major new organization—the National Farmers Organization (NFO), established in 1959—has developed. It operated as a bargaining association in competition with established cooperatives. Although NFO continues today, its influence has considerably weakened.

3. Cooperatives have maintained neutrality with regard to political parties, but they are assuming an increasing political action role to influence farm policy, macroeconomic policy, and the business climate.

4. Economic conditions, with a growing international dimension, have been the dominant force influencing cooperative structure and business strategies.

5. Larger cooperative size, whether achieved through internal growth or consolidation, has enabled major entry into the further processing of agricultural products.

6. Increasingly complex organizational structures have developed, including multiple subsidiaries, marketing agencies in common, joint ventures with cooperatives or other types of businesses, and a few international arrangements.

7. Although they have been involved in international trade for more than a century, cooperatives in the 1980s are now wrestling with alternatives for developing an international business strategy.

SUMMARY

Cooperative business enterprise developed in response to the social and economic upheaval brought on by the industrial revolution. Factory production disrupted the way people had lived and worked. Low wages and bad working conditions motivated people to look for ways to make life more pleasant. Some leaders also recognized the need for corrective action. This concern produced the building blocks of several economic, social, and political structures. Among them were socialism and communism, communal colonies, and self-help economic enterprises.

The idea of self-help collective action as an economic solution developed almost at the same time in several European countries. But basic operating rules guiding a cooperative to unprecedented success after its formation in 1844 in Rochdale, England, are credited with the greatest influence on the development of modern cooperatives.

Colonial United States survived largely on informal cooperation. The first recorded cooperative business, organized under the leadership of Benjamin Franklin, predated the Rochdale formation by nearly a century. Cooperatives have evolved as an integral part of the U.S. capitalistic private-enterprise system. General farm organizations played major roles in organizing cooperatives and bringing about a political climate favorable to cooperative growth. Cooperatives have matured to become a significant force in agriculture, developing integrated systems for operations and assuming an increasing role in influencing national agricultural policy.

DISCUSSION QUESTIONS

7-1. Cooperation has been characterized as both a young and an old idea. Explain.

7-2. What period in history, and what conditions, are cited as leading to the development of business cooperatives?

7-3. Compare the evolution and concepts of the Rochdale pioneers with those of Robert Owen.

7-4. What were some of the early forms of cooperation in the United States?

7-5. Who is credited with forming the first known U.S. cooperative, when was it formed, and what was its purpose?

7-6. Name three factors having major influence on the development of U.S. agricultural cooperatives.

7-7. What act clarified the antitrust law treatment of farmers' agricultural cooperatives?

7-8. Contrast the two major schools of thought that influenced the development of agricultural cooperatives.

7-9. Discuss the trends of agricultural cooperatives, concerning numbers, memberships, and business volume.

7-10. Summarize agricultural cooperative development of the past 40 years.

REFERENCES

ABRAHAMSEN, MARTIN A., *Cooperative Business Enterprise*. New York: McGraw-Hill, 1976.

CARR-SAUNDERS, A. M., P. SARGANT FLORENCE, AND ROBERT PEERS, *Consumers' Cooperation in Great Britain*. London: Allen & Unwin, 1938.

ERDMAN, H. E., and J. M. TINLEY, *The Principles of Cooperation and Their Relation to Success or Failure*. California Agr. Exp. Sta. Bull. 758, Univ. of California, 1957.

HOLYOAKE, GEORGE JACOB, *History of the Rochdale Pioneers*, 10th ed. rev. New York: Charles Scribner's Sons, 1893.

INGALSBE, GENE, *Cooperative Facts*. Washington, DC: USDA ACS CIR 2, 1987.

KNAPP, JOSEPH G., *The Rise of American Cooperative Enterprise: 1620-1920*. Danville, IL: The Interstate Printers & Publishers, Inc., 1969.

KNAPP, JOSEPH G., *The Advance of American Cooperative Enterprise: 1920-1945*. Danville, IL: The Interstate Publishers & Printers, Inc., 1973.

KRAENZLE, CHARLES A., AND KATHERINE C. DEVILLE, *Cooperative Historical Statistics*. Washington, DC: USDA ACS CIR 1, Sec. 26, 1987.

LAIDLAW, ALEX, *The Public, the Private, and the Cooperative Sector of the Economy*. Regina, Saskatchewan: Credit Union Central, 1976.

Part III
Economic Theory

8

*Economic
Justification*

Lee F. Schrader,
Purdue University

WHY A COOPERATIVE?

Why do farmers choose to organize and patronize cooperatives, and why does the public sanction or encourage the cooperative form of business? In a sense, the cooperative form of business needs no more justification than do the proprietorship, partnership, or corporate forms of business enterprise. The interesting issue is: Why do persons or firms at one level of a commodity system elect to extend their business into another level or levels of that system via a cooperative? A commodity sector or system includes the whole set of activities or functions from input supply to final users associated with the production and marketing of that commodity. Both terms ("sector" and "system") are used in the literature on this topic. We use the term *system*. Figure 8.1 illustrates the movement of products and general levels of activity in the corn commodity system.

Agricultural cooperatives usually extend farmers' businesses backward into input supply or forward one or more levels into marketing. Further, what is the impact of this integrative activity by farmers on others in the food and fiber sector? Although there are other motivations for

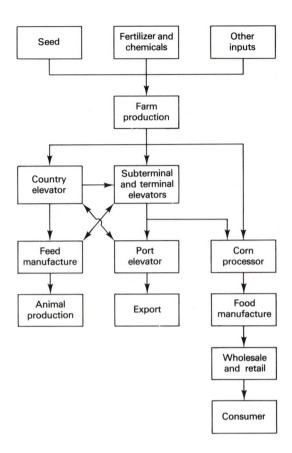

Figure 8.1 Illustration of an agricultural system.

cooperative activity, it is accepted that the primary motivation for farmer participation in a cooperative is to improve their well-being (usually income). Rational persons or well-managed firms would not voluntarily engage in activities that would be expected to leave them less well off. Economic growth is generally associated with increasing specialization and trade among economic units. This usually implies larger firms operating at each level of an economic sector, not the integration of activities from several levels into a single firm. The existence of economies of size up to some size, which is often larger than a typical proprietorship, encourages firms to expand horizontally. That is, if major economies of size exist, there is incentive to increase output of a given product or service rather than to expand by extending the firm into marketing or input production. Given this reasoning, one would expect a moderately sized farmer to invest available capital in expanding cotton production rather than in building a cotton gin, whether alone or as part of a group. Thus the choice by a group of relatively small farmers to invest capital (as a cooperative) in an operation at another level in the system rather than investing in the expansion of their farms begs for an explanation.

If markets for goods and capital were perfect, there would be no reason to expect any particular form of business organization (cooperative or other). The result of perfect markets would be a state in which prices would reflect marginal costs (the additional cost of producing an additional unit) throughout the system. Goods could not be reallocated to increase the satisfaction of market participants, given the existing distribution of income. Such a state is the end result of a competitive market system, but such a state has never occurred. If it existed, one could argue that there would be no economic reason for the cooperative business form. In other words, cooperatives exist because markets have failed to attain the competitive ideal. All reasons or justifications for special policies to promote cooperatives can be traced back to such failure. Markets fail for a number of reasons.

MARKET FAILURE

A primary characteristic of a competitive market is that there are a large number of active buyers and sellers on both sides of the market. But the production process for many products is such that plants must be large (and thus fewer in number) in order to produce at minimum cost. Farm equipment manufacture, fertilizer manufacture, petroleum refining, and food processing all represent cases in which technology dictates relatively small numbers on one side of a market. In these cases a large number of both buyers and sellers is not possible and a balance of buyers and sellers is unlikely. Power to price at a level other than cost is conveyed to firms on the small numbers side of the market. The oligopolist (one of few sellers) may price above cost to the disadvantage of buyers, and the oligopsonist (one of few buyers) may price to the disadvantage of sellers.

Even in cases where the economies of size are quite limited (e.g., retail fertilizer dealers or country elevators) the spatial dimension of a market may result in an imbalance of market power. The least-cost organization of firms marketing farm products and supplying inputs to farms may mean there will be only one firm in the best position to serve any given farm. The cost of assembly, conditioning, storage, and shipping of grain is illustrated in Figure 8.2. Note that the average cost per unit handled through an elevator decreases as the plant size increases (Chase, et al.). Also, the average cost of getting grain from farms to the elevator increases as the elevator size increases, because the grain must be assembled from a larger area. The plant size that minimizes the total cost of assembly and processing is often smaller than that which would minimize the cost of condition-

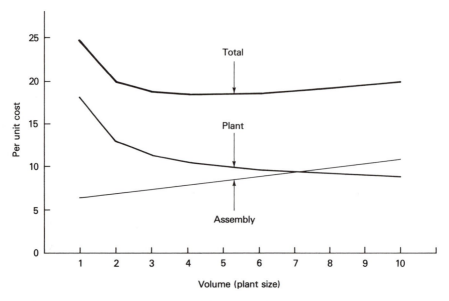

Figure 8.2 Costs of assembly and processing.

ing, storage, and shipping functions. More than one elevator at the same site would result in higher costs than those that would accrue if the same volume were handled in one facility. A cost-efficient system would have elevators serving nonoverlapping areas, and the individual farmer would have access to more than one buyer only by incurring added transportation costs. Even despite the appearance of a large number of elevators, there remains some possibility for prices to deviate from costs.

An efficient market also depends on the availability of information needed for efficient exchange and production. Having alternative trading partners is of no consequence unless the terms of trade available from each are known and understood. Information is available only at some cost, because it is not always in the best interest of buyers or sellers to facilitate the comparison of their prices or of any other aspect of their offers. Thus the cost of information, like the cost of transportation, may be the source of market imperfection.

Ideally, market prices provide all the information needed by firms to make appropriate decisions about what to produce (kind and quality), how much to produce, and what combination of inputs to use in production. A system of perfect markets would deliver the quantity and quality of goods at the time and place necessary to be of highest value to the ultimate consumer. However, the ability of market price to accomplish this task is complicated by a number of factors. In some cases a product attribute that is valued by a user can only be determined by inspection or testing that destroys the product. Some of these attributes may be determined and controlled only as part of the production process, and the user must therefore specify the production process to guarantee the quality of the product delivered. The need for such additional specifications increases the cost of market transactions, which may include inspection of processes and contract enforcement. These factors complicate the market exchange process and increase the likelihood of a failure of markets to attain the ideal.

Each of the problems cited above presents an incentive for firms to extend their activities into another level of the system and to internalize allocation decisions that might have been left to an open market. This internalizing may be accomplished by using contracts or by combining succeeding activities in the market channel into one firm. An operating cooperative represents a form of integration of farm production and other stages in the system. Bargaining cooperatives generally use contracts to deal with market failures. The sections to follow provide more examples of specific situations in which cooperative action might benefit farmers.

BENEFIT FROM ECONOMIES OF SIZE

Farmers who share in a gain in efficiency can enjoy a higher net income. As indicated in the preceding section, one may expect average in-plant costs to decrease as the size of operation increases for many of the input supply and product marketing firms serving farmers. Fixed costs of management are spread over greater volume, larger machines use less labor per unit, larger storage structures cost less per unit of capacity, and so on. At some point, as these plants get larger and larger, economies are offset by higher costs of product assembly or input distribution (Figure 8.2). But the least-cost size and location of these individual businesses is not consistent with the large number of farm product buyers and input sellers needed to ensure competition at the local level.

There probably is only one logically best buyer or seller for each farm location. This is often a rural condition common, particularly in areas of extensive agriculture such as Nevada, Wyoming, and North Dakota. In this case the buyer or seller is characterized as a spatial monopolist that has obtained monopoly power by locating in a market at a distance from competitors. Spatial monopolies usually exist when the size of the market in the firm's market area is too small, relative to economies of size, to support more than one enterprise and consumers are unwilling to travel to the next available source of supply—usually because costs are too high. The cost structure of providing services in many rural areas is similar to the utility industry. Costs of duplicating services (e.g., milk pickup and fuel delivery) are relatively high. Therefore, workable competition among two or more firms adds unnecessarily to the costs of a system. Cooperatives can serve in these situations without the fear of monopolistic profits being extracted.

Further, even if businesses are optimally located, the full benefits of their low costs are not always passed on to farmers, because the size of the market restricts the number of firms to one. These local monopolies or monopsonies have an opportunity to maintain margins above costs. For this reason, farmers may hesitate to encourage firms controlled by others to attain the least-cost organization if that leads to local monopolies. On the other hand, maintaining local competition might result in higher-than-minimum costs and margins, so farmers need another alternative to achieve low costs. A cooperative might provide a method of capturing the benefits of the more efficient organization in a way that investor-oriented firms (IOFs) would not. Operated at cost and controlled by farmers, it could bring the benefit of an efficient plant size to farmers without the adverse effects resulting from a loss of competition. With sufficient support, a cooperative can put in place the most efficient system with benefits for both farmers and consumers.

CAPTURE PROFITS FROM ANOTHER LEVEL

Farmers may perceive that firms at another level in their commodity system are earning larger returns to capital invested than are being earned at the farm level. This implies that the capital market is not efficient or that the nonfarm firms are able to prevent entry of competitors and thereby earn extraordinary (monopolistic) returns. One alternative is for farmers to invest in an existing firm at that level of the system that yields the desired higher returns. This may not be possible, however; and if it is, the asset value might already be inflated to reflect the higher earning level.

A farmer might also add the profitable function to the farm firm. Usually, however, the volume used or produced on a single farm is too small to match an efficient input supply or product processing operation. For example, a large egg producer may have sufficient volume to add grading and packing within the firm, but a hog producer would be unlikely to consider hog slaughtering as a feasible extension of his production operation. Another alternative is for like-minded farmers to organize a cooperative to initiate operations at the level where the above-normal profits are being taken. Entry of a cooperative enterprise could result in higher returns to the organizers' capital, probably in the form of prices more favorable to farmers.

The opportunity for such a gain arises because existing firms possess some market power and monopoly profits or because of differences in the opportunity cost of capital. Either market competition has failed to reduce profits of the higher earning firms or it indicates a failure of capital to leave production agriculture. In either case, entry of a cooperative enterprise may leave the system more efficient.

The capture of monopoly profits described in this section is closely related to the gain from economies of size discussed in the preceding section, in which it was implied that maintaining competition at a local level is likely to result in units that are below an efficient size. A gain from the capture of profits implies that margins may be high because of market power in the hands of firms serving farmers. These firms may be located and sized to be cost-efficient.

PROVIDE MISSING SERVICES

A commonly stated reason for existence of cooperatives is that they provide services (input or marketing) that otherwise would not be available. A new cooperative could provide a new service or buy a business that was not serving farmers well. Such reasoning, however, demands careful examination. If a real need exists, the filling of that need should provide a reasonable return for resources devoted to it. The fact that an entrepreneur does not see the opportunity to get such returns may be a signal that the need for the service is not real. To be of benefit, the service provided must have value which justifies resource use regardless of whether the providing organization is a cooperative.

It may be argued that unless there are other reasons for the service or product to be provided by a cooperative, an activity that cannot afford the ordinary investor a normal return would also be a mistake for a cooperative. To argue otherwise is to argue market failure, that the farmers opportunity cost of capital is less than considered normal by other firms, or that being a cooperative confers some advantage, such as lower taxes.

In the early 1970s, sugar beet growers in the Red River Valley of Minnesota and North Dakota formed a cooperative to acquire a sugar beet processing plant from an IOF corporation. The cooperative's organizers criticized the prior owners for being unwilling to expand, even though farmers wished to increase production of sugar beets because of their excellent return. Beet growers were represented by a bargaining cooperative which had some influence on the returns available to the beet processor. Instead of forming a cooperative, the beet growers might have perhaps allowed the former owner a higher return in order to induce expansion. However, the cooperative turned out very well for the organizing farmers, in part because sugar prices increased dramatically for other reasons shortly after the conversion to a cooperative.

ASSURE SUPPLIES OR MARKETS

The assurance of a source of supplies or a market for products is another reason closely related to but slightly different from those already mentioned for forming a cooperative organization. The primary difference is the consideration of risk. The question is not whether the service or product is available from an IOF but whether that source can be depended upon to place the needs of the farmer above those of all others. During the energy crisis of the early 1970s, domestic fertilizer prices were controlled at a level well under world prices. Much of the multinational IOFs' fertilizer output went to the high bidder, whereas farmer patrons had first call on the product of their cooperatives' output.

Access to a market is a major concern to farmers who produce perishable crops (such as fruits and vegetables) for processing. Farmers can lose their entire crop because of a loss of a market at a critical time. Therefore, membership in a cooperative may offer farmers more security than a year-to-year contract with an IOF processor.

The chance that a desired quality or quantity of service might not be available may also provide sufficient motivation for the cooperative patron to accept a lower rate of return on investment than that demanded by the outside entrepreneur. The degree of uncertainty is also related to the size and diversity of the firms involved. The small, limited-function, local firm probably has little alternative but to remain in business. On the other hand, the retail outlet owned by a large diversified firm may leave the farmer feeling much less secure about the continuity of the business segment he relies on. The element of control exercised by the user of a cooperative reduces this uncertainty.

GAIN FROM COORDINATION

The potential for gain from close coordination of inputs, production, and marketing appears to be large but mostly unrealized by cooperatives. Knutson found that what he called a committed integrated cooperative had the greatest potential for solving producer marketing problems. The common failure of markets to bring about close coordination is shown clearly in the cycles of price and production typical of many agricultural commodities. Periodically idle processing capacity and a failure to convey accurately to producers the quality desires of end users provide further evidence of market failure in these dimensions. It is difficult for market price to provide a clear signal to producers, especially when the quality and time dimensions of the products are complex. Futures markets, where they exist, provide some help in the time dimension but have not dampened the hog cycle. Another example of coordination failure is consumers desiring lean meat while the market appears to encourage the production of fat hogs.

Coordination of production and processing by means other than market transactions alone offers the possibility of increasing the value of production resources in several ways. If producers are committed to a particular processing operation, the need for buyers and the cost of search for raw materials is reduced. In addition, if the schedule of product delivery to processing is prearranged, scheduling of processing operations can be more precise and can be accomplished at a cost lower than it is if raw material arrival is more nearly random. A relatively constant and reliable flow of product allows the design of processing facilities to operate at minimum cost. There is also potential gain from production of crop or livestock varieties and sizes that allow lower-cost methods of processing or production of higher valued products.

Theoretically, businesses might achieve all these gains from coordination by using contracts just as well as they can by integration (establishing common ownership of the vertically adjacent stages). Indeed, contract coordination is common in businesses that process chickens (broilers), vegetables, and other products. But contract coordination can also be risky. Uncertainty regarding future events makes it very difficult to produce a long-term contract that will provide for all contingencies. Short-term contracts require frequent renegotiation and do not provide a firm basis for long-term investment. Also, since usually only a few processors deal with many producers, the fear of an imbalance of economic power may influence farmers to make decisions that achieve security for them.

Cooperatives have extended their operations backward into petroleum refining and fertilizer material production and forward from raw product processing into the manufacture of textiles and the marketing of consumer food products. But unfortunately, they have not exploited the potential for gain from coordination of farmers' and cooperatives' actions. Instead, the pattern of coordination between the member and the member's cooperative is generally similar to that of IOFs. For example, grain marketing cooperatives typically operate on a buy-sell basis similar to that of other types of grain marketing firms.

Gains in efficiency through cooperative coordination can be accomplished only with some decrease in the scope of decision making by the member. If the match between production and processing is to be enhanced, some direction from the processing level must be accepted by farmers. The member must be willing to commit a product to the cooperative and forego the option to pick the high day in the market or shop for the high bid at the last minute. The high value farmers place on independence has limited the extent to which cooperative coordination has been exploited. It may also be that a market failure is perceived where none exists.

BENEFIT FROM RISK REDUCTION

Uncertainty has already entered the discussion of cooperatives several times, and reducing individual uncertainty by risk sharing is clearly a part of cooperation. The extent of risk sharing varies depending on the cooperative. Risk sharing is the sole purpose of an insurance cooperative, for example, while operating cooperatives may share risks only by averaging unallocatable expenses. The process of sharing risks by an operating cooperative is illustrated by a feed truck that has a flat tire on the way to farm A. Common sense leads one to conclude that this is a cost of doing business in general, not a cost specifically attributable to the delivery to A. An IOF would probably take the same view. To the extent that any business bases price decisions on an average cost of performing a given function, there is also some risk sharing taking place.

Different levels of risk sharing are implied by the way in which patronage refunds are computed. Refunds based on earnings from a given product line involve less risk sharing across the broad group of patrons than calculating refunds separately only for farm supply and marketing patrons, the minimum separation required by tax law. At times, the result of usual cooperative practice is somewhat perverse. The members are, in effect, pooling their capital, but the returns to capital are allocated based on patronage. Thus the member whose crop failed also receives a smaller share of the cooperative's return to capital in that year.

Whether it takes the form of insurance, market pools, or costs averaged over lines of business, risk sharing ordinarily reduces the variation of returns to the individual member. A reduction in variation (risk) of returns, given the same average over time, is preferred if members are risk averse. The appropriate amount of risk sharing depends on the extent of risk aversion of members and the benefits of maintaining service at cost. In general, if variation in costs or returns results from the actions of an individual member, averaging costs or returns across members rewards behavior that leaves the group worse off. If the variation is not controllable by the individual member, averaging may increase the utility of the group through the reduction of uncertainty.

Risk sharing through cooperatives poses problems common to insurance programs. For example, there may be a tendency for adverse selection—that is, the most likely persons to participate may be the poorest risks. The poor marketer may be the first to participate in a marketing pool, the producer of low quality will be the first to join if there is averaging over quality, and so on. A second problem is called the moral hazard, which means that the averaging process reduces the incentive for individual participants to do their best. For example, when a marketing pool averages over quality classes that cannot be controlled by the grower, some growers may be less diligent in areas they can control because the price received (the average value) is influenced little by the quality of an individual lot.

But despite these problems, risk sharing is an important, valuable function of cooperatives, especially in light of recent transitions in farming practices. The trend toward specialized production to attain economies of size has resulted in the loss of some income stability provided by the diversity of crop and livestock enterprises on traditional farms. The extension of the farm business into processing via a cooperative enterprise may accomplish a degree of income stabilization, because profits from processing operations are often highest when production is high and farm prices (and profits) are low. Thus in a cyclical commodity system the total profit from production and processing may be more stable than the profits from either activity alone. A processing cooperative's net income would offset to some extent the variation in income from farming operations.

One must be aware, however, that this benefit from diversification is related to cycles of production and price. The argument does not apply in situations of general overcapacity such as the one that occurred during 1983–1987. Diversification of a farmer's capital investment into IOFs outside agriculture may be the more appropriate means to reduce the variation in a farmer's income due to the more general agricultural cycle.

MARKET POWER

In a prior section in this chapter we discussed the situation in which farmers organize to displace firms that earn above-normal profits as a result of their market power. In such a case, the entry of a cooperative tends to decrease the market power of other businesses in the commodity system involved. A cooperative may in fact be organized to acquire and exploit market power for the benefit of its members. Bargaining cooperatives and marketing cooperatives that either command a dominant share of a commodity at some level in the system or have been able to establish a brand with some degree of consumer loyalty may have attained some market power. Gaining some degree of monopoly power may be translated into more favorable terms of trade for the members than would be possible without the cooperative. The Sapiro school of cooperative thought advocates this approach.

Farmers' potential to influence the terms of trade by using market power appears to be large compared to the potential of the reasons for cooperation discussed previously. However, the ability to exercise such power is limited. Undue price enhancement is prohibited by the Capper-Volstead Act. The more serious limitation is the difficulty of exploiting market power by an open-membership organization. Cooperatives need some control of quantity in order to take advantage of market power. This is why a number of cooperatives such as Ocean Spray (cranberries) have found marketing orders helpful in maintaining the degree of control needed to manage a marketing program.

You should also be aware that gains from use of market power often come at the expense of other participants in a commodity system. The gains in efficiency discussed in the preceding sections, while initiated to benefit the cooperative organizers, operate to the advantage of consumers as well. Unfortunately, the use of market power to benefit the cooperative's members is less benign, often occurring at consumer expense. If there are other monopolistic elements in the system in which the cooperative operates, the acquisition of power by a cooperative may offset some of the undesirable actions of the already existing less-than-competitive elements. Little can be said about the impact of the accumulation of economic power at one level of a system when other levels are not competitive. Benefits would be expected to flow to the members of the cooperative in position to exercise some influence on terms of trade.

THE COMPETITIVE YARDSTICK

Some students of cooperatives consider all the roles of or reasons for cooperatives, except the acquisition of market power, as a part of the competitive-yardstick role. The competitive-yardstick role means simply that a major reason for cooperation is to maintain competitive and efficient systems to supply inputs and marketing services for farmers. Its benefit to farmers and society in general is in providing a measure by which the performance of other firms can be judged. The cooperative, operating at cost (including normal return for capital invested), provides an idea of what is reasonable for both its members and for others. Thus the benefit of doing business in the open, and in effect, of keeping others diligent, accrues to members and nonmembers alike.

If a competitive system is maintained, firms are forced by competition to use methods and to operate at sizes that minimize per-unit cost. Prices will be forced to levels that allow only a normal rate of return to the necessary capital and managerial effort. A cooperative that is successful in the yardstick role will find that the going price charged by all businesses is equivalent to pricing at cost when all costs, including a return to capital invested, are considered.

FOR THE INDIVIDUAL

In previous sections of this chapter we have explained the advantages that cooperatives present to farmers and the economy in general. You also need to be aware of special consideration that people must make when choosing how (if at all) they will interact with cooperatives.

There is no simple answer for the person who must decide whether to participate in the formation of a cooperative or to patronize an existing cooperative. The nature of the commitment that is necessary varies from providing initial capital for a new enterprise to making a decision based on present prices alone. The latter behavior involves virtually no commitment. Business from these persons may be important to the cooperative's existence even though they fail to recognize the competetive yardstick benefits. A farmer can ignore the value of patronage refunds (if any) yet still take advantage of the investment of those who have gone before. People need not consider the value of having the cooperative in the competition yardstick role. Nevertheless, this customer business may be important to the cooperative's existence. A capital-short young farmer or an operator near retirement may want to choose this customer role.

The person should also keep in mind that the transaction price between a patron and a cooperative is of little consequence if the cooperative follows strictly the business-at-cost principle. The final cost or return is not known until the books are closed for an accounting period and net income is distributed (or allocated) based on patronage. For example, if the net income of a cooperative is distributed based on patronage by major product lines, the cost per ton of fertilizer to a patron includes the initial transaction price less the value of any patronage refund associated with the business. The net cost includes allowance for any return to member-patron capital invested in the cooperative. The value of the patronage refund includes the amount paid in cash plus the present value of the retained portion to be received at some time in the future. To determine the short-term value of cooperative patronage for a given product, you must compare the IOF product price to the cooperative's initial price adjusted first for the present value of the patronage refund (including taxes) and then for the alternative yield of the member-patron's capital invested in the cooperative.

An evaluation of the long-term benefits must include the discounted value of a future stream of benefits which may depend, in part, on current patronage. The long-term view also includes the value of the business filling the yardstick and market access roles. However, this "What if the cooperative were not there?" question may be very difficult to answer.

SUMMARY

Farmers have elected to organize and use cooperatives extensively to extend their businesses into both input supply and product marketing levels of their commodity systems. By so doing, they can make their farming operation and commodity system more efficient and attain power to influence the terms of trade in their own favor. The first benefit implies that market operation without cooperatives has failed to use or allocate resources efficiently. Cooperatives represent an alternative in certain situations when other approaches result in market failure. If the market were efficient, however, there would be little special reason for cooperatives—although cooperative business would not be excluded.

Markets fail for a number of reasons. Economies of size (large plants that experience lower unit costs than small plants) are typical of businesses serving farmers. The larger plant's cost advantage results in a small number of firms dealing with a large number of farmers distributed over wide areas, leaving the individual farmer in a position to deal with only a very few sources of supplies or markets. This increases both the imbalance of numbers and the potential for above-normal returns to the few, below-normal returns for farmers, and higher prices for consumers.

The existence of an efficient market depends on the availability of information. Information has a cost, and larger operations will usually be able to afford more than the smaller ones. Thus an imbalance of information also increases the potential for distortion of resource allocation.

It may also be difficult for market price alone to provide clear messages from the ultimate consumer through all transactions to the producer. Consequently, inefficient allocation of resources or products may result.

Each of these market distortions provides an incentive for firms to integrate the functions on both sides of a poorly performing market into a single firm. A cooperative represents the extension of the farm firm to include other stages in the system.

The primary reason for cooperation may be to attain the benefits of economies of size in providing inputs or marketing services to farmers. Members who control a cooperative enterprise need not fear exploitation by their own cooperative even though it may appear as a monopolist or monopsonist. Similarly, if the firms presently serving farmers are able to earn higher-than-normal profits because of an imbalance of power, these firms might be replaced by a cooperative to the advantage of its members.

A cooperative may be organized to provide new or missing services. If no entrepreneur can be interested when a real need (willingness to pay) exists, a cooperative may be the answer. The assurance of a supply or a market is a closely related issue. A cooperative whose objective is to serve the member will not desert a market for temporarily higher returns elsewhere.

A cooperative may be organized to coordinate the flow of supplies to farmers and of farmers' products to processing. There is a substantial potential for gain from reducing the magnitude of production and price cycles and from more uniform use of processing capacity. The cooperative's success depends on firm commitment and surrender of some scope of individual decisions by individual members. However, this level of commitment to a cooperative program is unusual.

The sharing of risk is a possible advantageous characteristic of cooperative action. The degree of sharing varies from averaging expenses over a single line of business to operating a single seasonal marketing pool.

Cooperation to achieve market power represents a concept wholly different from the others mentioned. It means attaining the ability to influence the terms of trade in favor of the cooperative members. It is the deliberate unperfecting of a market. Public policy allows cooperatives to move in that direction, although it warns them not to "unduly enhance prices." This sanction of joint marketing efforts by farmers represents a recognition of the farmers' typically disadvantageous position in the market and the difficulty that a cooperative has in capturing and retaining gains from a position that would mean market power for an IOF.

DISCUSSION QUESTIONS

8-1. Why would forming a cooperative be of very limited advantage to a farmer if the market the farmer deals in is already highly competitive and products are undifferentiated?

8-2. Explain why the firms dealing with farmers are likely to have some market power even though there are many firms dealing in the same product in a state.

8-3. How would it be to farmers' advantage to have their cooperative establish a consumer brand on processed farm products?

8-4. Why would a grower of vegetables for processing probably be more concerned than a soybean producer about maintaining access to a market?

8-5. Explain why the farmer must make a firm commitment to the cooperative in order to realize gains from coordination of production and marketing.

8-6. How could close coordination between hog farmers and a cooperative slaughter plant reduce processing costs?

8-7. Which involves greater risk sharing: patronage refunds by product line or a single rate of refund for all products marketed?

8-8. Give an example of a situation in which the averaging of costs (or returns) would give an incentive for undesirable behavior by patrons.

8-9. Which of the reasons for organizing a cooperative does not operate to the benefit of the consumer?

8-10. How would you persuade a new farmer to patronize a cooperative that is performing the yardstick role?

9

Economic Theory and Its Application to Supply Cooperatives

Brian H. Schmiesing,
South Dakota State University

INTRODUCTION

A sound economic perspective is essential for effective cooperative decision making. Cooperative management, boards of directors, and patrons must make decisions on the goals or objectives of their cooperatives. These objectives, as well as the strategies for implementing them, will affect the cooperative's viability as a business organization and its ability to serve its patrons. Also, members may alter their own business objectives and strategies to adapt to the strategies being implemented by the cooperative.

Cooperative management should also understand the economic implications resulting from differences between cooperatives and investor-oriented firms (IOFs). If cooperatives are different from other businesses, what are the implications of these differences to patrons, boards of directors, and management? If differences exist, are these differences desirable for overall society?

You will need some background in elementary microeconomic theory, particularly for chaps. 9 and 10, to follow the details of the rationale presented in this section. Otherwise, you should be able to comprehend the conclusions and basic concepts.

BACKGROUND ON ECONOMIC THEORY FOR COOPERATIVES

Why Theory?

Theory forces discipline into the arguments concerning the conduct and performance of cooperatives. Conduct is the range of business methods, strategies, and policies that cooperatives use in responding to their business environment. Performance is the result of cooperative conduct.

Although economists frequently use prices and output levels to measure cooperative performance, industry conduct is also an indicator of cooperative performance. For example, the fertilizer industry used to sell fertilizer with predetermined chemical compositions to farmers. This aspect of industry conduct changed when local cooperatives began selling blended fertilizers. Farmers could specify the nutrient content tailored to the specific growing conditions and crop. Cooperatives forced a change in industry conduct.

Assertions about cooperative performance must be consistent with accurate assumptions about cooperative conduct and about the business environment confronting the cooperative. Strategic business decisions based on mythology rather than reality will result in the eventual failure of the cooperative as a business entity (Caves and Peterson). For example, during the 1970s regional grain cooperatives attempted to expand their market share in the international grain trade. However, they were mistaken in assuming that the future business environment would be conducive to their expansion. Because of their poor competitive position relative to the multinational grain companies, among other difficulties, a number of regional grain cooperatives either failed or exited from the grain merchandising entirely during the 1980s.

Theory can help us develop realistic, workable relationships between cooperative principles and assertions about how cooperatives should operate. Patrons, boards of directors, and management are concerned about the applicability of these principles to the actual operation and performance of the cooperative. Hopefully, theory can assist in clarifying this relationship.

The lack of a coherent understanding of cooperative theory hampers the development of decision aids. The computer has brought about a major revolution in the mathematical modeling of IOF decisions. But before cooperatives can apply similar models for themselves, they must develop specific theories of behavior and specify the objectives they want to pursue.

Cobia et al. in their analysis of equity redemption, provide an example of how cooperative theory, cooperative principles, and modeling can be combined to provide useful information for management decisions. In 1979 the U.S. General Accounting Office (GAO) made recommendations for mandatory equity redemption and/or dividend payments on equity. By modeling specific equity redemption programs, Cobia et al. were able to indicate the level of cooperative financial performance required to meet specified equity redemption programs and thus to evaluate the feasibility of the GAO recommendations.

Of course, any theory is only as good as what is put into it. Theories are based on assumptions, and logic is used to derive conclusions from assumptions. Therefore, it is essential that assumptions are valid. The most important ones are identified in this chapter. By knowing the assumptions of various models, you can better determine the weaknesses and strengths of the analysis.

Development of Cooperative Theory

Although cooperatives have existed in U.S. agriculture for over a century, formal economic theories of cooperatives to a great extent have evolved since the 1940s. The cooperative economic models used today are largely based on theoretical developments that occurred after World War II. Vitaliano (1978) provides a comprehensive and excellent review of the development of this literature.

A central debate in cooperative theory is about the very definition of a cooperative and the appropriate objectives for a member-user organization. How independent are the decision processes of the cooperative and member-patrons? (Hereafter we will use *patron* to refer to patrons who are also members of the cooperative.) What is the interaction between management, boards of directors, and patrons?

Cooperatives have characteristics that create a decision-making process uniquely different from that of IOFs (Garoyan). First, benefits are tied to the use of the organization, not to the level of equity investment (Caves and Peterson). Further, members of the cooperative board of directors are both stockholders and users, while IOF boards of directors generally only include, at the most, stockholders and managers.

In 1946, Emelianoff portrayed cooperative activity as individuals pursuing self-interests without a central authority to coordinate the system (Vitaliano, 1978). He did not perceive the cooperative as a firm, because it generated wealth for its membership and lacked control by a central authority. Subsequent literature, however, recognized cooperatives as separate firms having distinct decision making units (Vitaliano 1978).

Given recent concern over managers controlling cooperatives and their boards of directors, Emelianoff's concept of cooperatives now appears to have been rather idealized. He ignored the segmentation and specialization of the roles of patrons, the board of directors, and management.

During the 1950s the traditional theory of the profit-maximizing firm was modified to account for the unique features of cooperatives. For example, Helmberger and Hoos developed the model that is frequently used to analyze processing cooperatives. These models were based on concepts derived from the economic theory of the individual firm. Theorists directed their attention toward altering the assumptions of the basic models in order to discuss specific issues such as open versus closed membership (Youde and Helmberger). Although a number of insightful articles were written during the late 1960s and throughout the 1970s, cooperative theory received little attention from professional economists. However, interest and attention increased during the 1980s (Carson; Caves and Peterson; LeVay; Lopez and Spreen; Royer; Sexton, 1984 and 1986; Staatz; Vitaliano, 1983; Zusman). Although part of the stimulus for this recent increase was the need to clarify previously developed models, the theoretical developments in game theory and in other areas of economic theory provided new approaches for analyzing cooperatives. Internal group decision processes, rather than the conduct and performance in specific market structures, became a major area of inquiry. The internal politics of cooperatives were no longer assumed away.

The economic theory presented in this and the following chapter is limited primarily to the firm and game theory approaches. These represent a major segment of the current economic and organizational theory of cooperatives. The firm approach is based on the assumption that organizations have a predetermined objective, such as maximize profits or net income. In actuality, however, cooperative managers, boards of directors, and patrons may have a broad set of objectives they are attempting to achieve (Ladd). For example, a cooperative may simultaneously have the objectives of obtaining adequate net income, achieving maximum operational efficiency, maintaining and expanding facilities, and increasing sales volume.

A weakness in firm theory is that it cannot be used to analyze the internal decision-making processes of IOFs and cooperatives. For example, it cannot be used to evaluate how and what determines the conduct and performance of a multibillion dollar cooperative (Rhodes). Game theory is a better approach to at least partially analyze this issue. Also, economists have attempted to develop alternative theories that more realistically describe how organizations make decisions (De Alessi). However, before we can use and understand these new approaches, we must first understand the basics of the firm and game theory.

BASIC ECONOMIC MODELS OF SUPPLY COOPERATIVES

In this chapter the term *supply cooperatives* refers to a broad range of cooperatives whose patrons purchase goods and services. Examples of this type of cooperative are rural electric cooperatives, rural water cooperatives, farm supply cooperatives, petroleum cooperatives, and the Farm Credit System.

In this section we use basic economic models of cooperatives to partially answer four cooperative management and organizational questions: (1) Can and do cooperative business objectives differ from IOFs? (2) Are the identified cooperative objectives complementary or in conflict? (3) How can cooperatives achieve the identified cooperative objectives? and (4) Does the structure of the industry affect the ability of cooperatives to achieve their objectives? We examine these questions for supply in this chapter and for and marketing cooperatives in chap. 10.

Basic Model of IOFs

Although alternative objectives such as sales maximization may exist, IOFs are usually assumed to maximize the value of the firm to its investors by maximizing net income (Ferguson and Gould). Net income is defined as total revenues *(TR)* minus total costs *(TC)*.

Consider the fertilizer supply IOF that sells fertilizer to farmers on a per-ton basis (Figure 9.1). The firm allocates net income of the IOF to investors, based on their investment (number of shares held) in the fertilizer company. Stock in the IOF is purchased from either a previous investor or when the company issued stock.

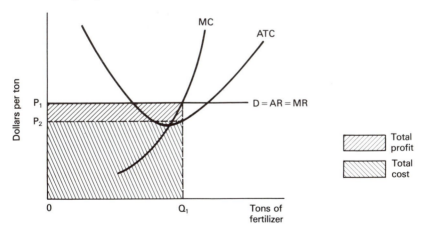

Figure 9.1 Net income point for an investor oriented firm in a perfectly competitive industry.

In a fertilizer supply IOF, marginal cost *(MC)* is the cost added to the firm's total cost *(TC)* resulting from the production of one additional ton of fertilizer. Marginal revenue *(MR)* is the revenue added to the firm's total revenue *(TR)* resulting from the sale of one additional ton of fertilizer. To achieve net income maximization, economists have thus derived a decision rule for IOFs: marginal cost *(MC)* must equal marginal revenue *(MR)*. The rationale in a competitive market is that the cost of producing an additional unit *(MC)* eventually increases and that *MR* is constant. As long as *MR* is greater than *MC*, the IOF can increase total net income by increasing output. In other words, the sale of an additional ton of fertilizer exceeds the cost of providing it. If *MR* is less than *MC*, the firm can increase total net income by decreasing fertilizer output because the reduction of sales will reduce costs by more than revenues.

The implication of this decision rule for IOFs' pricing and output decisions depends on the market structure of the industry (Boyle; and Caves). Market structure is the number and relative size of the competing firms in the industry. In a perfectly competitive market structure, there are many buyers and sellers of a homogeneous product such as nitrogen fertilizer.

If the IOF is competing in a perfectly competitive market structure, the IOF must accept the price *(P)*, because the IOF does not have sufficient market share to influence the market price. Restricting output will not increase the price paid for the product, because other sellers will have the product available. Further, if the IOF raises the product's price, buyers again will purchase the product from competing firms. In both cases the competing firms are willing to sell at the market price. Therefore, the IOF can sell *all* its output at P_1 (Figure 9.1). Because the demand curve is horizontal, average revenue *(AR)* equals *MR* and P_1. To maximize net income, the IOF will equate *MC* and *MR*. The output that the IOF markets will be Q_1 and the price will be P_1.

Total net income for the firm equals the difference between the IOF's average revenue *(P1)* and average total cost *(P2)* multiplied by the quantity marketed (Q_1). In other words, net income = $Q(P_1 - P_2)$.

However, a perfectly competitive market structure rarely, if ever, exists. In fact, the IOF will probably operate in a market where it must lower its price in order to sell a greater quantity of fertilizer. For example, farmers may have only three to five fertilizer companies competing for fertilizer sales in the local market. These firms probably devote considerable management effort to differentiating their fertilizer products from those of competing businesses. That is, they make their product appear different or decrease the ease with which buyers may substitute a product from another firm. This product differentiation can be achieved by altering service quality, product delivery procedures, availability of credit, application services, brand identification, or advertising. If a firm differentiates its products, it will not lose nor retain all its customers by changing prices.

A market for a differentiated product implies that the demand (D) for output is downward sloping (Figure 9.2). In this market AR and MR are not equal as they are in a perfectly competitive market structure. Rather, the demand curve is the AR curve. The MR curve is below the demand curve because the IOF must reduce its price on *all* the units sold, not just on the additional unit sold. Because of the loss of revenues on the other units, the MR received for selling an additional unit is less than the price for which it was sold. The IOF or cooperative will maximize total revenues when MR is zero. At levels of sales below this level, additional revenue can be added to total revenue, while at levels above this level, total revenues will actually decline because marginal revenue is negative.

This type of market structure is called monopolistic competition. Although the IOF has competitors, the IOF's production or pricing decisions will alter the price for the product. [The IOF will also follow the rule that MR should equal MC in order to maximize net income in this type of market structure. The net income maximizing sales level is Q_1 with a price of P_1 (Figure 9.2.).]

The steepness of the demand curve's slope for an individual firm is dependent on competition or the availability of close subsitutes. As com-

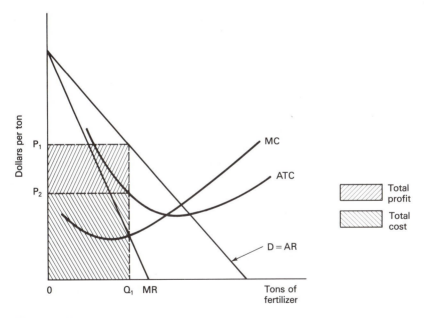

Figure 9.2 Net income maximization point for an investor oriented firm in a monopolistic competitive market.

petition increases, the slope of the demand curve will flatten. Diagrams used in this and subsequent chapters contain demand curves with rather steep slopes. This was done to represent spatial monopolists in some parts of the country and to dramatize the price differences that result when cooperatives pursue objectives other than maximizing net income. Price differences between the objectives will not be nearly as dramatic with less steep demand curves. Adjustments will be shifted to the quantity demanded rather than through the prices. In actual situations, management must direct efforts toward identifying the competitive structure of their industry and characteristics of the demand curves confronting the firm.

A Model of Supply Cooperatives

Now assume that the business organization form of the fertilizer company has been changed from an IOF to a supply cooperative. The supply cooperative is assumed to have the same cost structure as the IOF. This means that the supply cooperative is as technically efficient as the IOF, and both firms have identical average total cost and marginal cost curves.

Possible Cooperative Objectives Several possible objectives have been proposed for supply cooperatives. The cooperative can attempt to (1) have average revenue equal average total costs, to (2) minimize the net price paid by patrons for the product, or to (3) maximize the total returns to the patrons as a group (Vitaliano, 1978). Objectives (1) and (2) are discussed in this chapter. Those interested in objective (3) should read the article by Sexton (1983). Are any of these the same as net income maximization? With one exception they are different. A detailed answer to this question follows.

As discussed previously, a cooperative is a special type of firm that links returns with use rather than investment and which distributes these returns as patronage refunds. For the purpose of current discussion, we assume that the total net income of the cooperative is returned to the patrons as cash patronage refunds. The patronage refund per ton of fertilizer equals the price per ton minus the average total cost per ton. The total cash patronage refund to patrons will equal total net income.

According to standard economic analysis, the average total cost curve includes all costs, including a normal return for capital. Therefore, when average total cost equals average revenue, the cooperative will still have some funds available to distribute to equity and debt capital. These funds are sufficient for retaining the invested capital within the cooperative.

Who Establishes the Cooperative's Objectives? Although the patrons own the cooperative, they delegate to the board of directors the responsibility of supervising the cooperative's management and operation. The board of directors and management establish goals and objectives. In the following analysis, assume that the board of directors and management establish one objective and use a specified pricing policy to achieve the objective. (This is consistent with the assumption that the cooperative is a separate firm from the patrons.) In addition, all patrons pursue their own objectives. Patrons will use either the market price or a price adjusted for the patronage refund in their purchase decisions.

The cooperative initiates a pricing and patronage refund policy to achieve a specific cooperative objective and the patrons respond to the implementation of the firm's strategy. Whether a specific cooperative objective will actually be achieved depends on the response of the patrons.

Analysis of Specific Cooperative Objectives Assume the cooperative attempts to operate just as the IOF and sells only Q_1 to its patrons by charging a price P_1 (Figure 9.3). The patronage refund to the patrons is $P_1 - P_4$ (objective 1, Table 9.1). The actual net price to the patron is P_4 or P_1 minus the patronage refund. Is this objective consistent with cooperative principles, the objectives of the patrons as a group, or the objectives of individual patrons? Are there other pricing strategies that are more effective in achieving these objectives?

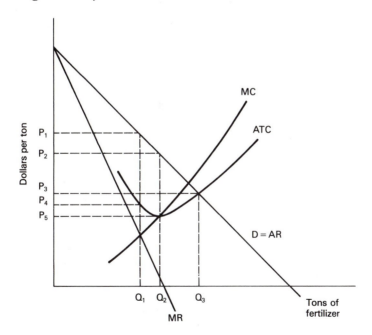

Figure 9.3 A supply cooperative with analysis of different cooperative objectives.

TABLE 9.1 Summary of potential cooperative objectives, decision rules, output prices, patronage refunds and net prices presented on Figure 9.3.

Objective[a]	Decision rule	Output	Price paid	Patronage refund	Net Price paid
1. Maximize net income like an IOF	$MC = MR$	Q_1	P_1	$P_1 - P_4$	P_4
2. Minimize net price paid by patrons	$MC = ATC$	Q_2	P_2	$P_2 - P_5$	P_5
3. Operate at cost or break-even	$ATC = AR = P$	Q_3	P_3	$P_3 - P_3$	P_3

[a]Two other objectives can be pursued: sales maximization, where $MR = 0$, and patrons surplus maximization, where $MC = P = AR = D$.

The cooperative principle of business at cost implies that returns are associated with use. If a cooperative exists in a monopolistic competitive industry, the cooperative manager may perceive this principle to mean that average total cost should equal average revenue. In this situation, the cooperative will supply Q_3 to its patrons and will not have any net income beyond a normal return for distribution (Figure 9.3). This output level is higher than the net income-maximizing level for the IOF, but at a lower sale price (P_3) (objective 3, Table 9.1).

If the cooperative's objective is to minimize the net price charged, that minimum price occurs when ATC is at a minimum or $MC = ATC$. Marginal cost is greater than ATC at sales levels above Q_2; this implies that the ATC of supplying the product to the patron can be reduced by reducing sales. At sales levels below Q_2, the ATC of supplying the product can be reduced by increasing sales. The net price equals $P_2 - P_5$. Patrons will have the lowest net price when only Q_2 is sold (objective 2, Table 9.1).

Unstable Equilibrium Reality may be defined as those events to which an individual pays attention. For the cooperative patron, the crucial question is whether to pay attention to initial prices or to expected net prices. This issue has important pricing policy implications for management and directors attempting to implement the three cooperative objectives specified earlier.

Patrons who base their purchase decisions on initial prices, rather than on expected net prices, probably either do not expect to receive patronage refunds or else believe that the present value of patronage refunds will be low or even negative. For example, if a cooperative allocates a low percentage (e.g., 1%) of its sales prices as patronage refunds, or if the present value of patronage refunds is near zero, patrons will probably ignore patronage refunds in their purchasing decision. Instead, they will respond to initial prices charged, and the quantity they demand will be consistent with these prices. For example, if a cooperative charges a price of

P_2, then patrons will buy Q_2 (Figure 9.3). Therefore, patrons will receive a cash patronage refund of $P_2 - P_5$, and the cooperative will achieve its objective of minimizing net price.

However, if patrons respond to expected net price (adjusted for an expected patronage refunds), an instability is introduced into the system. Again, assume that the board and management are attempting to sell at an output level consistent with a minimum net price. This output level creates an unstable equilibrium for the cooperative because the net price is less than what the membership is willing to pay, as indicated by the demand curve. Thus patrons have an incentive to buy more fertilizer than the amount needed to achieve minimum net price, since they know it is actually cheaper than the initial price charged. However, if the cooperative is achieving the objective of making ATC equal to AR, this instability will not exist.

Achieving Minimum Average Total Cost

What will patrons do if their objective is to obtain a product at the minimum ATC and the cooperative has been operating at high production levels that prevent this minimum ATC? The membership has two basic alternatives.

First, an economic incentive exists for a group of the cooperative's membership to form a closed membership cooperative. By excluding part of the membership, the new cooperative can shift the demand curve to the left or from D_1 to D_2 (Panel 1, Figure 9.4). The cooperative will then achieve stability when the demand curve intersects the minimum point of the ATC curve. Realization of cooperative objectives 2 and 3 converge at the same level of sales. There is no incentive for the patron to shift to other businesses, since the cooperative is achieving the minimum net price for the product.

The second alternative is to alter the cost structure of the business. The cooperative can expand its physical plant and shift the cost structure to the right. Assume that the cooperative shifts the marginal and average total cost curves from MC_1 and ATC_1 to MC_2 and ATC_2. By expanding facilities the cooperative may be able to shift the minimum ATC point to the right and achieve an even lower net price P_2 and larger output Q_2 for the patrons. Since successful cooperatives tend to expand their productive capacity and business volume, cooperatives probably select this second alternative more frequently.

Are Supply Cooperatives Different?

The decision rules that govern the management decisions of IOFs are relatively few and consistent. The primary IOF organizational objective is generally to maximize net income. This results in the basic decision rule

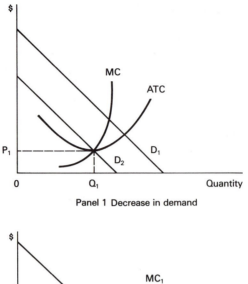

Panel 1 Decrease in demand

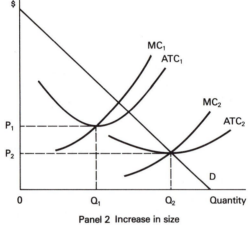

Panel 2 Increase in size

Figure 9.4 Two approaches for achieving a stable long-run equilibrium for a supply cooperative.

to produce at the quantity for which additional revenue is the same as additional costs, or $MC = MR$. On the other hand, the decision rules for cooperatives are considerably more complex.

A number of alternative objectives, including net income maximization, have been advocated for cooperatives. However, these objectives can cause problems for cooperatives. Instability results when management attempts to achieve any objectives other than a normal return on investment, because patrons who do not understand or agree with other objectives will pressure the cooperative to change prices or output. The main benefit of supply cooperatives in our society is that incentives exist for the cooperative to have sales levels beyond those which result in maximum net income but at a lower price level. Only under perfect competition in long-run equilibrium, or when $MR = P = ATC = MC$, will there be equality in output and price levels when the results of net income maximization and meeting cooperative objectives converge.

In a competitive industry, cooperatives and IOFs are expected to achieve essentially the same level of financial performance in the long run. Both types of firms must be earning a rate of return sufficient to retain invested capital. In long-run equilibrium, cooperatives and IOFs will both operate so that $MR = P = ATC = MC$. Thus it seems that making comparisons between cooperatives and IOF businesses in established competitive industries would be a justifiable practice.

However, the issue becomes more confused when net income maximization and possible cooperative objectives result in IOFs' and the cooperatives' achieving different prices and quantities (Figure 9.3). For example, supply cooperatives in industries with downward-sloping demand curves and increasing average total costs will merchandise products at lower price levels and greater levels of output than an IOF will. This implies a lower rate of financial return for a supply cooperative than for a competing IOF.

Cooperative managers usually emphasize the importance of maximizing financial performance in order to compete successfully with and be respected by IOF managers. However, patrons often prefer to receive lower prices rather than to improve financial performance. This management-patron conflict over price levels may be perceived as just the normal tension in buyer-seller relations, but the conflict can become serious in a supply cooperative. Further, because the cooperative's consumers also are on the board of directors, the conflict and tensions exist in the boardroom as well as in merchandising activities.

Some people argue that the patronage refund is a substitute for lower prices, but this is not totally true. Patrons know that the cooperative's projected net income is only an estimate, while the prices they pay are a certainty. Patrons' desire for lower prices and their misunderstanding of the cooperative principle of business at cost can eventually push a cooperative to allocate financial returns at such a high level that the cooperative can neither retain nor attract capital. As a result, the cooperative also might be unable to meet other objectives, such as equity redemption.

Excess Capacity

The underlying assumption of the previous analysis is that the cooperative's business volume is too large, in relation to the size of its specified physical plant, to achieve minimum ATC. Sexton and other economists have argued that cooperatives frequently confront the opposite problem—insufficient demand for achieving the minimum point on the ATC curve (Sexton, 1983; Cotterill).

When a firm has excess capacity, it cannot achieve the point of minimum average total cost at any demand level (Figure 9.5), because marginal cost is below-average cost at all levels of sales. In this case cooperatives could use average cost.

The IOF will maximize net income by producing the amount necessary for marginal cost to equal marginal revenue. The IOF will charge P_1 and sell Q_1. The IOF objective results in the restriction of output and higher prices compared to that of cooperatives. In other words, the market power of the firm reduces the welfare of the consumer. Obviously, this is not an acceptable strategy for a cooperative. Instead, the cooperative might require that $ATC = AR$, following the average cost pricing strategy of charging P_2 and selling at Q_2. This price is lower than the IOF's price, and the quantity supplied is larger.

Beyond making pricing decisions, cooperatives confronted with excess capacity must examine the feasibility of other strategic solutions to the problem. When a cooperative is competing with other cooperatives in the same market, the cooperatives may merge to achieve economies of size and to reduce excess capacity. For example, assume that three farm supply cooperatives each own a bulk fuel truck operating at 30% capacity. A

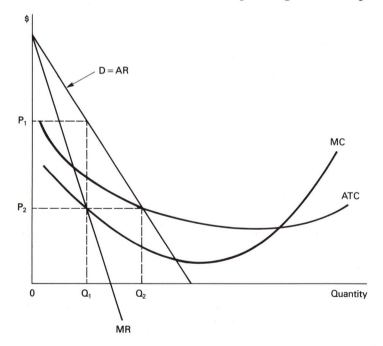

Figure 9.5 Closed membership supply cooperative operating in the declining portion of its cost function.

merger may enable the cooperatives to sell off at least one truck and excess fuel storage capacity, as well as to reduce labor expenses. The cooperatives could thus achieve economies of size by expanding quantity demanded and reducing total industry capacity. On the other hand, if the cooperative is in a weak strategic position, the best decision may be to exit from the industry and sell its capacity to an IOF competitor. Alternatively, if the cooperative is in a strong strategic position, its best decision may be to buy the capacity of its IOF competitors.

Nonmember Business

Supply cooperatives operating in the declining cost portion of their cost function have a cost incentive to expand nonmember purchases in order to increase the total quantity sold to reduce the average cost of providing the product to members (Figure 9.6). By expanding nonmember purchases, a cooperative using an average cost pricing strategy can reduce the price charged to patrons from P_1 to P_2. If the cooperative's objective is obtaining the minimum net price for members, nonmember business will be expanded to the point where average total cost achieves this minimum.

However, the underlying assumption is that the cooperative does not transfer potential net income associated with nonmember business to the

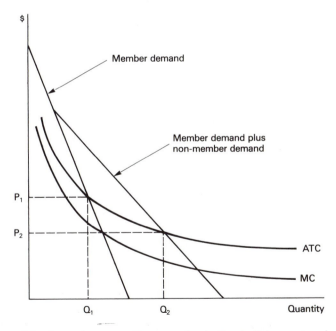

Figure 9.6 A supply cooperative operating in the declining cost region of its cost function and non-member demand is available.

membership, because such transfers, if allowed between members and nonmembers, would alter the incentive system within the cooperative. The larger the proportion of business done with nonmembers, the greater the incentive of the members to use the nonmember business as a profit center. (See the sections on dividends and unallocated equity in Chapters 13 and 14.) Increasingly, the cooperative will operate like a net income-maximizing firm as the nonmember business becomes a more significant proportion of the business. To avoid this, the U.S. Internal Revenue Code and a number of state laws have specific provisions regulating the level of nonmember business. This is particularly true for marketing cooperatives because the Capper-Volstead Act specifies that the cooperative must do the majority of its business with members.

When a cooperative attempts to expand nonmember business, its actual pricing objective will probably be to either maintain or decrease the net price paid by members. Nonmember business motivates a cooperative to act increasingly like an IOF in its pricing strategies. Therefore, the asserted societal welfare gains for concentrated industries in terms of increased output and lower prices for consumers are decreased (or weakened). This is not a concern in competitive industries, because competing firms limit the ability to overcharge nonmembers for products and services provided. Cooperative membership policies, on the other hand, must be evaluated carefully for their implications to cooperative conduct and performance (Youde; Youde and Helmberger). Policies in the best interest of member-owners may not always achieve what society desires from cooperatives as business organizations.

SUMMARY

Cooperative theory is important because it helps to determine price, output, and membership policies logically and consistently. The formal theory of cooperatives has largely developed since World War II. Cooperatives were not always perceived as separate entities with centralized authority, but during the 1950s economists altered the traditional theory of the firm to recognize the unique features of cooperatives. During the 1980s additional theoretical developments enabled economists to analyze the economic incentives involved in cooperatives decisions. These developments are important tools for cooperative managements, whose objectives are often more complex than those of IOFs.

The theory developed for an IOF selling a product in a competitive and monopolistic competitive market is based on the assumption that the IOF's objective is to maximize net income. The decision rule for achieving maximum net income is that marginal revenue should equal marginal cost. This decision rule implies that the IOF will expand output only if the additional revenue exceeds the additional cost of expanding output or will contract its output level if the additional revenue is less than the additional cost of production.

Cooperatives, however, often have more objectives than maximizing net income. Consequently, they often use decision rules different from those used by an IOF in order to deal with unique cost and demand situations. We have identified three potential objectives for a supply cooperative: (1) maximizing net income, (2) minimizing the net price, and (3) breaking even. Each objective requires a distinctly different decision rule for determining the appropriate pricing strategy. Only in competitive long-run equilibrium will an IOF and *all* the cooperative strategies result in identical prices and output. Therefore, the management issues confronting supply cooperatives, as identified by theory, are different from the issues confronting IOFs and imply a different type of conduct and performance.

A frequent problem for cooperatives is an insufficient business volume to achieve the minimum of the average total cost curve (i.e., because of excess capacity). Potential solutions to this problem are merging with other cooperatives, exiting from the industry, acquiring IOFs, or expanding nonmember business. If a supply cooperative distributes patronage refunds only to members, it will depend increasingly on nonmember business and thus increasingly assume the conduct and performance of an IOF.

DISCUSSION QUESTIONS

9-1.   A local supply cooperative conducts a survey of its patrons and discovers that its patrons ignore cooperative patronage refunds when making purchase decisions. How does this affect the cooperative's ability to achieve specific objectives through its pricing policy? What happens when patrons consider net prices in their purchase decisions?

9-2.   Patrons assert that their local supply cooperative should have the same rate of return on investment as ConAgra, Pillsbury, and other agribusinesses have. The patrons also argue that the cooperative should sell inputs at prices equal to those of other local competitors. What assumptions about the cooperative's cost and market structures must actually be true if the patron's assertions are consistent? Are these realistic assumptions?

9-3.   What are the possible objectives of a supply cooperative? Do IOFs have the same objectives? If differences exist between supply cooperatives and IOFs, why do these differences exist?

9-4.   A cooperative has insufficient demand for achieving the minimum point on the average total cost curve. What are two strategies a cooperative can use to remedy this problem? What are the advantages and disadvantages to each of these strategies? Be sure to state your assumptions and their impact on your analysis.

9-5. A supply cooperative's board of directors wants to determine whether they should expand the amount of business volume with nonmembers. What demand and cost conditions must exist for the board of directors to decide in favor of expanding nonmember business? What assumption did you make about the cooperative's primary objective?

9-6. One cooperative does not distribute patronage refunds to nonmembers and conducts approximately 45% of its business with nonmembers. Net income associated with nonmember business is distributed to members on a patronage basis. Assume that another cooperative with an identical situation distributes patronage refunds to both members and nonmembers. Is the first cooperative a true cooperative? Do you feel both cooperatives should be taxed and regulated by government in the same manner? Defend your answer.

9-7. Identify two types of situations in which you observe supply cooperatives operating. Does the competitive or monopolistic competitive model best describe the situation confronting the cooperative? Defend your position by considering the assumptions of the two models. Are there any weaknesses common to both models as discussed in the chapter?

9-8. Why do economists advocate the study of theory for improving the business decisions of cooperatives and IOFs? Do you agree with this perspective? Defend your answer.

9-9. In the long run, is there any difference between the expected conduct and performance of a cooperative and that of an IOF in a competitive industry? Is there any justification for the existence of cooperatives in a competitive environment? Assume that long-run equilibrium occurs and that average revenue = average cost = marginal revenue = marginal cost. Are your conclusions the same for a monopolistic competitive industry in the long run? Explain why or why not.

REFERENCES

BOYLE, STANLEY E., *Industrial Organization: An Empirical Approach.* New York: Holt, Rinehart and Winston, 1972.

CARSON, R., "A Theory of Co-operatives," *Can. J. Econ.* 10(1977):565.

CAVES, RICHARD E., *American Industry: Structure, Conduct and Performance,* 5th ed. Englewood Cliffs, NJ: Prentice-Hall, 1982.

CAVES, RICHARD E., and BRUCE C. PETERSON, "Cooperatives' Share in Farm Industries: Organizational and Policy Factors," *Agribusiness* 2(1986):1.

COBIA, DAVID W., ET AL., *Equity Redemption: Issues and Alternatives for Farmer Cooperatives.* Washington, DC: USDA ACS RR 23, Oct. 1982.

COTTERILL, RONALD W., ED., *Structure and Strategies in the Food Retailing, Consumer Food Cooperatives.* Danville, IL: The Interstate Printers & Publishers, 1982.

DE ALESSI, LOUIS, "Transaction Costs, and X-Efficiency: An Essay in Economic Theory," *Am. Econ. Rev.* 73(1983):64.

FERGUSON, C. E., AND J. P. GOULD, *Microeconomic Theory,* 4th ed. Homewood, IL: Richard D. Irwin, 1975.

GAROYAN, LEON, "Developments in the Theory of Farmer Cooperatives: Discussion," *Am. J. Agr. Econ.* 65(1983):1096.

HELMBERGER, PETER G., and SIDNEY HOOS, "Cooperative Enterprise and Organization Theory," *J. of Farm Econ.* 44(1962):275.

LADD, GEORGE W., *The Objective of the Cooperative Association, Development and Application of Cooperative Theory and Measurement of Cooperative Performance.* Washington, DC: USDA ACS, ACS Staff Report, Feb. 1982.

LEVAY, C., "Agricultural Co-operative Theory: A Review," *J. Agr. Econ.* 34(1983):1.

LOPEZ, RIGOBERTO A., AND THOMAS H. SPREEN, "Co-ordination Strategies and Non-member Trade in Processing Cooperatives," *J. Agr. Econ.* 36(1985):385.

RHODES, V. JAMES, "The Large Agricultural Cooperative as a Competitor," *Am. J. Agr. Econ.* 65(1983):1090.

ROYER, JEFFREY S., *Cooperative Theory: New Approaches.* Washington, DC: USDA ACS SR 18 (Jul 1987).

SEXTON, RICHARD J., "Economic Considerations in Forming Consumer Cooperatives and Establishing Pricing and Financing Policies," *J. Consum. Aff.* 17(1983):290.

SEXTON, RICHARD J., "Perspectives on the Development of Economic Theory of Cooperatives," *Can. J. Agr. Econ.* 32(1984):423.

SEXTON, RICHARD J., "The Formation of Cooperatives: A Game-Theoretic Approach with Implications for Cooperative Finance, Decision Making and Stability," *Am. J. Agr. Econ.* 68(1986):214.

STAATZ, J. M., "The Cooperative as a Coalition: A Game-Theoretic Approach," *Am. J. Agr. Econ.* 65(1983):1084.

VITALIANO, PETER, "The Theory of Cooperative Enterprises—Its Development and Present Status," *Agricultural Cooperatives and the Public Interest,* ed. B. W. Marion, NC 117 Monograph 4, College of Agr., Univ. of Wisconsin-Madison, 1978.

VITALIANO, PETER, "Cooperative Enterprise: An Alternative Conceptual Basis for Analyzing a Complex Institution," *Am. J. Agr. Econ.* 65(1983):1078.

YOUDE, JAMES, "Cooperative Membership Policies and Market Power," *Agricultural Cooperatives and the Public Interest,* ed. B. W. Marion, NC 117 Monograph 4, College of Agr., Univ. of Wisconsin-Madison, 1978.

YOUDE, JAMES, AND PETER HELMBERGER, *Membership Policies and Market Power of Farmer Cooperatives in the United States.* Res. Bulletin 267, Univ. of Wisconsin-Madison, Aug. 1966.

ZUSMAN, PINHAS, "Group Choice in an Agricultural Marketing Cooperative," *Can. J. Econ.* 15(1982):220.

10

Theory of Marketing Cooperatives and Decision Making

Brian H. Schmiesing,
South Dakota State University

INTRODUCTION

Marketing cooperatives differ from supply cooperatives because of their linkage with patrons. Patrons purchase their farm supplies from their supply cooperatives and sell their output to marketing cooperatives. Patrons' output is an input in marketing cooperatives' production processes. Marketing cooperative patrons thus desire to increase prices for their output through their sales to a cooperative. Consequently, supply cooperatives and marketing cooperatives differ in the theory used to explain their relationship with member-patrons.

In this chapter we will explain how marketing cooperatives contrast with investor-oriented firms (IOFs), especially in the way they conduct business and in the way their performance is affected by differences in market structure and cooperative objectives. Also, we will explain how management issues confronting the patrons, managements, and boards of directors of marketing cooperatives are different from supply cooperatives.

Finally, we introduce you to the implications of game theory to cooperative decision making. Game theory can be used to analyze situations in which two or more members of a group are in conflict. Rather than analyzing cooperative conduct and performance when specific market structures and cooperative objectives exist, economists should pay attention to factors that determine how decisions are made *internally* in the cooperative.

Again, you will need to have been exposed to elementary microeconomic theory to follow the details of the rationale presented in this chapter. Otherwise, you should be able to comprehend the conclusions and basic concepts.

In this chapter, as in Chapter 9, we will use basic economic models to partially answer four management and organizational questions: (1) Can and do marketing cooperative business objectives differ from IOFs? (2) Are the identified cooperative objectives complementary or in conflict? (3) How can marketing cooperatives achieve the identified cooperative objectives? and (4) Does the structure of the industry affect the ability of marketing cooperatives to achieve their objectives?

MARKETING COOPERATIVES

Consider a marketing cooperative that processes sugar beets. This cooperative pays prices and patronage refunds in dollars per ton. It sells its product to buyers, who are neither producers of sugar beets nor patrons of the cooperative. Therefore, unlike the supply cooperative, there is no direct linkage between the demand for the final product and the patronage refunds.

The linkage between the patron and the marketing cooperative involves the patron's product, which the cooperative uses in its production process (Helmberger and Hoos; LeVay; Lopez and Spreen; Vitaliano). The *net revenue surplus (NR)* available for payment to patrons equals the cooperative's total revenues minus total costs associated with processing and marketing the product. However, this total cost figure **DOES NOT** include the payments made to producers for their sugar beets. Average net revenue *(ANR)* equals the net revenue surplus divided by the quantity of sugar beets marketed.

The *ANR* is derived by subtracting the adjusted average cost *(AAC)* from average revenue *(AR)*. The *AAC* curve equals average total costs *(ATC)* minus the average price paid for the sugar beets *(APSB)*. It does not matter whether the final product demand curve is downward sloping or horizontal; the demand curve *(D)* indicates the *AR* for a specific quantity of processed sugar beets.

The shape of the *ANR* curve depends upon the relationship between *AAC* and *AR*, (Panel 1, Figure 10.1). When *AAC* is greater than *AR*, as represented by the demand curve, the *ANR* is negative (Panel 2). If the quantity of sugar beets processed is less than Q_a or greater than Q_b, *ANR* is negative. In other words, there is no revenue available for the cooperative to pay producers for their sugar beets after it subtracts the other processing costs.

A positive *ANR* implies that producers can be paid for their sugar beets. A cooperative patron can receive the payment in the form of a price and/or patronage refund. The *ANR* is at a maximum when the difference between *AR* and *AAC* is at a maximum. The amount that producers are actually paid depends upon whether the processor is an IOF or a cooperative and upon the cooperative's objective.

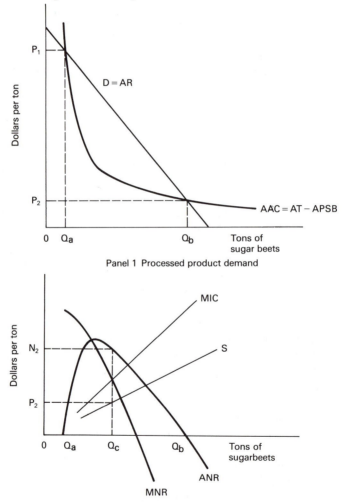

Panel 1 Processed product demand

Panel 2 Input demand

Figure 10.1 Analysis of net income maximization by an IOF when buying an input.

IOF Net Income Maximization

Assume that the objective for the IOF is to maximize net income from the sale of sugar beet products. To maximize net income, the IOF must generate a marginal net revenue from the sale of sugar and by-products that is equal to the marginal cost of purchasing sugar beets.

The ANR curve represents an IOF's and a marketing cooperative's demand curve for sugar beets. Each point along the ANR curve indicates the average price that producers could receive for their sugar beets, assuming that all the net revenue surplus is used to pay producers. Marginal net revenue (MNR), which is derived from this ANR demand curve, is the additional net revenue surplus that results from processing one additional ton of sugar beets. For ANR to increase, MNR must be greater than ANR (Panel 2, Figure 10.1). The maximum ANR is achieved when MNR equals ANR. When ANR decreases, MNR is less than ANR. The supply curve indicates the average price that must be paid to obtain a specific level of production from sugar beet producers. If a processor does not have sufficient market share to affect the input's price, the supply curve is a horizontal line. This implies that the average cost of purchasing the input is constant and is equal to the marginal input cost (MIC). In such a situation, the processor lacks monopsony power in the purchase of inputs.

An alternative assumption is that the processor is a monopsony or the sole outlet for the product. The aggregate supply curve (S) is the horizontal summation of the sugar beet producer supply curves. Therefore, the supply curve (S) indicates the marginal cost of producing additional sugar beets. Since the individual producer's supply curves are upward sloping, the aggregate supply curve is also upward sloping (Panel 2, Figure 10.1). To purchase an additional ton of sugar beets, the IOF must pay a higher price. The marginal input cost (MIC) of purchasing an additional ton of sugar beets is greater than the price paid by the IOF.

The IOF will maximize net income by equating $MIC = MNR$ at the input purchase level of Q_c. The price the IOF must pay to producers to obtain Q_c is indicated by the supply curve. At P_2 producers are willing to supply Q_c to the IOF processor. The IOF's net income per ton will equal the difference between ANR and S, or $N_2 - P_2$. Total net income equals this difference multiplied by the quantity of sugar beets processed.

IMPLICATIONS

What Should The Cooperative Maximize?

At this juncture, as in Chapter 9, we must make assumptions about the cooperative's organizational objectives and the patrons' response to the cooperative's implementation strategies. The cooperative can attempt to break even, to maximize net income like an IOF, or to maximize the net price received by the patrons. The cooperative patrons can ignore the patronage refund in their production decisions or calculate a net price paid that includes the patronage refund.

Assume that the marketing cooperative is as cost-efficient as the IOF processor and is a spatial monopsony. As before, assume that the cooperative distributes all its net income as cash patronage refunds to the patrons. However, unlike a supply cooperative, the patronage refund increases producer revenues per ton rather than decreasing input costs, because the marketing cooperative adds patronage refunds to the prices it pays to its patrons. For example, if the cooperative attempts to achieve the objective of net income maximization, it will allocate a patronage refund equal to ANR minus price paid or $N_2 - P_2$ (Figure 10.2). The patrons will receive a net price defined as the patronage refund plus the price paid to the producer.

The cooperative can achieve the break-even objective when the producers' aggregate supply curve S intersects the ANR curve (Figure 10.2,

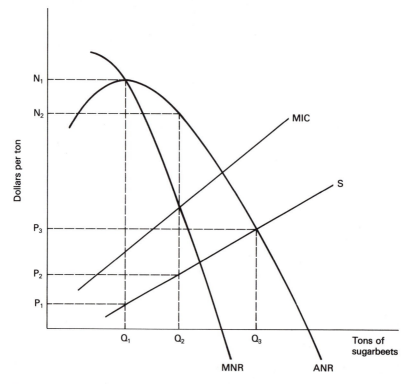

Figure 10.2 A marketing cooperative with analysis of different cooperative objectives.

Table 10.1). At this production level the net price received equals the price paid to producers, and producers lack an economic incentive to alter their output level from Q_3.

At any level of production beyond Q_3, the marginal cost of producing the additional output exceeds the price received P_3. At production levels below Q_3, the net price exceeds the marginal cost of producing additional output; thus individual producers have an incentive to increase output. However, as producers increase their output levels, they will decrease the net price they receive for their sugar beets.

Table 10.1 Summary of potential cooperative objectives, decision rules, output, prices, patronage refunds, and net prices presented in Figure 10.2.

Objective[a]	Decision rule	Output	Price received	Patronage refund	Net price received
1.Maximize net price received by patrons	$MNR = ANR$	Q_1	P_1	$N_1 - P_1$	N_1
2.Maximize net income like an IOF	$MNR = MIC$	Q_2	P_2	$N_2 - P_2$	N_2
3.Operate at cost or break-even	$ANR = S$	Q_3	P_3	$P_3 - P_3$	P_3

[a]Two other objectives can be pursued: sales maximization for the cooperative, by equating marginal revenue for the processed products to zero, and patron surplus maximization, where $MNR = S$.

Finally, the cooperative may want to maximize the net price received by the patrons. This occurs when ANR is at a maximum or when $MNR = ANR$. The producers then receive a price of P_1, and the patronage refund equals $N_1 - P_1$. Although producers may want to receive the maximum net price possible from their cooperative, this objective conflicts with that of net income maximization for producers. At the maximum net price N_1, the marginal cost of producing sugar beets is considerably below this price level. Producers can thus expand their net income by expanding their sugar beet production. If producers could sell an unlimited quantity at the maximum price, the highest price would be consistent with producer net income maximization. Producers would expand production to equate their marginal cost of production to the net price received. However, this is not possible because increased production beyond the maximum net price quantity implies a lower average net revenue.

Are Marketing Cooperatives Different from IOFs?

The objective of maximizing net income like an IOF can be achieved if producers make production decisions based on price received rather than on net price. If the cooperative pays the price of P_2, producers will only supply Q_2. The production decision will be based on P_2, not N_2.

However, if producers make production decisions based on net price received, output level is no longer stable. At Q_2 the marginal cost of producing an additional ton of sugar beets is only P_2, while the net price received is N_2 for that output level. Producers have the incentive to increase output because they can make a profit by expanding output. For example, if only one producer increases output, and the cooperative accepts the additional output, ANR will decline only slightly; therefore, it is profitable for the individual to overproduce. However, if all the patrons expand production, they will eventually end up expanding production to the $ANR = S$ level.

Herein is a major difference between cooperatives and IOFs. The IOF will maximize the value of the firm and restrict the production level. On the other hand, cooperative patrons have an incentive to increase production individually, even though as a group they are better off to restrict output. Thus one of the arguments in favor of cooperatives is that marketing cooperative patrons will expand production beyond the IOF level.

If a marketing cooperative obtains market power to increase consumer prices and to increase net revenue surplus for producers, the producers will receive a higher net price. Producers will then expand production levels, which will cause prices at the retail level to drop. Clearly, cooperatives cannot achieve the same type of market power as an IOF, because patronage refunds encourage output increases by patrons. On the other hand, if a marketing cooperative can restrict the expansion of output by producers and/or prevent other firms from adding processing capacity, this benefit of expanded output for consumers will not be realized; however, the cooperative will be increasing the net price for its producers.

The Significance of Instability

The difference between the objectives of $MNR = MIC$ and $ANR = S$ is not always a significant issue. If the ANR curve is relatively flat or horizontal for a fairly broad range of output levels around the producers' supply curve, the impact of the instability decreases.

If ANR is horizontal, then $MNR = ANR$, and the three cooperative objectives (Table 10.1) result in identical prices. This situation exists in industries whose demand curve for a specific cooperative is horizontal and whose average total cost curve flattens out rather than rising sharply as volume increases. Midwest local grain elevators are probably an example of such a situation. Local grain elevators sell in a competitive market, in which the sales of a single elevator are not large enough to affect grain prices at the destination markets. Therefore, the demand curve for the elevator's grain is horizontal. Also, economies-of-size studies have shown average total costs to approach a horizontal level as the volume of grain handled increases. In the short run the average costs will eventually rise very sharply after a firm begins operating at full capacity. The long-run average costs are not going to rise as sharply because the firm can adjust the capacity of its physical plant.

The need for production coordination by cooperatives and patrons is the greatest for industries in which the demand for the processed product is rather steep. This causes the slope of the *ANR* curve to be steep. Small changes in quantities marketed can cause major changes in the processed product prices. Total returns to producers will decline rapidly for a relatively small increase in producer output levels. Therefore, producers need to enforce the necessary production coordination.

Enforcement of Coordination

Output coordination can be enforced in two ways. First, government can restrict production through acreage reduction programs, marketing quotas, marketing orders, grading systems, or other regulations. Second, cooperatives can coordinate production levels through production restrictions, specific penalty schemes, or patron education (Lopez and Spreen).

The most direct way of controlling production is to issue quotas to each member of a closed membership cooperative. (A closed membership cooperative allows members to market their products only to the cooperative and restricts the admission of new members.) The quota may be in the form of acreage or production restrictions. If the cooperative imposes quotas, it also must impose penalties for overproduction.

If the cooperative imposes monetary penalties, they need to exceed or equal the benefits gained from over production. In other words, the penalties should ensure that overproduction will be unprofitable. For example, if producers lack an alternative outlet, the simplest penalty is to refuse to process their excess production or to buy the product at a significant discount.

If quota certificates are used to obtain the desired level of output, the cooperative can also fix the total number of certificates it allows. For example, sugar beet cooperatives control the number of acres planted and then allow the members to buy and sell the limited number of certificates.

However, this approach causes a dilemma for new cooperative members. the price of the market certificates is determined by competitive bidding and reflects the producers profit from growing the crop. If the cooperative is successful and raises patron prices, there is a positive economic value in belonging to the cooperative. As producers bid for certificates, this benefit is capitalized into the price of marketing rights, and the initial owners of the certificates benefit from the price increase. But the certificate represents a cost to the new members for marketing with the cooperative. Therefore, certificate cost will represent a barrier to entry for the new patrons. However, if the cooperative does not expand productive capacity, a producer group may be able to form a new processing cooperative to add processing capacity.

Alternatively, cooperatives might attempt to educate patrons about the need for cooperation in achieving specific price and net income goals. However, education does not eliminate the economic incentives for overproduction by the patrons. The cooperative must have methods for enforcing production discipline even among enlightened cooperative patrons.

One of the principal problems with using only the education approach is that the cooperative inevitably must deal with the free-rider problem. Assume that a cooperative institutes an education program and convinces part of its membership to restrict production. Although all patrons will receive higher prices, the price will not be as high as what would have been accomplished if everyone had participated. And unfortunately, the most profitable producers will be those who overproduced and exploited the benefits resulting from the reduced production of other patrons. These producers had a so-called free ride because of the sacrifices made by other patrons. This is not to imply that cooperative education is useless, but rather that any effective education program must be accompanied by disciplinary actions and controls.

Industry Level Implications

In the long run, cooperatives and IOFs operating in a competitive input industry will produce the same level of output, will pay the same price for the producers' output, and will lack the ability to influence the price received by the producer. Therefore, the supply curve for these firms is horizontal, and marginal input cost is equal to the price paid for the input.

Both types of firms eventually recover total variable and fixed costs, and no incentive exists to enter or exit the industry. This long-run equilibrium is achieved when $ANR = MNR = P_1 = S = MIC$ (Figure 10.3). At this output level, the marketing cooperative also achieves the maximum price for the product. All the discussed cooperative decision rules result in optimal prices of P_1 and output of Q_1. Therefore, in the long run, cooperatives are not confronted with the instability discussed previously.

For example, in a competitive industry, excess net income will eventually disappear. An excessive net income level is above a normal long-run net income level and is sufficiently high to attract additional capacity into the industry. As marketing cooperatives and IOFs enter the industry and expand processing capacity, prices for the processed products will decline and excess net income will disappear.

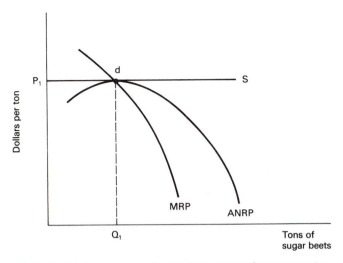

Figure 10.3 Stable long run equilibrium for a processing cooperative.

Therefore, one might expect that the economic incentive for cooperative membership would dissipate in the long run, because a cooperative member could obtain the same price from an IOF as from the cooperative. However, as discussed previously, the justifications for cooperatives are much more extensive than simply the price paid for the product.

Furthermore, some analysts question whether long-run equilibrium is ever really achieved in agricultural markets. Because agriculture is a spatially dependent industry, the markets for products are regionalized. Although numerous processing plants of a specific type may exist nationally, producers in a specific region may have to depend on a single processor as the only market for their product. Markets that appear to be competitive on a national basis may actually have local markets with considerable market concentration. Also, industry entry barriers may make long-run competitive equilibrium impossible.

In such markets, marketing cooperatives are confronted with an upward sloping supply curve (Figure 10.4), and thus have an incentive to achieve the maximum price for their producers in the long run. Cooperatives can achieve this objective either by altering their processing capacity or by altering the supply curve. If a marketing cooperative is confronted with a producer supply curve S_3, the cooperative and its patrons have an incentive to increase producer production capacity or to reduce the cooperative's processing capacity, because an increase in producer production capacity shifts the supply curve to the right toward S_1.

If, as a result, the cooperative is confronted with too much producer production capacity, as indicated by S_2, the cooperative has two alterna-

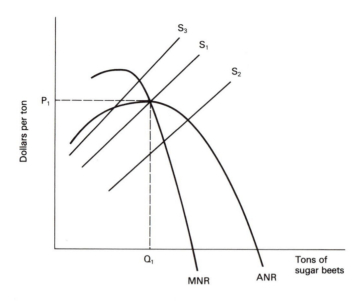

Figure 10.4 Stable and unstable equilibriums for a marketing cooperative.

tives: it can reduce the producer production capacity, or it can expand the capacity of the processing cooperative. A reduction in producer productive capacity causes the supply curve to shift left toward S_1.

GAME THEORY AND COOPERATIVES

Thus far, our discussion of the various economic models has not included an analysis of the internal choice process of cooperatives. The results of the internal decision process—namely, specific cooperative objectives—have simply been assumed. It is now helpful to discuss group choice issues, especially because cooperative managements, boards of directors, and patrons are paying increasing attention to this issue. As farms become less similar in size, supply cooperatives are under increasing pressure from large farms to institute price discount schedules for large-volume purchases; small-farm patrons are unable to take advantage of these volume discounts. Also, farms are becoming increasingly specialized in specific commodity complexes. Instead of the typical farm of the 1940s which included different livestock enterprises and field crops, the typical farm of the 1980s may have only field crops or may specialize in producing only hogs, rather than hogs, cattle and turkeys. Thus, intensified fighting among commodity groups occurs in some diversified marketing cooperatives, because the patrons' self-interests have become less homogeneous.

Game theory involves the study of situations in which two or more members of a group are at least in partial conflict (Chiang). For example, cash grain farmers in a processing cooperative may want to expand the cooperative's soybean processing division, while the dairy farmers may prefer to expand the dairy processing unit. Since the cooperative probably does not have enough equity capital to expand both divisions, the two farmer commodity groups are at least in partial conflict. Game theory examines how and what factors determine the cooperative's decision to expand only one of the processing divisions.

The two major game categories are games of chance and games of strategy. Games of chance are games in which no skill is involved (i.e., betting on coin tosses). Games of strategy involve deliberate choices or courses of action that imply specific outcomes. The application of game theory to cooperatives involves games of strategy.

In this section we present the recent game theory research and implications developed by Sexton and Staatz. Those interested in more specific applications of game theory to cooperatives can read the referenced articles by Staatz and Sexton.

Cooperative Games

Many cooperative choices involve group decisions that can be conceptualized as "cooperative" games. These groups of individuals can gain from acting jointly. Successful cooperative action, however, requires that the group members communicate and bargain, as well as compromise, in order to determine rules for the allocation of costs and benefits. Once they have agreed on allocation policies, the individuals must then make and keep binding commitments to follow these policies (Staatz).

For example, when farmers act jointly to build a processing plant, they must determine who will pay and benefit from the processing plant. Producers may be required to sign a contract which requires them to deliver their products to the cooperative and which specifies how much and when the cooperative will pay the producers.

Although cooperative operations can be perceived as group actions by patrons, the actual coalitions formed may consist of individuals other than cooperative patrons. Creditors, managers, employees, and other cooperatives may all be involved in the formation of a coalition to influence actual cooperative decisions.

If the stakeholders change, the feasible coalitions may also change. Because of the rapid change in the structure of agriculture, many cooperatives are concerned about whether or not they can maintain their traditional coalitions. There is some question whether cooperatives can continue to serve all farmers effectively or whether they must concentrate on a specific category of farms, such as large or small farms.

Cost Allocation

The basic insights gained from game theory can assist in understanding how cooperatives must deal with these issues, especially cost allocation. Cost allocation is particularly important to cooperatives as they attempt to implement price discount schedules based on volume or deal with conflicts between commodity groups within the cooperative.

Assume that a service cooperative is attempting to allocate the costs of providing a specific service to patrons. The farms of patrons are heterogeneous in their cost and size characteristics. Also, assume the following:

1. The total costs of any coalition of producers attempting to produce the service is less than or equal to costs of producing the service individually. An economic incentive exists for some producers to form a cooperative.
2. The farmers have three choices: (a) purchasing the cooperative's service, (b) purchasing the service from a competing firm, or (c) forming a new coalition of dissatisfied producers who leave the original cooperative.
3. The demand for the service by one patron does not affect another patron's demand for the service. The patrons decide to do business with the cooperative based on their own self-interest, not on what is best for the group.
4. Some of the costs in producing the service are joint costs or, in other words, costs that cannot be allocated to serving a specific patron. Because the costs cannot be specifically allocated, flexibility exists in how these costs are allocated among the patrons.

Given these assumptions, the board of directors and management must determine how to allocate the costs among the membership. Feasible allocation schemes may include as recipients all of the cooperative's current patrons or smaller coalitions (Staatz). The possibility exists that no allocation scheme is feasible and the cooperative will thus dissolve.

Several factors affect the allocation scheme selected by a board of directors and management. Because bargaining involves uncertainty, single members or coalitions of members attempt to influence the cost allocation by threats and counterthreats. In evaluating the validity of this posturing by the membership, we must identify (1) how the costs to the other members are affected if a coalition or single member exits from the cooperative and (2) what service costs the coalition or single member can obtain outside the cooperative.

Let us consider the price discount issue and assume that large farms with large transactions represent a major proportion of the business volume of a local cooperative. The withdrawal of their business implies a major reduction in business volume. If the supply cooperative is operating in the region of increasing average total cost (a rare situation), the cooperative may actually move closer to minimum average total cost. Thus the threat of withdrawal is not a major concern to the smaller farms. But if the cooperative is operating in the declining-cost region of the average total cost curve (a much more common occurrence), the loss of a major proportion of the business volume represents a major threat to smaller farms in terms of higher average total costs.

Another feature of the larger farm or a coalition of larger farms is the ability to achieve economies of size. A large farm may be able to buy direct from a chemical manufacturer and receive volume discounts from the manufacturer. Because of these two features, large farms can attract lower-cost alternatives from other agribusiness firms than can small farms. Because of these abilities to affect other members' costs or to establish lower-cost coalitions, owners of larger farms have a relatively strong bargaining position in the cooperative decision-making process.

Doing business with both larger and smaller farms also increases the uncertainties involving the bargaining process. If a cooperative has a relatively homogeneous membership, it can determine quite accurately the strength of its specific coalitions. Board members and management will have a fairly realistic perspective of the estimated costs of members exiting from the cooperative.

Cooperative Instability

As discussed earlier, a cooperative will also fail to achieve minimum net price for members and thus become politically unstable if it operates at sales levels beyond minimum ATC. This problem occurs when a coalition of members can lower costs by forming a new cooperative with a lower sales level. This problem also confronts diversified cooperatives if some divisions generate net income and others do not. If the members patronizing successful divisions do not patronize the losing divisions, internal conflicts will develop. Some will thus advocate divesting unprofitable divisions or forming a new, more profitable cooperative. However, these strategies may be offset by cost savings in the business function areas, such as finance or marketing of the diversified cooperatives.

Fairness of Allocation

Game theory models are based on the recognition that a large number of feasible cost allocations are possible and that the board and management decisions on cost allocation will be somewhat arbitrary. Concepts of fairness, self-interest, and cooperative principles may all influence decision makers in determining appropriate cost allocation.

Changing The Game

Cooperative management, boards of directors, and patrons all have the potential to change the institutional rules governing the cooperative.

Management can change operating procedures and policies, boards of directors can alter policies governing management, and patrons can alter bylaws and articles of incorporation.

In addition, state and federal laws may be altered to affect the bargaining power of specific coalitions. For example, in the past many states required a two-thirds vote of the membership to approve a cooperative merger. However, because of the difficulty in convincing membership coalitions to favor the mergers, a number of states have recently passed laws enabling cooperative mergers to be approved by one-half of the membership. This kind of alteration in a voting system can greatly affect the relative bargaining power of specific coalitions within cooperatives.

The cooperative can also attempt to alter patron perceptions about the benefits of cooperative membership or about the relative importance of long-run versus short-run gains. Another alternative is to alter the product choices available to patrons. For example, soil testing and fertilizer can be sold as a package or as separate products.

DYNAMIC MARKETS AND FUTURE THEORIES

The cooperative theory we have discussed covers the comparative statics of cooperatives and IOFs in stable business environment—the theory of the firm and game theory models presented implicitly assume stable markets with fixed technology. A major frontier of cooperative theory will be the analysis of potential decision rules for cooperatives competing in international markets and/or in markets with rapid rates of technological change. These will make coalitions even more unstable and require shorter investment payback periods. The future role of cooperatives will thus be highly dependent on their ability to develop business strategies for effective competition in such markets.

SUMMARY

When patrons sell their output to a marketing cooperative, the cooperative uses this output as an input in its production process. Cooperative patrons are thus input suppliers to the marketing cooperative. They choose to patronize the cooperative rather than an IOF because they desire a higher price for their products, and a positive patronage refund increases the net price they receive.

IOFs do not distribute patronage refunds. Net income is distributed to investors based on investment rather than patronage. Thus, to maximize net income, they will simply purchase inputs from producers until the marginal net revenue from selling the processed products equals the marginal input cost of buying the input. If an IOF firm is competing for producer production in a competitive input market, the price paid for the input is equal to marginal input cost. If the IOF's input purchases affect the input's price level, the marginal input cost will be higher than the input's price.

Marketing cooperatives often must use a different strategy, however, because they often have two other possible organizational objectives besides net income maximization. If patrons ignore the patronage refund in their marketing decisions, the cooperative can accomplish these goals through its pricing policy. However, if patrons include positive patronage refunds in their production decisions, cooperative patrons will expand output.

Marketing cooperatives may attempt to coordinate producer production levels by using production restrictions, specific penalty schemes, and education. To be effective, any penalties imposed must exceed the benefits of overproduction to the producer. Educational programs must also be accompanied by disciplinary actions, or else production coordination efforts will be hampered by free riders.

Because cooperatives operate according to group decisions, game theory can provide a conceptual base for analyzing cooperative decision making within the firm. Cost or revenue allocation schemes developed by cooperative boards of directors and managers will reflect the viability and strength of various coalitions associated with the cooperative.

DISCUSSION QUESTIONS

10-1. How is a marketing cooperative different from a supply cooperative in terms of objectives, the relationship between the patron and cooperative, and the role of patronage refunds?

10-2. Why can the average net revenue (ANR) curve be considered the demand curve for the producers' product? What decision rule does the IOF use to maximize net income? What is the underlying logic of this decision rule?

10-3. What are the three potential marketing cooperative objectives? How do these objectives differ in terms of their resulting decision rule, producer input level, price received, patronage refund, and net price received?

10-4. Which potential marketing cooperative objectives lead to an instability if producers use a net price in their production decisions? Is this instability good or bad, in terms of conduct and performance, from the perspective of the producer? of society?

10-5. What are the basic methods a cooperative can use to enforce coordination into the cooperative system? What are the advantages and disadvantages of each method?

10-6. In long-run competitive equilibrium, will producers selling to a cooperative receive a higher net price for their production than producers selling to an IOF?

Defend your position graphically. Does your answer change if the market is not competitive? Why?

10-7. How is game theory different from the standard theory of the firm?

10-8. A local marketing cooperative processes three vegetables: sweet corn, peas, and beans. Given the processing and marketing costs, sweet corn is twice as profitable as peas and beans. Assume that each producer produces only one of these vegetables. The sweet corn producers want a separate patronage refund based on the profitability of each crop, while peas and bean producers want a single patronage refund for the whole cooperative. Based on game theory, what factors will determine the possible outcomes of this conflict?

10-9. At the coffee shop a local cooperative manager is informed that the five largest farms in the cooperative's trade area are going to form a buying club to purchase their inputs. The manager does an account analysis and discovers that 50% of the cooperative's business is associated with the five farms. You have been retained as a consultant to advise the manager. Develop a plan of action for the manager and specify possible solutions to the problem.

10-10. You are on the board of directors of a financially strong local cooperative attempting to merge with another cooperative of approximately the same size. The second cooperative is in a weak financial condition. It has little owner equity and has been unprofitable in three of the past four years. Your manager has indicated that the cooperative has an 80% chance of becoming as profitable as your current operation and a 20% chance of continuing to be unprofitable. However, to accomplish this improved performance, a major business reorganization is necessary. The second cooperative's board of directors wants your cooperative to accept its stock at par value and to double the number of board members so that all board members can retain their positions. Based on concepts of fairness, self-interest, and cooperative principles, what would your position be on these two issues?

REFERENCES

CHIANG, ALPHA C., *Fundamental Methods of Mathematical Economics*, 3rd ed. New York: McGraw-Hill, 1984.

HELMBERGER, PETER G., and SIDNEY HOOS, "Cooperative Enterprise and Organization Theory," *J. Farm Econ.* 44(1962):275.

LEVAY, C., "Agricultural Co-operative Theory: A Review," *J. Agr. Econ.* 34(1983):1.

LOPEZ, RIGOBERTO A., and THOMAS H. SPREEN, "Co-ordination Strategies and Non-member Trade in Processing Cooperatives," *J. Agr. Econ.* 36(1985):385.

SEXTON, RICHARD J., "The Formation of Cooperatives: A Game-Theoretic Approach with Implications for Cooperative Finance, Decision Making and Stability," *Am. J. Agr. Econ.* 68(1986):214.

STAATZ, J. M., "The Cooperative as a Coalition: A Game-Theoretic Approach," *Am. J. Agr. Econ.* 65(1983):1084.

VITALIANO, PETER, "The Theory of Cooperative Enterprises—Its Development and Present Status," *Agricultural Cooperatives and the Public Interest,* ed. B. W. Marion, NC 117 Monograph 4, College of Agr., Univ. of Wisconsin-Madison, 1978.

Part IV
Marketing

11

*Product
And Pricing
Strategies*

David W. Cobia,
North Dakota State University

Bruce Anderson,
Cornell University

MARKETING CONCEPTS

Cooperative marketing is satisfying the wants and needs of customers, whether those customers be farmer members or consumers. It involves all the business activities between member-patrons and their cooperatives, as well as the activities that direct the flow of goods and services from cooperatives to final consumers. Marketing includes identifying the wants and needs of customers and the goods and services required to satisfy these needs, making sure those products are available when and where customers want them, and communicating the facts about these goods and services so that customers know what they are buying. It also involves pricing the products to maximize their value to member-patrons. This definition applies to the marketing strategies of all cooperatives. This chapter applies equally well to supply and service cooperatives as to marketing cooperatives. However, Chapter 12 pertains mainly to marketing cooperatives.

The primary objective of a cooperative is to enhance the economic well-being of member-patrons. In traditional marketing management there are four major marketing policy areas considered to fulfill this objective: product, place (channels of distribution), price, and promotion.

These are commonly referred to as the 4 P's. Cooperatives employ unique operating methods in each area as a result of the patron-owner relationship and by government statutes and regulations. So, besides dealing with standard marketing decisions, cooperatives also must determine policies (e.g., pooling, contracting, and bargaining) for paying for producer products.

Considerable variation may exist in the way cooperatives implement their marketing strategies, as long as the economic principles of cooperatives are preserved. That is, when net income is generated, statutes require distribution of patronage refunds to maintain business at cost. The finance chapters contain information about the mechanics and terminology associated with patronage refunds.

Consistent with the goals of this book, these marketing chapters will emphasize marketing strategies unique to cooperatives. Individuals interested in general marketing management strategies should consult textbooks such as McCarthy and Perreault. For information on a market-level agricultural approach, see the book by Kohls and Uhl. Branson and Norvell as well as Rhodes treat both firm- and market-level concepts of marketing from an agricultural perspective. For market structure concepts, see Caves and Section III of Marion.

Alternatives

Alternative marketing strategies employed by farmer-owned cooperatives can be divided into those that leave basic strategies to the individual and those that employ group marketing strategies such as pooling or bargaining. Cooperatives using the former must also decide whether to pay at delivery or use a delayed-payment plan. We have represented these alternative strategies in Figure 11.1.

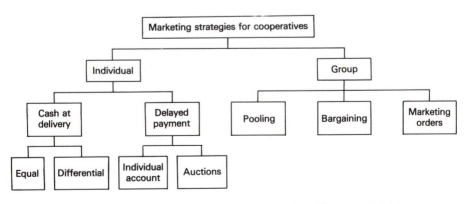

Figure 11-1 Alternative marketing strategies employed by cooperatives.

In addition to pricing and payment strategies, cooperatives must decide on product line or degree of diversification (the number and variety of products and services). Would the cooperative serve members best by concentrating on one product or by spreading fixed costs and expertise over many products? The degree of vertical integration is another consideration. To what extent should the cooperative integrate backward to sources of supplies or forward to export market and final consumer products?

This chapter focuses on the two major components of a marketing program: a product strategy, and pricing strategy. Chapter 12 includes several topics unique to marketing in cooperatives. They include payment policies, marketing contracts, pooling, bargaining, marketing orders, and international trade.

PRODUCT POLICY

A product strategy involves policies that provide a customer orientation and indicate what products a cooperative will carry.

Product Line Considerations

Cooperatives should carefully determine the appropriate mix of goods and services that they will offer their members. Some issues of concern are breadth of product line, the handling of single or multiple products, and the positioning of products with respect to quality, performance, and image that will best serve customers.

Any firm considering expansion of product line should consider such issues as spreading of risks over more products and operating more efficiently by utilizing the seasonal demands in labor and plant capacity. On the other hand, they must also avoid spreading management, financial, and other resources too thin.

Cooperatives should also determine whether it will really be advantageous for them to expand in the first place. Several arguments favor a single-product or specialized cooperative. Such a cooperative, for example, has more uniformity among its patrons. This leads to several advantages. For example, business at cost (no group subsidizing another) is easier to carry out, and uniform pricing practices are more practical. As a result, member education and advertising and promotion strategies are simplified. Board members can also more readily grasp and deal with problems because they can specialize in a narrow product or service category.

On the other hand, a wider product line and service offering is convenient to patrons. Improved economies can also be realized since additional product lines can share some of the same fixed costs. Further, multiple-product cooperatives may be better able than single-product cooperatives to move into a void created by the exit of a traditional supplier or marketer. Finally, board members benefit by exposure to a wider variety of opportunities and contacts.

Farmers depend on several inputs and services for their farming operation. Many farmers desire a business that can supply most, if not all, of their input needs. As a result, most supply cooperatives carry more than one input commodity. In contrast, farmers may be selling only one farm product and require the marketing services just for that one commodity. In addition, the required marketing functions vary between commodities. For example, the marketing of milk is considerably different from the marketing of grain.

The relative importance of these issues varies, depending on the circumstances. Several questions can be used to evaluate their relative importance.

1. Will a new product or service duplicate adequate services already provided by another cooperative in the geographic area? If so, members may be better off by working through the other cooperative to avoid unnecessary duplication.

2. Is there a reasonable assurance that net income generated will repay the investment and provide adequate returns to patrons?

3. Will the expanded product line fulfill a genuine need of members, one that they are willing to pay for and patronize?

4. Will the product complement the existing management and physical plant? This should be reflected in calculating return on investment.

Evaluating these factors is a responsibility of the board. Appropriate answers will probably be different in different parts of the country, and these answers will change with time.

Consumer Product Marketing

An issue related to breadth of product line is forward integration, meaning the extent to which cooperatives should take products closer to the consumer. Several U.S. cooperatives are involved in marketing consumer products. Some of these organizations started as bargaining cooperatives, but over time evolved into vertically integrated organizations that sell branded products to food wholesalers, retailers, and institutions.

The general marketing strategies of these cooperatives are no different from those of other cooperatives or even investor-oriented firms (IOFs). However, cooperatives that market consumer products face several additional challenges. A few of these challenges are discussed in this section. They include the following: (1) developing an effective customer-oriented organization, (2) determining the appropriate role of new-product development in the product life cycle, and (3) establishing a strong and well-respected brand name.

A Customer Orientation   To be successful at marketing, a cooperative must have a customer orientation. This is true whether the cooperative is a supply or service cooperative and its customers are member-patrons or if it is a marketing cooperative and its customers are processors, wholesalers, retailers, and final consumers. The motivation for a customer orientation is to increase sales, capture higher marketing margins, and pay higher returns (patronage refunds) to members. But these benefits do not occur automatically. To market consumer goods successfully, a cooperative must offer unique products that customers want or need. This is a process that generally involves incorporating real and perceived services into common commodities. Doing it well requires experienced, creative, and customer-oriented marketing personnel, who often command salaries higher than many cooperatives are accustomed to paying. However, the added benefits can be well worth these additional costs, because cooperatives that properly market to consumers can command a premium price for their products and provide more or better services to members.

Premium prices are possible only if customers view the cooperative's products and services as unique and valuable ones for which they are willing to pay a premium. Thus cooperatives that market products must first research and understand current and potential customers and then develop products that meet customers' ever-changing wants and needs. Good marketing makes a cooperative distinctive by offering members and customers products and services with a little something extra.

Although the process sounds simple, it often is a more difficult one for cooperatives than for other types of organizations, because agricultural cooperatives place primary emphasis on satisfying their members, as owners rather than customers. Consequently, there are likely to be conflicts that will arise in both supply and service cooperatives, as well as marketing cooperatives.

For example, today's customers of marketing cooperatives demand extremely high-quality products, but high quality entails additional costs, and some members who worry about those costs do not understand why lower-quality products cannot be shipped and paid for. Others cannot understand why the cooperative should dictate cultural and production practices. As a result, it is often difficult to please customers without losing crucial member support.

Consider the following illustration. A few years ago a marketing cooperative with a well-recognized brand name unilaterally banned its growers from using a particular pesticide which was beginning to receive media attention. Although the pesticide had not been banned by state or federal health agencies, the cooperative banned it to increase the confidence of customers and to avoid the possibility of future negative publicity. Many members at first objected to the ban. However, since that time, public concern about the pesticide has increased, and the cooperative is pleased to assure its customers that the pesticide is not used on their products.

Another conflict sometimes arises when a cooperative limits the amount of a product it will take from its members. Most brand-name cooperative products have an extremely high-quality image, allowing the cooperative to charge premium prices. But the cooperative cannot charge premium prices if it floods the market with its products. This means that the cooperative may need to allocate and limit production. Such a system may not be practical in some types of cooperatives such as grain cooperatives.

Another conflict can result from a related issue. To charge premium prices, the cooperative may try to make the product attractive to certain groups of consumers through its product offerings, positioning, and advertising. But sometimes members find it difficult to understand and appreciate the objective or product image the organization is trying to create among the consumers it is trying to reach. One example is a supply cooperative that tries to attract suburban customers, which increases store traffic and creates inconvenience for members. The purpose of such a strategy would be to increase sales and profits for the cooperative.

Nevertheless, agricultural cooperatives must remain committed to achieving a customer orientation whether those customers be members or not. Even though this commitment almost certainly implies that conflicts with member interests will increase, many cooperatives have found that the increased returns to members are worth the effort. Successful cooperatives have learned that if they can keep customers happy (and therefore buying their products), it is relatively easy to keep members happy by paying above-market prices and/or paying significant patronage refunds.

The Role of New Product Development The product life cycle is a very real concern for marketing, supply, and service cooperatives. It is also a phenomenon over which cooperatives have control.

All products go through different phases over time (Figure 11.2). In the introduction phase a product is introduced, and it struggles as product, pricing, distribution, and advertising problems are worked out and fine-tuned. Then if the product satisfies customer wants and needs, it will enter the growth phase and experience a rapid increase in sales and profits. But eventually, the increase in sales will slow down and the product will enter the maturity phase, during which it will probably approach the normal growth rate in the population. Finally, substitute products that are preferred by customers will probably arrive on the market and erode the original product's sales—this period is called the decline phase. These four phases constitute the product life cycle, a phenomenon experienced by most products unless aggressive marketing efforts are used to prevent maturity and decline.

The primary long-term strategy used to prevent maturity and decline is new product development and finding new markets for old products. This involves developing, testing, and introducing new products to meet the ever-changing wants and needs of customers. In addition, these new products must outperform the substitutes offered by closely related industries. The potential impact of successful new product or market development is shown by the dashed lines in Figure 11.2. Although no comparative studies are available, observation suggests that agricultural cooperatives are not overly aggressive in developing new products and markets, although there are some outstanding exceptions.

Why have cooperatives not been more aggressive in new product and market development when it is the primary way to increase the demand for agricultural products? There are no solid answers to this question, but we can speculate that the most likely reason has to do with the immediate costs versus the uncertain benefits of new product and market development, as well as with the probability of success and the likelihood of competitors copying one's success.

Let us analyze the case of new product development. New product development requires a moderate to significant amount of resources.

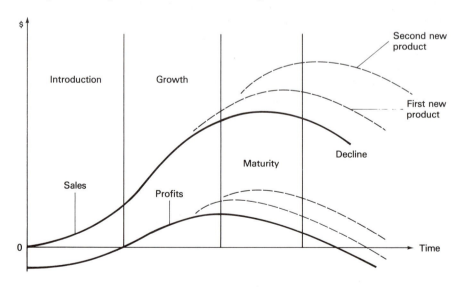

Figure 11-2 Product life cycle.

Moreover, success is not guaranteed. Although many estimates exist, let us say that only three of 10 new products are likely to become profitable. If a cooperative is active in new product development, it will bear the costs of developing and testing all 10 new products, but only three will contribute to earnings. This means that there is significant risk involved, especially for agricultural markets that have relatively low profit margins.

Moreover, cooperative members often cannot understand why only three of 10 new products are successful, especially since, as they argue, all 10 are costing them money. Although an organization should strive for above-average performance, members should realize that there will almost always, under any circumstances, be some new product failures. However, many members see these failures only as an indication that new-product development does not work.

But let's suppose that three of 10 products are successful and the members agree to continue developing them. What happens then? If a cooperative does develop a highly successful product, competitors will probably copy the product quickly and thus drive profit margins down. Moreover, since competitors have a tendency to copy only successful products, the original product developers must compete at a disadvantage against other firms that have not incurred the costs associated with unsuccessful products.

As a result of the behavior described above, firms dealing in mature markets with narrow profit margins have a tendency to let others develop new products and new markets and then copy them when they are successful. However, a problem arises if all firms in the industry adopt such a strategy, because then no one will be active in new product or market development. Product life cycles will continue their normal course through the maturity and eventually the decline phases. Clearly, then, both strategies can be risky. Therefore, members, directors, and management may need to reconsider whether it is in members' best interest to be active in new product or market development by assuming the associated risks and pursuing the potential rewards, or whether they should instead abstain from product development and allow the product life cycle to run its normal course.

Brand Names To market consumer products successfully, cooperatives also need to develop and use a recognized and well-respected brand name. Several well-known cooperative brand names are Sunkist, Welch, Tree Top, Blue Diamond, Norbest, Land O'Lakes, and Ocean Spray, just to name a few. For supply and service cooperatives, the firm's name and corporate image plays the same role as brand names in marketing cooperatives. A brand name can benefit marketing and new product development efforts. It gives the cooperative recognition and an image among consumers. The characteristics of this brand umbrella are referred to as a brand franchise. If a cooperative does not use a brand name, competitors can easily copy and accrue the benefits of its innovative marketing efforts. In fact, without a brand name or corporate image, it has little incentive to engage in new product or market development and other marketing activities, because consumers will probably not associate the product or activity with the cooperative.

Some cooperatives marketing consumer products do not have any well-established brand names at all, and many other cooperatives have brand names that are recognized only in limited geographical areas. As cooperatives expand, they often must decide whether to invest in and build on their current brand or corporate names or to purchase well-recognized brand or corporate names from other organizations. Current evidence suggests that most cooperatives think it is easier and more cost-effective to build on their existing brand names. Although the value of an existing well-recognized brand name is difficult to determine, it may be more economical to purchase a new brand name rather than to build on the existing brand name.

PRICING POLICIES

Cooperatives must set prices for products, supplies, and services and determine what to pay farmers for their products if they operate on a cash at delivery or buy-sell rather than a delayed payment basis. Pricing policies are critical because they dictate the distribution and timing of benefits.

Theoretical concepts, particularly those illustrated by cost and demand curves, are developed in the theory chapters and are used to explain concepts and to justify conclusions in this section. However, by reading this chapter, those not familiar with these theoretical concepts still should be able to understand the conclusions and intuitive reasons behind those conclusions. The demand curves used in this chapter represent member demand for supplies and marketing services. Thus concepts for both marketing and supply cooperatives can be shown in the same graph.

Price Level

Common pricing strategies employed by business in general include full cost, skim, market penetration, follow competition, and marginal cost strategies. These strategies may be relevant for cooperatives dealing with nonmembers, but cooperatives usually follow three basic approaches. They use price maintenance, offer favorable prices, or operate at cost plus a margin.

Price maintenance is a pricing policy designed to generate relatively large margins or margins as large as competitive pressures will allow. This means using prevailing market prices or charging relatively high prices for supplies and paying relatively low prices for products marketed. This approach is represented by maximizing net income (P_1 in Figure 11.3) or by minimizing net price (P_2). Favorable pricing means passing the benefits immediately to patrons in the form of low prices for supplies and high prices for products. Favorable price policy is characterized by minimum initial prices to patrons being equal to average costs or P_3. However, in most agribusiness industries, cooperatives have excess capacity; therefore, the demand curve crosses a declining average total cost curve (P_2, Figure 11.4). Thus a favorable price yields the lowest possible net price. (The more generic terms of *favorable prices* and *price maintenance* are used to avoid the awkward repetition of referring to the situation in both supply and marketing cooperatives.) The difference between the price maintenance price (P_1) and average cost is returned as a patronage refund (P_1-P_4) (Figure 11.3). If favorable pricing (P_3, Figure 11.3) is used, there are no patronage refunds to distribute because patrons already received benefits as favorable prices.

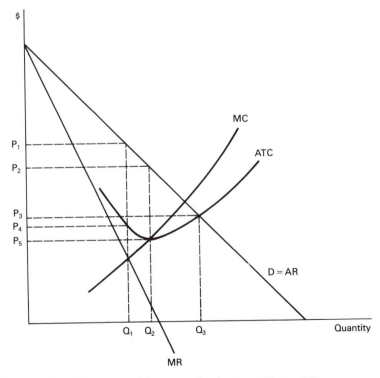

Figure 11.3 Alternative pricing strategies (same as Figure 9.3).

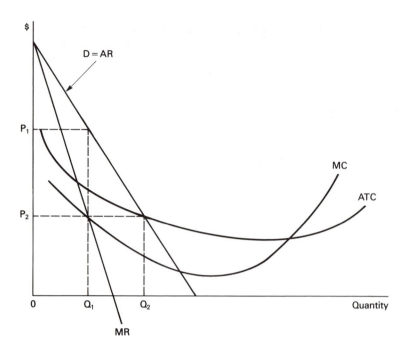

Figure 11.4 Cooperative operating in the declining portion of average cost curve (same as Figure 9.5).

Favorable price and price maintenance strategies are both legitimate cooperative policies and both achieve business at cost. The choice depends on the cooperative's objectives, methods to generate equity capital, cost structure, competitive environment, and ability to forecast costs. However, most cooperatives in the United States follow a price maintenance policy for one or more of the following reasons:

1. Management fears that competitors will retaliate if the cooperative offers favorable prices. Incurring the vengeance of competitors may force the cooperative to operate below cost and could ultimately result in the failure of the cooperative.

2. Large net income provides funds necessary to generate and service equity and to facilitate growth of the cooperative.

3. Management's stature is enhanced. Most annual reports highlight net income as evidence of cooperative success.

4. Free-loading nonmembers do not benefit if they are not given patronage refunds.

5. Costs are often unpredictable and price maintenance provides a margin for error.

On the other hand, some cooperatives have chosen to use favorable prices for one or more of the following reasons:

1. Management may wish to disturb the price structure in the market. This may be a major objective of the cooperative to begin with.
2. Favorable prices may be used as a strategy to encourage more producers to patronize the cooperative and become members.
3. Immediate benefits in the form of favorable prices are popular among members (Cobia and Navarro).
4. Occasionally, costs are predictable, allowing a cooperative to price at cost without financial risk.

Reduced cash flow, lower net income, and the possible drain on equity position are the most serious consequence of favorable prices. Patrons must understand that growth and equity redemption will of necessity be slow or absent.

A cooperative following a third price policy sets its margins according to cash flow needs, ignoring competitive influences. This policy generally can be followed only when the cooperative has either a dominant position in the market or a lower cost structure than competing firms have. If these conditions exist, cooperatives can set prices without much concern for competitor reaction.

Differential Pricing

Differential pricing means using discounts and premiums. Discounts are price reductions on supplies and services, and premiums are price increases on products marketed. Differential pricing is practiced in several industries. For example, airlines use peak versus off-peak pricing, telephone companies offer evening versus day rates, and utilities distinguish between industrial and residential rates.

Cooperatives may offer discounts used by other businesses, such as cash and season discounts. These discounts have few or no unique features among cooperatives and are generally accepted by members. Therefore, we will not discuss them further. In this section we focus instead on the contrast between average-cost pricing (often referred to as equal pricing) and differential prices based on volume per transaction (often referred to as equitable prices) and on differences in demand.

For price differentials to be successful, a cooperative must separate patronage into categories. Objective criteria for categories are based on cost and demand differences. Patrons may be segregated by volume, location, time, method of payment, and type of patron. The program must also be accepted by a majority of patrons.

Instead of offering quantity price breaks, cooperatives have traditionally used average-cost pricing. This is still the dominant approach, but cooperatives are finding that they must be more aggressive in order to compete in today's markets. Average-cost pricing is basing the price of a product for all patrons on the average total cost. This policy was appropriate as long as the size and enterprise combinations of farms were relatively uniform. But the changing size distribution of farms (more small, part-time, and very large farms), the economies of size associated with servicing different-size farms, and the presence of excess capacity in many agribusiness industries often has led to differential pricing by IOFs as a competitive marketing tool. Consequently, cooperatives often have lost valued patrons. Nonetheless, many directors of farmer cooperatives still do not want to change their policies, finding it difficult, if not impossible, to discuss or even consider the issue rationally.

Opposition to Differential Pricing Members of farmer cooperatives continue, for the most part, to oppose differential pricing, even though the practice and rationale justifying it is not new (Rust). Feelings against this practice seem to be particularly strong in certain geographic regions. Several reasons for this opposition have surfaced. Farmers might explain those reasons as follows:

1. Small farmers feel that they are always being discriminated against. We joined the co-op to avoid this. It just isn't fair. Large-volume patrons should not receive both a price break and a patronage refund.
2. Cooperatives are a people-oriented business. Treating different groups of members differently is therefore against the spirit of the cooperative movement.
3. Uniform pricing is a basic cooperative principle.
4. Equality is necessary to comply with state and federal statutes.
5. Differential pricing is a nuisance. The additional member education and record keeping is not worth the cost.

These ideas probably stem in part from commitment to the one-member, one-vote principle and from a feeling of frustration in fighting monopolistic powers on every side. Members join cooperatives to capitalize on economies of size not available to them as individuals, and differential pricing appears to them to defeat that purpose.

Objections to differential pricing reflect a misunderstanding. Equal treatment or uniform pricing is not a legal requirement. Different prices can be used as long as they are based on costs or on competitive pressures. Equal treatment is not a cooperative principle. In fact, strict adherence to equal prices violates the principle of business at cost because members who are less costly to service should probably be charged less.

Managers face the dilemma of losing business to rivals that offer differential prices, on the one hand, and of angering members opposed to differential prices, on the other. Some managers have been known to offer favorable prices secretly to keep the business operating effectively, while others accept and abide by the wishes of the board.

Less of this sub-rosa behavior by managers is now evident. Forced by competition to face the disadvantage in their traditional pricing policies, some boards have come to accept and even embrace the practice of differential pricing. It can be defended or justified because costs of providing the service or demand are different by class or category of patronage. If differential pricing is not practiced, valued patronage may leave the cooperative and remaining members will experience higher average costs as a result.

Equal Margin Pricing Equal margin pricing reflects differences in the cost of providing a service to different classes of patronage so that the net margin (price less costs) of each category is equal. For example, if it costs 4 cents/gallon less to make a 2,000-gallon delivery than to make a 200-gallon delivery, a 4-cent/gallon discount is appropriate for the larger delivery. From a theoretical perspective, the system will be more efficient if prices accurately reflect the costs of resources employed. Members may be motivated to change the scale or method of operation in a way that reduces resources employed by their cooperative.

Ideally, then, patrons, their cooperative, and society will all be better off. (If they are not, compliance with principles and theoretical concepts is meaningless.) All members, including those not receiving a price break, benefit from differential prices. Suppose that a cooperative operates at Q_1 and is confronted with competitors who are bidding away large volume members or those located in fringe areas (Figure 11.5). If the cooperative refuses to offer quantity premiums to keep this patronage, its volume will fall to Q_2. Average costs will increase from P_1 to P_2. Thus the remaining small-volume members will be worse off because now the cooperative's margins will have to absorb the higher costs. Both the small- and large-volume members need the larger combined volume of each other to keep average costs down. For this reason, small-volume members should, in their own self-interest, encourage quantity discounts and premiums based on the cost of providing the service.

The situation illustrated in Figure 11.5 occurs when both groups of members share joint fixed costs, such as physical plant and administrative expense. If there are no joint fixed costs, both groups may be just as well off forming separate cooperatives. In some cases, differences in operations and philosophies among various groups of members are so great that they overshadow common elements and have led to the creation of separate cooperatives. If two organizations exist for two types of members, the characteristics of members can be more uniform, and single prices can be more easily implemented.

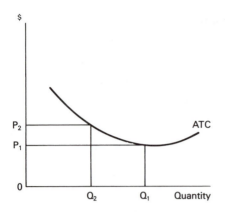

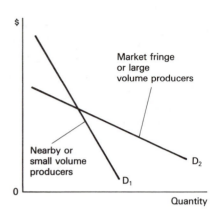

Figure 11.5 Impact of a change in volume upon average cost.

Figure 11.6 Different demand categories of patronage.

A major concern for cooperatives is that differential prices be based on cost. Therefore, management must periodically evaluate relevant costs to ensure that price differences accurately reflect associated cost. Otherwise, one class or category of patronage may end up subsidizing another. For example, if the cash discount for early payment is 10% when actual saving from cash payment is 15%, cash patronage is subsidizing credit patronage.

Demand Differences Demand for services by patrons will vary because of differences in the number of competitors and the vigor of resultant competition. Competitors will seek the patronage of some groups of patrons (e.g., large-volume and market-area fringe patrons) more than others. Therefore, the demand for services by large-volume, market-fringe patrons is more elastic than the demand by the small-volume and nearby patrons (Figure 11.6). Elasticity of demand is the measurement of patron responsiveness to a change in price. Patrons with a more elastic demand (D_2) are more responsive than those with inelastic demand (D_1) to a change in price. Consequently, it takes less of a price change to motivate them to move their patronage from one firm to another, because they have more alternatives to choose from.

For example, suppose that patrons are scattered over the market area of a grain marketing cooperative. The net farm price to members is the posted price minus the farmer's delivery costs. Because delivery costs depend on distance, producers located near the cooperative will receive a higher net price than those farther out. Therefore, there will be less competition for those near the elevator. Their demand is less elastic (D_1, Figure

11.6), while the demand of those on the market fringe (D_2) would be more elastic, because the net farm price offered by all competing elevators, including the member's cooperative, is about the same (D_2).

The demand curve of large-volume patrons may be more elastic for two reasons. Many agribusiness firms, particularly those eager for increased volume, will offer favorable prices to large-volume producers to help cover fixed costs. Large-volume producers are also sometimes in a position to capitalize on the transportation economies of large trucks, thus reducing the per-unit/mile delivery cost. This raises their net farm price and makes them more responsive to the prices of neighboring firms competing for their business.

Usually, a firm's net income is increased by decreasing the price to patrons with more elastic demand (P_2, Panel 1, Figure 11.7) and by increasing the price to patrons with more inelastic demand (P_1). As long as the additional revenue is greater than the additional cost, net income will increase relative to that achieved by uniform prices (P_3), but total output may not increase. However, this price discrimination analysis for a profit maximizing IOF does not apply to cooperatives. Pricing so that $MC = MR$ in each market (D_1 and D_2) yields a greater revenue with no increase in volume and therefore with no change in costs. As a result, net income is greater than it would be if the cooperative used an average price (P_3), of $MC = MR_1 + MR_2$ (Panel 1, Figure 11.7). This result does not benefit a cooperative's patrons because, after patronage refunds are distributed, patrons are no better off. Volume is the same with or without discriminatory pricing, and average costs are still the same. Therefore, after patronage refunds are deducted, the net price is the same with or without differential pricing.

But there is an advantage for many cooperatives to respond to demand differentials. A common scenario is characterized by a cooperative with excess capacity (resulting in marginal costs that are constant in the relevant operating range) and a policy of not responding to the more elastic demand of some producers, D_2. The normal price is P_1 and the volume is Q_1. If the cooperative changes its policy by responding to D_2, any favorable price (e.g., P_2) offered to D_2 patrons results in increased volume, lowering average fixed costs from C_1 to C_2, and benefiting all patrons (Panel 2, Figure 11.7). Panel 2 is the same as Panel 1 except that the marginal revenue curves (MR) were deleted and the average total cost (ATC) curve was added.

Price differences based on demand rather than cost place cooperatives in a dilemma. Even though all patrons may benefit from lower average costs resulting from increased volume, price differences are not based on the business-at-cost principle. But this dilemma can be resolved by returning differential patronage refunds to the two groups of patrons.

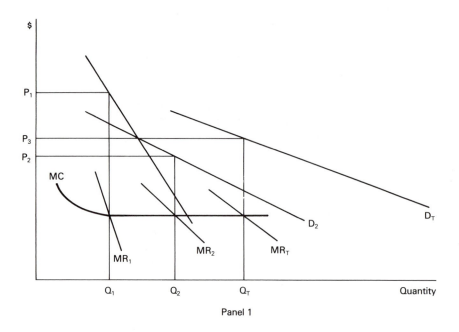

Panel 1

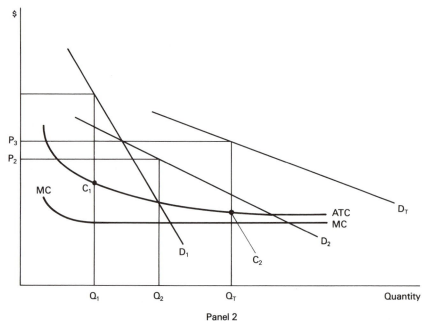

Panel 2

Figure 11.7 Differential pricing with different demand curves and constant marginal cost.

A cooperative must issue differential patronage refunds to maintain the business-at-cost principle if differential prices are based entirely on different demands and if costs of servicing the two groups are equal. In such a case, patrons receiving favorable prices should receive correspondingly lower patronage refunds. D_1 patrons should pay P_1 and receive a patronage refund of $P_1 - C_2$ (Panel 2, Figure 11.7). D_2 patrons should pay P_2 and receive a patronage refund of $P_2 - C_2$. The eventual cost to both groups is the same or C_2. In order for this scenario to operate, patrons must respond to prices (P_1 and P_2) rather than to the eventual decrease in cost or C_2. The net advantage to D_2 patrons is the present value of the $P_1 - P_2$ favorable price. D_1 patrons benefit from a lower average cost of $C_1 - C_2$.

Legal Implications Differential pricing is legally justified if there are differences in the cost of providing different services or if differential pricing is used as a defensive move to meet the prices of competitors (Greer). Equal margin pricing satisfies the first requirement, and the rationale given for responding to D_2 satisfies the second. Concern over making legal justification for different prices is unnecessary as long as differential patronage refunds ultimately result in uniform prices and business at cost.

Number of Categories To benefit from using price differentials, cooperatives must also carefully determine the appropriate number of categories and the brackets. Taken to the extreme, an excessive number of categories would result in a different price for each transaction, and the costs of administering such a policy could easily exceed any benefits. Categories should continue to be added only as long as the costs of administering an additional category are equal to or greater than the benefits. The number of categories and the brackets should also be based on physical factors that determine costs. A 7,000- to 8,000-gallon bracket on fuel deliveries correspond to the capacity of a tanker truck, for example.

When are Differential Prices not Justified? Despite all their benefits, quantity discounts and premiums are not advisable in some situations, especially if costs and/or demand for possible categories are the same. This is often the case in some marketing cooperatives when delivery is made by the patron and transaction costs are inconsequential. In other cases, the nuisance of keeping additional records and the extra burden of keeping members informed may exceed the benefits. If the cooperative has a spatial monopoly, it does not need to worry about the loss of large-volume members.

Cooperatives should also not use differential pricing if, as is often the case, the majority of members oppose it. Even if large-volume members leave, the cooperative will be obliged to continue with single prices as long as a majority of members insist on them. However, it must be remembered that the remaining members will eventually end up with higher costs. Hopefully, an education program about the realities of the competitive environment can change opposing members' feelings.

Netting

Netting is a policy of establishing prices so that the losses of one division are absorbed by net income from another division. In effect, the cooperative pools income and losses from its various divisions. Cooperatives use netting in cases when they need to spread risk, to absorb developmental costs associated with a new product, or to extend market area.

Other less noble examples also exist. In one case the board of a local supply cooperative voted to keep an automobile service bay open, even though they knew it was losing money. In that case, netting was not appropriate because members not using the service subsidized those that did. Netting often occurs because cost analysis is not done carefully enough for management to realize what is occuring. Unless a cooperative has special needs like the ones mentioned above, netting violates the business-at-cost principle and should be avoided as a deliberate policy. However, clearly no blanket rule can be mandated. Each situation must be decided on its own merits.

DISTRIBUTION AND PROMOTION STRATEGIES

Distribution and promotion strategies are other common concerns of marketing management. Distribution strategy refers to the organizational or institutional linkages (such as wholesalers and retailers) that perform marketing functions. To the extent that organizational linkages are unique among cooperatives, they are introduced in chap. 3 and Part VII.

The nature of cooperatives precludes use of any misleading advertising when dealing with members. Some argue against any advertising to members unless it is strictly informational. Authors of cooperative literature have generally not detected unique advertising and promotion strategies for cooperatives or discussed promotion to members under a separate topic of communication or member education. This book continues that tradition. See Chapter 19 for a discussion of such topics as promotion methods that are unique to cooperatives.

SUMMARY

Marketing involves satisfying the wants and needs of customers. Cooperatives face unique challenges and opportunities in developing and adopting strong marketing programs. The purpose of this chapter has been to explore some of the unique aspects of cooperative marketing. Particular attention was devoted to a cooperative's product strategy and pricing strategy. Cooperatives must evaluate the trade-off between the benefits of more uniformity among its members and specialization associated with a narrow product line with the benefits of sharing fixed costs and single-stop service associated with a broad product line. A few cooperatives have benefited from premiums associated with marketing consumer products. But the customer orientation required for this strategy often conflicts with the perceived primary emphasis traditionally placed by cooperatives on satisfying farmer member needs. Pricing strategies, involving price level and volume discounts and premiums, dictate how benefits are distributed. Price maintenance, use of prevailing market prices or relatively high margins, provides cash flow for the cooperative, enhances management image, and avoids the free-rider problem. Favorable prices, on the other hand, pass benefits immediately to patrons, encourages increased patronage, and may affect the price structure of the relevant market.

Equal pricing (charging the same price regardless of quantity) has been a traditional cooperative practice. The appearance of more large farms and increased competitive pressures have led to differential or equal margin pricing (offering discounts and premiums that reflect benefits from large volume or differences in demand). This sometimes controversial practice, if properly administered, benefits all members, is legal, and is in harmony with cooperative principles. In chap. 12 we examine special issues related to cooperative marketing.

DISCUSSION QUESTIONS

11-1. Define marketing.

11-2. What are the four major components of a marketing strategy?

11-3. What is unique about cooperative marketing?

11-4. List six brand names found in supermarkets that are owned and used by cooperatives. What advantages might accrue to a cooperative and its member-patrons from establishing a recognized brand?

11-5. Outline a presentation that you would make to a board of directors of a supply cooperative to convince them to adopt differential prices for members receiving tanker load deliveries of fuel.

11-6. What is the dilemma faced by cooperatives that are confronted with demand curves of different elasticities, such as members located nearby and those in the market-area fringe?

11-7. Can you reconcile netting with the business-at-cost principle? Why or why not?

11-8. Most agricultural supply cooperatives are organized as multicommodity cooperatives in that they handle several types of farm inputs. In contrast, most agricultural marketing cooperatives are single-commodity cooperatives. What might be the major reason(s) for this contrast?

11-9. Select a product and develop a brief marketing program for presentation to the cooperative's board of directors. The marketing program should include a product strategy, a distribution strategy, and a pricing strategy as well as an advertising and promotion strategy.

REFERENCES

BRANSON, ROBERT E., AND DOUGLAS G. NORVELL, *Introduction to Agricultural Marketing.* New York: Macmillan, 1983.

CAVES, RICHARD E., *American Industry: Structure, Conduct and Performance,* 5th ed. Englewood Cliffs, NJ: Prentice-Hall, 1982.

COBIA, DAVID W., AND LUIS A. NAVARRO, *How Members Feel about Cooperatives.* Dept. of Agr. Econ. Res. Rep. 86, North Dakota State Univ., 1972.

GREER, DOUGLAS F., *Industrial Organization and Public Policy,* 2nd ed. New York: Macmillan, 1984.

KOHLS, RICHARD L., AND JOSEPH N. UHL, *Marketing of Agricultural Products*, 6th ed. New York: Macmillan 1985.

MARION, BRUCE W., ed., *Agricultural Cooperatives and the Public Interest.* NC 117 Monograph 4, College of Agr., Univ. of Wisconsin-Madison, 1978.

McCARTHY, E. JEROME, AND WILLIAM D. PERREAULT, JR., *Basic Marketing,* 8th ed. Homewood, IL: Richard D. Irwin, 1984.

RHODES, V. JAMES, *The Agricultural Marketing Systems*, 3rd. ed. New York: Wiley, 1986.

RUST, IRWIN, W., *Providing Equitable Treatment for Large and Small Members.* Washington, DC: USDA FCS Info. 21, Dec. 1961.

12

Special Topics for Marketing Cooperatives

David W. Cobia,
North Dakota State University

INTRODUCTION

There are several topics that apply mainly to marketing cooperatives created by the owner-patron relationship of cooperatives that do not fit neatly within the standard marketing management framework. Therefore, in this special topics chapter we explore those issues. The first two topics include payment methods and marketing contracts. The second group of topics include marketing agencies in common, cooperative bargaining, and marketing orders. These are alternative forms of group action. The last section analyzes the role of agricultural cooperatives in international trade.

PAYMENT POLICIES

Payment policies apply primarily to marketing cooperatives. Cooperatives may pay cash for commodities on delivery or delay payment until costs and income have been determined.

Cash at Delivery

Cash at delivery means that cooperatives pay a cash price for and take title to products delivered by patrons. These products are then processed to a greater or lesser extent and sold in the market at the most advantageous price possible. Net income remaining after expenses is refunded to patrons. Most grain and oilseed marketing cooperatives operate this way.

There are several advantages of cash-at-delivery policies. First, farmers know at once what they will get; uncertainty is removed. Second, the cooperative reaps more goodwill and generates additional extra business as a result. Third, cash-at-delivery practices require somewhat less bookkeeping than delayed payment schemes. Fourth, the policy forces management to be on top of prices and costs, because it must constantly be alert to changes in order to post realistic prices. Finally, members generally have more control over marketing strategies

There are also some disadvantages as well: Cooperatives assume additional risk because they take title. (However, many of these risks can be transferred or offset by hedging or forward contracting.) In addition, considerably more working capital is required and members must create their own marketing strategies. Also, some errantly say that cash payment is not in agreement with cooperative principles.

Operating on a policy of cash payment at delivery is popular when producers have several marketing alternatives, especially when futures markets exist. This policy is more comparable to the operations of an IOF and therefore facilitates direct comparison between cooperatives and IOFs.

Delayed Payment

When a delayed payment policy is used, farmers deliver their commodities to the cooperative and may or may not retain title. The cooperative sells the commodities at the most advantageous prices. Expenses are then deducted and the patron receives the residual amount. Delayed payment is desirable because the cooperative's exposure to risk and financial requirements is at a minimum, particularly if it does not take title and thus avoids risks of ownership.

Moreover, the policy is clearly in harmony with the business-at-cost principle—one year's business does not subsidize another's—and it substantially reduces working capital requirements of the cooperative, because only partial payments are advanced to patrons before accounts have been closed. Patrons also like the system because they receive the same price that their products bring in the marketplace.

Members sometimes do not like the system because they do not know what they will receive until several months after delivery. Other disadvantages include the reluctance of individuals to join cooperatives with delayed payments and the difficultly of educating members about methods of delayed payment.

Cooperatives can establish delayed-payment practices by using individual accounts, commission sales, auctions, and pools. In the following sections we explain each of these practices.

Individual Account If cooperatives establish individual accounts, deliveries by each patron are handled and traced separately, and products from other producers are not commingled. Livestock trucking associations often use this approach. It is desirable because producers receive what their individual products bring in the marketplace and because it comes nearest to complying with the business-at-cost principle.

However, in many situations this method is so unwieldy that it is not economically feasible—the extra administrative expense of keeping track of every transaction could more than offset any benefit from cooperative activity. Even if it is feasible, relatively little market power can be exercised by cooperatives, because sales are fragmented.

Commission Sales Commission sales are similar to individual accounts in that each lot is handled separately. However, the cooperative sales agency, generally a creation of a federation of local cooperatives, handles products at a terminal market on a commission basis. These agencies typically handle products from both members and nonmembers. Because of the rapid decline or total disappearance of terminal markets, use of commission sales by cooperatives has also declined substantially.

Auctions Auctions are another vehicle that farmers can use to sell their products through cooperatives. In this case, payment is made shortly after the product is sold. Farmers may be attracted to auctions to take advantage of widely publicized competitive prices, low marketing costs, guaranteed payment, and reduced liability for marketing losses.

There are also some drawbacks to auctions because buyers must be physically present, because buyers may avoid auctions due to the loss of market power, and because only a limited amount of products can be sold per hour. Auctions are not well suited for large growers who find it advantageous to sell direct. Neither do auctions work for producers of semi-durable products such as apples, because a compelling reason for regular exchange does not exist.

One of the few farmer-owned auctions is Vineland Cooperative Produce Auction Association., Inc. (VCPAA) in New Jersey. This cooperative is the only survivor of 29 fruit and vegetable cooperative auctions that existed in 1949 (Harrington). However, it has reported consistent growth and financial strength. To join this cooperative, a $1.00 membership fee is required. Income is generated by a 3% commission on all sales. Auction sales account for 75% of the volume. The remaining income is generated from direct negotiations conducted at the cooperative between growers and buyers. Patronage refunds amount to about 1% of sales. The cooperative typically gives a 20% cash refund and operates on an eight-year revolving fund.

POOLING

Pooling is a distinctive cooperative practice. It is a delayed-payment scheme involving a signed marketing contract. It therefore incorporates the features of these two practices. Products of many producers are commingled and, after deducting expenses, the average net price received is paid to producers. Key elements of a pool are the sharing of risks, expenses, and revenues and the payment of an average price, with possible adjustments for product quality and for time and location of delivery.

Each cooperative pool has its own operating procedures. However, most have the following characteristics. Farmers sign marketing contracts (see the following section, "Marketing Contracts") with the cooperative that guarantees delivery of all or part of their production to the pool. The contract transfers all authority over marketing decisions (including timing, pricing, and further processing) to the cooperative and its professional management. An initial advance is paid to members upon delivery of the product. The advance is generally a percentage of the government support price or an estimated market price if no support price is available. One or more progress payments may be made as the product is sold out of inventory.

When all or most of the product has been sold, generally within 12 months of delivery, the pool is closed. A total value, including an estimated value of any remaining inventory, is determined for the pool. Operating and administrative expenses are allocated and subtracted. Any excess over previous payments is then distributed to patrons. This final payment results in zero net income for the cooperative or business at cost. A per-unit capital retain is generally withheld from payments to producers. These funds, ultimately returned in cash, provide equity capital necessary to finance the cooperative (see Part V).

Milk, fruit, vegetable, and nut cooperatives generally use marketing pools. Sunkist Growers, Sun-Diamond Growers, Ocean Spray Cranberries, American Crystal Sugar, National Grape Cooperative, and California Almond Growers Exchange are cooperatives among the list of top 50 cooperatives (Table 2.1) that use marketing pools. Grain and oilseed marketing cooperatives have tried pools with mixed results. One of the few successful pools for oilseeds is operated for soybeans by Riceland Foods of Stuttgart, Arkansas. Infrequently, supply cooperatives may pool orders for inputs and then shop for the best price. Pools are used in other countries such as Canada and Australia, to a greater extent than they are used in the United States.

Some publications discuss two types of pools—*seasonal* and *contract*. But contract pools merely involve purchase and sale or open-market transactions. Therefore, attention in this section is focused on seasonal pools.

Pools can be divided into two groups, *single* pools and *multiple* pools, depending on how they are organized. In a single pool all products from several producers are commingled and sold by the cooperative. The net proceeds, after deduction of expenses, are paid to producers. Each producer receives the average net price, usually with predetermined premiums and discounts for quality differences. In multiple pools, products may be segregated on the basis of grade, variety, time of delivery, and/or location. Each category constitutes a separate pool. Length of the pool depends in part on the marketing characteristics of the product. Some products (nuts, for example) are stored over a number of months; some (like tobacco and grapes), even for years.

Hammonds describes a pool operated by Calcot, Ltd. (a cotton cooperative headquartered in Bakersfield, California) as follows:

> Each member agrees to deliver all his cotton to the pools, with a yearly sign-out period from February 1 to February 15 for those wishing to terminate the agreement . . . . Each member agrees to deliver his crop steadily, as ginned. . . . All advance payments are made on the basis of U.S. government classifications. However, once the advances have been made, Calcot further classifies the cotton for sales purposes . . . according to six basic grade standards . . . a further separation [is] based on eligibility . . . for government loan programs . . . . The schedule for advance and progress payment is established each fall by the board of directors. Generally, the seasonal pool initial advance is made at delivery for an amount equal to that which the association can borrow on the cotton, either from the U.S. government or from lending institutions, minus a $3 per bale primary retain. Progress payments are then made during the marketing year as actual sales conditions warrant. A final settlement is generally made in July of each year for the equalized balance due to each pool, minus a 25-cent per bale secondary retain (Hammonds, p. 35).

The following illustration of a marketing pool (Table 12.1) has been adapted from Dunn (p. 8). For this pool, three members produce one commodity with three grades. Each grade is treated as a separate subpool. The deliveries of each member are tabulated by grade (as illustrated in the first section of Table 12.1). Producer Thor delivers 2,000 bu of grade 1, 4,000 bu of grade 2, and 8,000 bu of grade 3. Upon delivery, the pool member receives an *advance payment* of $1 per bu regardless of grade. The pool manager then sells the contents of the pool, obtaining the highest prices available. As the contents are sold, the pool receives money from which an interim payment of $0.50 per bu is made to pool members. When the pool's contents are completely sold, an average unit sales price for each grade is calculated on the basis of actual sales. Subtracting the per-unit advance and interim payment from the average price results in gross final payments perunit for each grade.

TABLE 12.1   Illustration of member pool receipts and payments

	Cooperative receipts from members (bu)			
		Grade		
Item	1	2	3	Total
Thor	2,000	4,000	8,000	14,000
Stressgaard	9,000	3,000	1,000	13,000
Loman	6,000	7,000	4,000	17,000
Total	17,000	14,000	13,000	44,000

	Payment & deduction schedule ($/bu)		
		Grade	
Item	1	2	3
Advance payment	1.00	1.00	1.00
Interim payment	0.50	0.50	0.50
Operating expense	0.05	0.05	0.05
Capital retain	0.03	0.03	0.03
Net final payment	0.57	0.45	0.30
Avg. price received by the pool	2.15	2.03	1.88

	Member payments		
Member	Payment	Calculation	Amount
Thor	Advance	14,000 x $1.00 =	$14,000
	Interim	14,000 x 0.50 =	7,000
	Final—grade 1	2,000 x 0.57 =	1,140
	Final—grade 2	4,000 x 0.45 =	1,800
	Final—grade 3	8,000 x 0.30 =	2,400
	Total		26,340
	Average price received: $1.8814/bu		

(Calculate payments for Stressgaard and Loman in the same way)

Source: Adapted from Dunn et al.

The pool costs $2,200 to operate. To cover these operating expenses, a $0.05 per bu deduction is taken from the average market price. In addition, a $0.03 per bu capital retain is withheld to cover the cooperative's need for capital. After all deductions have been made, the final net payment per bushel for each grade is determined. Producers receive a final net payment of $0.57, $0.45, and $0.30 for each bushel of grade 1, 2, and 3, respectively, that they delivered to the pool. In the "member payment" section of Table 12.1 an example of member payments to Thor is calculated.

Advantages and Disadvantages of Pools

Pools have three major advantages to producers—(1) an assured market, (2) shared risks, and (3) enhanced prices. These are important because opportunities for producers to sell their products in open markets have been and are continuing to decline. Increased vertical integration blocks out nonparticipants. Thus, signing up with a cooperative pool that has access to a given market may be the only way to gain access to markets.

Prices of agricultural products have been relatively unstable, especially in recent years. Farmers who participate in pools face less danger of having to sell at bottom prices, because all participants share the risk of some producers receiving a low price. Of course, they all share in the high prices, but this way no one alone has to bear the burden of a low price.

Enhanced terms of trade (primarily regarding price) are the major attraction of pools. Achieving better prices stems from the following conditions:

1. Marketing decisions are made by professional specialists who devote a majority of their time to seeking maximum prices for producers and to identifying and capitalizing on market trends, customer needs, and coordination of supplies with these needs. The cost of this expertise and focused effort is covered by a much larger volume than an individual producer could generate.

2. Improved prices can be achieved through control over a large supply. The cooperative is in a position to bargain for improved prices. Buyers are often anxious to enter preharvest agreements and even pay premiums for timely deliveries of guaranteed quantities and qualities. These factors often create more buyer competition, thus expanding demand.

3. Operating costs, including costs of assembly and of transactions with growers, can be reduced with a known quantity and control over pricing and distribution.

4. Selling a high-quality product or providing a unique service.

In spite of their benefits, pools are not adapted to all circumstances and not all growers want to participate in them. In the first place, producers lose control over marketing. Some producers may not wish to delegate the responsibility of marketing their products to a specialist; nor do they want to commingle the results and thus forgo some short-term opportunities. Commingling products may be disadvantageous to some producers, because the grading system may not reward superior quality differences nor accommodate their harvest schedule. Producers may also resist the necessary discipline required to fulfill quality, quantity, and delivery requirements of some pools. In addition, complexity and increased record keeping can lead to increased costs and confusion among members.

For pools to be successful, members must be committed. Commitment is typically created by expectations of higher-than-average returns and, in some cases, by having access to markets. Most successful pools have special features in addition to risk pooling and the use of marketing experts to enhance income. These features include special quality control, grading procedures, or delivery guarantees that make products more desirable to buyers. Other pool operating cooperatives (such as Riceland Foods, SunKist, and Ocean Spray) have integrated forward by marketing consumer-branded products.

MARKETING CONTRACTS

Effective commodity pools require marketing contracts. Contracts also are employed in bargaining and other marketing situations and can even be used by supply cooperatives.

Marketing contracts (also called member or marketing agreements) are written agreements between cooperatives and their members stating the rights and duties of both parties regarding how products will be marketed or purchased for a specified period of time. Contracts are typically linked with marketing cooperatives, but they are also used, although infrequently, as purchasing agreements. Contracts may in turn be tied to marketing pools, to individual accounts, to cash purchases, to commission sales, or to bargaining cooperatives. The cooperative agrees to a certain payment method and to other provisions necessary to market or acquire products to the best advantage of the member.

Legal Aspects

Agreement formalities between cooperatives and patrons range from nothing more than a purchase and sale to an agreement binding the cooperative to provide a service and requiring the member to patronize the cooperative. Contractual arrangements require cooperatives to define carefully the relationship they desire with patrons, and to spell out each party's rights and responsibilities under the agreement. Special consideration is given to contracts because the rules governing cooperative-patron relationships are so important and because contract-related provisions are contained in many cooperative statues.

The Uniform Commercial Code (UCC) in each state gives legal rules for sales and purchases, including those that occur between members and cooperatives. It also requires that certain terms be included to make the agreement enforceable. For example, suppose that a member signs a contract agreeing to sell and deliver a specified amount of grain to the cooperative at a stated price to be delivered in the future, and the cooperative agrees to its purchase. Several events may occur between the agreement date and the delivery date that might cause farmers not to deliver. They might, for example, find a better price elsewhere. When the cooperative attempts to enforce the agreement, the UCC will determine if it is a valid, enforceable contract.

Three issues are of interest in this brief review of common disputes. First, oral agreements are enforceable unless other forms are provided. However, the UCC requires a sales agreement for more than $500 be in writing. Second, although not all possible terms must be written, the agreement must contain enough information to alert the parties to their obligations. It would not be possible to enforce an agreement, for example, that did not specify how much was to be sold under the agreement. On the other hand, a specific price is not required as long as the agreement identifies a means (such as a market price quotation at a specific location) by which the price can be determined. Finally, a written agreement, as a rule, is binding only if it is signed by the participating persons. When a farmer agrees to deliver a commodity to the cooperative, the cooperative may write out a short agreement, signed by an authorized employee, and send it to farmers for their signature. If the farmer signs it, it becomes a binding agreement. If the farmer does not sign it within a specified time, the UCC claims that the farmer is still bound only if the farmer is considered a "merchant" of the commodity. States are split on whether a farmer is a merchant under those circumstances, and cooperatives must be careful to meet all conditions in their state for an enforceable agreement.

Cooperatives and members may also have contracts more detailed than the ones described so far. These agreements are usually marketing contracts under which farmers agree to market all or a part of their specified commodities through the cooperative. Most cooperative statutes mention this type of contract, although special laws are not required for it. Most statutes simply state whether the cooperative may or may not take title to members' products, recognizing the marketing and bargaining functions that cooperatives may perform. Many statutes also, however, place a limit (usually 10 years) on the length of time a marketing contract can exist.

Typical Provisions

Cooperatives should incorporate clearly stated provisions into the contract or into bylaws. Typical provisions include the following (from a marketing cooperative's point of view):

1. The percentage (all or part) of products the member must market through the cooperative.

2. An agreement that the cooperative will seek maximum net price for members. Management is typically granted authority to carry out marketing functions to accomplish these objectives.

3. The length of time the agreement is in force (e.g., one or three years). Conditions for renewal or termination often are included. The length of contract is largely determined by required investments and physical and marketing characteristics of the product. Commodities that are stored and marketed over a number of months or even years and that require substantial capital investments may require longer contracts.

4. The qualifications producers must meet for membership. First, they must qualify as farmers. In addition, they may have to qualify on the basis of their product, their location, and their ability to fulfill requirements of the marketing contract.

5. Authorization for per-unit capital retains to finance the cooperative. Per- unit capital retains are the prime source of equity capital.

6. Provisions for enforcement of the contract, penalties for noncompliance and a description of damages to be paid to compensate for breach of the contract. Damages are an estimation, commonly called *liquidated damages*, of the loss suffered. Many statutes also make it illegal for anyone to interfere with the cooperative's agreement with its patrons. Black argues that the effectiveness of contracts depends

> upon the promptness and thoroughness with which the initial breach of contract are handled. If the members and the Board turn their back on the initial contract breech, then the contract will be breached in greater frequency and severity. When this happens, the contract serves no useful purpose except to create distrust and confusion . . . . [I]f the initial breach of contract is promptly handled . . . then usually no other attempts at breaching the contract occur . . . . Action [must also be taken] . . . against the buyer. In most cases, the buyers are aware that they are purchasing a commodity that has been committed through the marketing agreement to a cooperative . . . . In some instances, the buyers of such committed products accept delivery to purposefully destroy the cooperative. Prompt legal action by the co-op discourages other buyers from encouraging members to breach their marketing agreements (Black, p. 229).

7. The method of payment for the product.

8. Control of product quality. Growing, harvesting, and delivery conditions may be specified to meet requirements of buyers and to minimize costs.

9. Control of quantity. This provision attempts to balance the quantity supplied with market requirements at prices that compensate growers for their inputs.

The ability to specify these types of requirements is beneficial for both cooperatives and farmers. Increased use of specification buying has moved contracts to the forefront as a method of achieving coordination in meeting ultimate consumer needs.

Purpose of Contracts

Patrons receive many benefits when their cooperatives use contracts. Members like contracts especially because they assure a market for their products. Contracts also give producers an important advantage in securing credit, because loan officers recognize that a contract reduces risk and provides a market at reasonable prices. Further, enforced contracts help solve the free-rider problem.

Cooperatives themselves can also improve their business by using contracts. First, they can reduce or eliminate costs associated with annual solicitation of patronage. Also, cooperatives can offer forward contracts with quantity, quality, and schedule guarantees; and they can acquire supplies and capital at more favorable rates and prices because of the image of reliability created by producer contracts.

Contracts enable cooperatives to plan for the best combination of resources to minimize costs. By using contracts, cooperatives can reduce procurement, assembly, and delivery costs because there are, for example, no soliciting costs. (These savings may be offset by broken contracts and legal fees.) They can also carry out processing and other activities at minimum costs because their volume is known (Figure 12.1). If a firm does not know its volume, it must incorporate flexibility into its operation to minimize risk. This normally results in a flatter cost curve (AC_1) or lower costs (C-D) over a wider volume. However, if volume is known, the cooperative can achieve a lower cost (E) at the known volume. But if volume changes, say from A to B, then average costs would be higher on AC_2 than on AC_1. Possible trade-offs include equipment rental and part-time, inexperienced

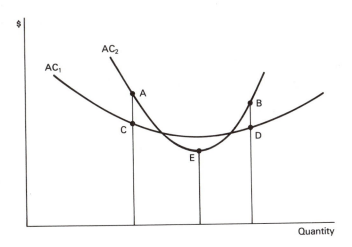

Figure 12-1 Depiction of potential average costs for variable volumes (AC_1) and for known volumes (AC_2).

labor for fully utilized purchased equipment and experienced, full-time employees.

Contracts are of legal importance to both cooperatives and patrons because they restrain outside interference, because they guarantee that either side can collect for damages incurred by a breach of contract, and because they record the rights and duties of both cooperatives and their members.

Problems with Contracts

Contracts are not, however, appropriate for all situations; and even when they are appropriate, cooperatives need to make efforts to minimize their potential weaknesses. For example, some potential members may be hesitant to make the type of binding commitment that the contract requires. One of the greatest drawbacks for farmers who sign contracts, particularly for those producing bulk commodities such as grain and oilseeds, is the loss of freedom to use other marketing alternatives. In addition, a cooperative's management may become careless because it may take the nearly guaranteed volume for granted. Cooperatives must keep in mind that contracts are no substitute for efficiency. No contract will hold members together indefinitely unless it produces expected benefits.

Contracts cannot be relied upon simply because they can be enforced by the courts. Even IOFs are loath to sue farmers for breach of contract because of the invariable loss of goodwill that results from law suits.

MARKETING AGENCIES-IN-COMMON

Marketing agencies-in-common, federated cooperatives that act as marketing agents for their members, have been created for two situations. The motivation is different in each case, but in both situations the agencies spread fixed costs and/or enable different cooperatives to share otherwise unavailable resources. In the first situation, the agency markets homogeneous products such as beet sugar and its by-products. For example, the three sugar beet cooperatives in the Minnesota–North Dakota area have formed Midwest Agricommodites, with offices in Corte Madera, California, to market their molasses and beet pulp. Two of these cooperatives market their sugar through North Central Sugar Marketing with offices near Minneapolis. In this case, the cooperatives' contacts, skill, and reputation for reliability and timeliness in industrial markets are critical.

In the second situation, marketing agencies-in-common are used to spread the otherwise astronomical costs of establishing consumer-recognizable brand names for differentiated products. Economies of mass advertising rates have been well established. Benefiting from the Norbest brand of turkeys, for example, are six independent turkey processing cooperatives located in five states from Oregon to North Carolina. Member cooperatives sign a contract with Norbest, Inc. of Salt Lake City to use

Norbest's marketing services and brand name. The contract is much like those used between producers and cooperatives. Per-unit capital retains or *authorized deductions* are Norbest's major source of equity capital. They use both pooling and individual accounts.

BARGAINING

Bargaining cooperatives or associations are organizations of producers of one or more commodities that represent their members in negotiating for the best terms of trade, including price, in light of economic realities. Labor unions negotiate for improved wages and working conditions, whereas farmer bargaining associations negotiate for prices and conditions of sale.

To bargain is to negotiate over the terms of trade. Emphasis is placed on terms of trade rather than just on price, because other considerations such as payment provisions, delivery point, and quality may more than offset a favorable price. Therefore, any fair agreement should specify arrangements for these details; otherwise, misunderstanding and adversary relationships are created.

Bargaining power is the ability to influence to one's own advantage the terms of trade in any transaction or series of transactions. It is also the ability to influence any of the aspects that would enhance selling or purchasing activity, such as product development, quality standards, government regulations, and expanding demand. Organizations created to carry out this function for a group of farmers often are called *bargaining associations*, but they could just as well be called *bargaining cooperatives*.

Scope

Many bargaining cooperatives negotiate for one crop. Others bargain for two or more crops. The most recent data on the relative importance of bargaining cooperatives were collected in 1978 for fruit and vegetable cooperatives (Biggs) and in 1978 and 1980 for dairy cooperatives (Stafford). Fruit and vegetable bargaining cooperatives that negotiated contracts are concentrated in the Pacific states. Of the 75 bargaining cooperatives, 34 reported consummation of contracts with processors; 41 did not. Either the processors could not handle the volume, grew their own volume, or simply refused to bargain. Of the total of 435 dairy cooperatives in 1980, 181 engaged in bargaining activities. Thirty-three of them bargained and operated processing plants. They accounted for 54% of milk receipts. The balance (148) had no processing facilities and accounted for 21% of milk receipts.

Bargaining Procedures

The typical bargaining procedure is as follows. Producers sign individual contracts with IOF buyers, the provisions of which have been determined by negotiations between the bargaining cooperative and the buyer. In their purest form, bargaining cooperatives neither take title nor handle the product. They simply negotiate the contracts between growers and purchasers. Producers deliver their products directly to buyers who take title. Sometimes the cooperative may provide additional services such as verifying grades and weights. As a bargaining chip, cooperatives also may arrange for improved efficiencies; for example, they will schedule varieties planted to avoid harvest gluts, or they will provide incentives for improved quality. To further enhance the association's bargaining position, cooperatives may acquire processing and merchandising facilities of their own as an alternative outlet; they do this primarily as a threat to processors during negotiations but also to handle surpluses, which then are usually sold for lower-valued uses.

Bargaining procedures of fruit and vegetable cooperatives were reported by Biggs (1982). Negotiation issues were selected by the board of directors by 65% of the fruit and vegetable bargaining cooperatives, by membership at large (29%) and by a committee of members (6%). These groups generally consider relevant economic factors such as production costs, carryover, competition, and general economic conditions. The views of members regarding which issues to be negotiated are sought regardless of which group makes the final selection.

Negotiations themselves were handled by a producer committee in 59% of the bargaining cooperatives, by a manager in 20%, by a board of directors in 18%, and by a hired negotiator in 3%. If the negotiating party was unable to reach an agreement within the guidelines previously agreed upon, they would go back to the membership (or to the board, in the case of the professional negotiator) for further instructions or approval of the processor's counterproposal.

Strategies employed in and benefits achieved by the bargaining process depend on sources of gain, characteristics of the industry, legal environment, and organizational and personal characteristics. Readers interested in a comprehensive and yet pragmatic approach to bargaining should consult Bunje's book *Cooperative Farm Bargaining and Price Negotiations*.

Sources of Gain

Benefits or gains from the bargaining process arise from (1) improved operating efficiency, (2) a third party or outside group such as consumers or government, and (3) the opponent. Efficiency can be enhanced by such activities as gathering and distributing information to reduce risks and by helping to coordinate product flow, particularly at harvest, in order to avoid harvest glut. Gains from improved efficiency can be shared between the buyer and the bargaining cooperative. Benefits from a third party, such as consumers or government, can be achieved through effecting favorable legislation or promotional campaigns. Gains extracted from the opponent are obviously difficult to achieve, and adversary relationships may result in the process.

Under What Conditions Is Bargaining Likely to Be Successful?

Several elements are required for successful bargaining. The most important is control over supply. This implies considerable producer discipline with no close substitutes of the product available. Farris observes that

> to be successful in such a confrontation, a substantial amount of membership control and group discipline are required, such as fairly complete control over supply or an ability to alter the legal and economic environment or rules of trade. Moreover, this kind of power will need to be held on a long-term basis, not easily subject to erosion from competing producers, competing products, or counter measures by the opponent (Farris, p. 87).

The market structure characteristics that make control over supply most likely are barriers to entry, lack of close substitutes, geographic concentration of production, and a relatively small number of producers. Examples of barriers to entry are the high capital cost for dairy, geographic production restrictions for cranberries, and the length of time to bring new walnut trees into production. If terms of trade are improved over other alternative employments without barriers to entry, other producers will enter the market, increase aggregate supply and cause havoc with bargaining efforts. If substitutes are available, consumers may switch to them and thus erode many of the benefits of bargaining, as happened in the butter and cotton markets. Geographic concentration of relatively few growers facilitate organizational efforts.

To capture control over supply, producers must also be organized. Creation of a bargaining cooperative representing all or at least a bulk of the producers immediately changes market structure. Before organizing, processor(s) face a large number of producers whom they could bid off against one another. But when a bargaining cooperative exists, processors are confronted by an organization capable of restricting their access to supplies.

Protective legislation also helps bargaining efforts. For example, the state of Michigan attempted to overcome the free-rider and uncontrolled-supply problems by passing the Michigan Agricultural Marketing and Bargaining Act (PA 344). This act establishes a state board that evaluates petitions for accreditation of bargaining groups as the exclusive bargaining agent for all producers, whether or not they are members. Accreditation was normally extended if producers voting in favor of the bargaining unit represented more than 50% of all producers as well as more than 50%

of all production of the commodity in question. All producers were required to pay a service fee to the bargaining unit, and the bargaining cooperative was empowered to represent all producers, regardless of producer membership status (Thomas J. Moore, letter to author, May 13, 1987).

In 1985 the U.S. Supreme Court handed down a decision which held that except for the exclusive agency section, all provisions of the Michigan act, such as good-faith bargaining and arbitration, were legal. However, it also held that a cooperative could not coerce a producer to participate in a bargaining effort against his will. This, the court maintained, violated the Federal Agricultural Fair Practices Act. Federal and state statutes protect farmers from coercion, intimidation, or boycotting by handlers and others who exercise their right to join and be represented by a bargaining cooperative (Marcus).

Finances

Revenue received by bargaining cooperatives comes from one or more of the following sources: check-offs, service charges paid by processors, annual dues, and membership fees. Processors forward the check-offs withheld from member payments to the cooperative, and service charges are then paid directly by processors—they are not deducted from producer payments. Justification for this practice is that the processor is relieved of the burdens of soliciting producer contracts. Annual dues are determined by a per-acre or per-ton basis. Membership fees are typically one-time charges made only when members join the cooperative. Check-offs were used by 94% of the fruit and vegetable bargaining cooperatives, service charges by 21%, and membership fees by 24%. Annual dues were not used at all in 1978 but had been used by four cooperatives in 1957.

Limited success from bargaining has been achieved by cooperatives. Either gains are modest or the bargaining effort breaks down because of free riders or because of increased production burdens created by incentives for improved profitability of production. Favorable prices result in increased production from both members and nonmembers. However, this is as it should be. First, it is illegal for cooperatives to unduly enhance price. Second, it would be hard to justify public policy protection from antitrust laws for organizations that themselves engaged in predatory practices to acquire monopolistic profits. One way to improve the effectiveness of bargaining is to use marketing orders.

MARKETING ORDERS

Marketing orders are institutions, enabled by government legislation, that farmers may use to promote collective orderly marketing of selected commodities. Cooperatives play an important role in their creation and maintenance. Marketing orders often help cooperative bargaining efforts and reduce the free-rider problem. This discussion will focus on the interface of cooperatives and marketing orders. A more comprehensive treatment of marketing orders is found in Branson and Norvell or McBride.

Enabling legislation for marketing orders flows from the Agricultural Marketing Agreement Act of 1937. This act is "enabling" because it does not create marketing orders, but it makes their creation legal and establishes the institutional framework and procedures for their creation. Many agricultural commodity producers (including milk, fruit, vegetable, and nut) can qualify.

State and federal legislation allows marketing orders (1) to stabilize erratic markets of selected farm commodities and to promote reasonable prices for them and (2) to provide consumers with an adequate supply of wholesome food. A marketing order is created only after producers have requested it and shown that it is administratively and economically feasible. Hearings are held to determine if more orderly marketing is needed, if production areas (or consumption areas, in the case of milk) are independent, and if an order is not against the long-run interest of consumers. To be effective, marketing orders must have two or more markets with different price elasticities, a perishable commodity, some barriers to entry, organized and knowledgeable leadership among producers, and a geographic concentration of producers.

If conditions warrant a marketing order and producers approve by a two-thirds majority, marketing regulations are drawn up which control specified marketing functions. Only handlers (not producers) are regulated. They are required to submit reports to the administering agency regarding compliance with the order and product flows. Marketing order regulations include factors such as quality, market flow for fruits and vegetables, minimum and blend prices for milk, blend prices for producers, research and promotion check offs, elimination of unfair trade practices, and gathering and reporting of market information. Marketing orders for milk are administered by an appointee of the Secretary of Agriculture. Orders for other commodities are administered by a committee elected by growers. Operational costs are covered by check-offs deducted from the prices received by farmers.

Several marketing orders are currently in operation. The United States is blanketed with 43 federal milk marketing orders that account for about 80% of the country's milk. Fruit and vegetable orders are scattered throughout the United States but are concentrated in California. Several are also active in Florida, Texas and Washington.

Active support by cooperatives representing a majority of producers is an essential element for successful marketing orders. Legal statutes require that marketing orders be initiated by and voted upon by producers.

Without the organizational efforts and leadership of cooperatives, this would be difficult if not impossible.

Marketing order statutes confer cooperatives with specific powers and functions that in most cases give them control over whether the order continues or not. These powers and functions include the following:

1. *Block voting.* This means that all members of cooperatives cast their votes as a unit. Nonmembers vote individually. Thus cooperatives may have complete power to veto the marketing order. Block voting does not, however, give cooperatives dictatorial power or control. Regulations must be approved by respective state or federal agencies, who are responsible for preserving the long-run interests of consumers.

2. *Power to make payments to producer members.* For example, cooperatives receive payments for milk from handlers and then pay producer members the uniform price.

3. *Authority to offer producer members selected services such as grading and market information.* The market administrator's office handles these functions for nonmember producers.

The last two functions do not give cooperatives direct control, but do enhance the cooperative's role and provide it additional revenue to carry out its functions.

Although statues do not require it, cooperatives, almost by default, perform the following tasks:

1. Coalesce the common feelings and frustrations of producers

2. Act on producer concerns, make official requests for improvement and changes in the order to the Secretary of Agriculture or state agency, and help compile necessary supporting data and rationale for the proposal

3. Present testimony at public hearings regarding advisability and specifics of proposals

Producers who are not members of cooperatives usually have very little influence on the outcome because they are generally relatively few in number and unorganized. However, they are not likely to be disadvantaged because their interests generally parallel those of cooperative members.

Cooperatives have been relatively more successful when bargaining efforts are initiated under the umbrella of a marketing order. For example, dairy cooperatives have received considerable publicity for their success in obtaining over order premiums.

INTERNATIONAL TRADE

Expanding international trade by cooperatives has been hailed as a key to augmenting benefits to members. The percentage of U.S. agricultural production exported has increased from 2% in 1940 to nearly 20% today. With the dramatic explosion of exports in the mid 1970s came renewed recommendations that cooperatives become more actively involved in foreign markets.

But exporting is also fraught with risks. Complexities and hazards arise out of an array of financial, insurance, transportation, and currency exchange transactions, resulting in an environment that can be especially intimidating to the uninitiated.

Common risks are magnified in the export market. For example, exporters face unstable exchange rates, unstable political structures, competition operating under different sets of rules, culture shock, and our own unstable export policy. Although several agencies such as the Central Bank for Cooperatives and the Foreign Agricultural Service can provide help and guidance, successful exporting cooperatives must take the initiative and assume some risks.

Despite these negative factors, opportunities exist for cooperatives to enter and expand exports in selected markets, particularly markets in non-bulk commodities such as almonds and seed potatoes. In 1985-1987, U.S. cooperatives exported $3.4 billion or 11.7% of total agricultural exports (Table 12.2 and Figures 12.2 and 12.3) (Bunker and Kennedy, 1987). Total dollar volume was greatest for grains and feeds, including sugar beets, citrus, and potato by-products, but cooperatives accounted for the greatest market share of U.S. exports in nuts and fruits.

Basic management techniques of international trade are explained in such references as the book by Cateora. Relevant institutional and technical aspects, as they apply to cooperatives for major commodity areas, have been prepared by the ACS, (Bunker and Kennedy, 1984). In this section we

TABLE 12.2   Cooperative shares of U.S. agricultural exports, 1985.

Commodity groups	Number of co-ops	Co-op volume (thousands)	Co-op share of U.S. (%)
Animals	14	$ 27,933	0.9
Grains & feeds	15	1,736,860	14.9
Oilseeds	10	398,861	8.7
Cotton	5	413,172	25.0
Fruits	29	377,602	31.8
Vegetables	7	37,688	6.4
Nuts	3	279,011	40.8
Other	14	124,820	[a]
Total	87[b]	3,395,948[b]	11.7

Source: Bunker and Kennedy (1987).
[a]Not available.
[b]Does not total. Some co-ops exported in more than one commodity group. The U.S. volume does not add up, due to rounding.

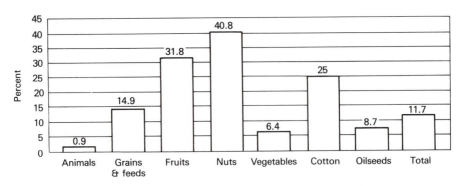

Figure 12.2 Cooperative share of U.S. Agricultural exports, by commodity group, 1985. *Source*: ACS USDA.

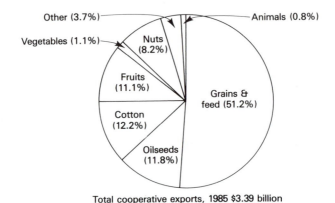

Total cooperative exports, 1985 $3.39 billion

Figure 12.3 Cooperative exports by commodity groups as a total percent of cooperative exports, 1985. *Source*: ACS USDA.

discuss separately bulk and nonbulk commodities because the comparative advantages of each are quite different for cooperatives.

Nonbulk Commodities

Cooperatives' greatest opportunities for expansion appear, at least in the short run, to be in promoting the export of specialty crops and value-added commodities such as nuts, processed fruits and vegetables, and broilers. Cooperatives exporting nonbulk commodities face the same problems that other firms face regarding management techniques and resources.

Bunker and Kennedy (1984) outline the exporting activity of 23 fruit, vegetable, and nut cooperatives (Figure 12.4). These 23 cooperatives accounted for 31% of the fruit, 2% of the vegetables, and 35% of the nuts exported in 1980. By catering to the unique economic and cultural needs of a particular nation, by developing a reputation for consistent quality, and by

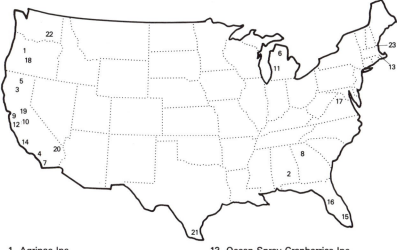

1. Agripac Inc.
2. Anderson's Peanuts
3. Blue Anchor Inc.
4. Calavo Growers Association
5. California Almond Growers Exchange
6. Cherry Central Cooperative Inc.
7. Gold Crown Macadamia Association
8. Gold Kist Inc.
9. Guild Wineries and Distilleries
10. Lindsay Olive Growers
11. Michigan Blueberry Growers Association
12. Naturipe Berry Growers
13. Ocean Spray Cranberries Inc.
14. Oxnard Frozen Foods Cooperative Inc.
15. Pioneer Growers Cooperative
16. Seald-Sweet Growers Inc.
17. Shenandoah Apple Cooperative Inc.
18. Stayton Canning Company Cooperative
19. Sun-Diamond Growers of California
20. Sunkist Growers Inc.
21. Texas Citrus Exchange
22. Tree Top Inc.
23. Welch Foods Inc. (Headquarters)

Figure 12.4 Locations of cooperative exporters of fruits, vegetables and nuts. *Source*: ACS USDA.

being a reliable supplier, cooperatives can develop the rapport and confidence necessary to penetrate and expand selected markets.

Another organizational step almost universally recommended is the formation of joint export entities. This may involve coordinated efforts with existing governmental and trade institutions that have been created to help companies of all kinds to export U.S. products. These institutions include state export development agencies, the Foreign Agricultural Service, the Department of Commerce, and Central Bank for Cooperatives.

Joint efforts with other cooperatives also have intuitive appeal, because cooperatives can in this way share management and other resources, as well as capitalize on the economic features of each cooperative.

Bulk Commodities

Cooperatives marketing bulk commodities (e.g., food and feed grains or oilseeds) face a unique set of challenges to their export efforts. These challenges are created by the structure of trading in these bulk commodity markets. The few large private firms competing in this arena (e.g., Cargill Inc., Continental Grain, Dreyfus Inc., and Bunge) all have tradition, experience, and a global network of information and operations behind them. As a result, cooperatives have a disadvantage that cannot easily be overcome because of their linkage to their patrons. IOF exporters also have major synergistic advantages made possible by their economies of multiple origin, information, and risk-spreading capabilities. These economies are created by worldwide sources and on-site facilities in several importing and exporting countries.

Economies of Multiple Origin IOFs can originate commodities from any of several countries and even from several ports in some countries. They can also then capitalize on slight differences in commodity prices and ocean freight rates from any available origins. Volatile prices on freight rates work to the disadvantage of cooperatives, however, because they are locked into U.S. origination points. Cooperatives' bulk exports are also obstructed when ports are blocked by strikes, physical disasters, or political embargoes. IOFs, on the other hand, can fulfill commitments in spite of these problems by originating their commodity sales from other countries.

Economies of Information IOF exporters are intimately involved in every significant market. They know what is wanted and what is available. They know the impacts of weather, as well as economic and political events on demand and supply, often before anyone else. Because of their experience, these firms are also able to anticipate changes before they become public information. For example, merchandisers at Cargill (a company that operates oilseed crushing plants and feed mills in a few countries) know firsthand the coming changes in demand for feed grains and supplies of oilseed products. Companies such as Cargill are famous for specializing in centralized information systems that digest volumes of information.

Economies of Risk Management Instability is a ruling force in export markets. Not only are prices continually changing, but ocean freight rates, exchange rates, political relationships, and weather all compound to make exporting a risky endeavor. IOF exporters can be self-insured by spreading their risks over a large number of transactions.

These three economies also have a synergistic effect; they complement one another. On the other hand, cooperatives, even large federations, are severely limited in comparison. For example, during the 1980 embargo, cooperatives could not make direct export shipments, but IOFs simply acquired grain from other country origins to fulfill their forward commitments. Strikes at a cooperative's major port also delay exports unexpectedly and thereby give cooperatives a reputation for being unreliable suppliers.

Cooperatives have made efforts to link with other cooperatives or IOFs (domestic and foreign) in order to achieve these economies. But in the past such arrangements have led to problems ranging from fraud by management to a conflict of interest by participants, the latter being particularly damaging to the success of the combined cooperatives. When there is a choice, domestic cooperatives have a vested interest in exporting U.S. commodities. Consequently, an international organization that includes U.S. cooperatives may not be as free to change its country of origin when the need arises. Even when domestic regional cooperatives form a joint exporting activity, participating parties often sell their commodities through other channels to achieve short-run gains.

It may be unrealistic to expect cooperatives to become a dominant force in the export market of bulk commodities. Even so, they can play a significant role. Despite their limitations, they have carved out niches for themselves in the exporting market and today account for about 15 to 20% of the exports of food and feed grains and 9% of the oilseeds. The continued presence of cooperatives in the export business is healthy for competition, gives farmers direct access to these markets, and assures farmers that undue monopolistic profits are not being extracted by IOF exporters.

Because of extremely narrow margins, limited risk-spreading ability, and the risks of closed ports, farmers may be better off investing in other endeavors. Unsuccessful attempts involving substantial losses by individual regional cooperatives, as well as by combined groups of cooperatives and other firms, are stark evidence that cooperatives do have a comparative disadvantage in exporting bulk commodities.

Several large U.S. cooperatives are attempting to overcome these comparative disadvantages by investing in A.C. Toepfer International through InTrade. A.C. Toepfer is an multinational merchandising firm, and InTrade is a consortium of 11 German, French, Dutch, Canadian, and U.S. cooperatives. The U.S. founding cooperatives include Gold Kist (Georgia), Land O'Lakes (Minnesota), Citrus World (Florida), and Agway (New York). Indiana Farm Bureau Cooperative Association and Harvest States (Minnesota) joined this group later. Even with InTrade these cooperatives have recognized and are battling the same kind of problems mentioned above. William Gaston, President of Gold Kist, stated:

> One of the hang-ups some of the co-ops have . . . is that they feel their role is to market only the grain or other commodity produced by their members. InTrade, on the other hand, has developed the philosophy that if we are going to be in the world's commodity trade center, we have to buy and sell from all sources, to all sources—every day . . . .

InTrade takes the position that if market conditions suggest that Brazilian beans are going to move to Europe or Japan anyway, at the expense of U.S. soybeans, we would rather know about it and participate in that movement. Hopefully, that way we can make some profit from that activity that would be returned to our members.

Toepfer is in this business—all over the world—every day . . . . We don't want to cut ourselves off from the knowledge of that movement or the profit potential—or downside risk—of that movement (Haffert, p. 32).

Banks for Cooperatives

Cooperatives received a boost in their exporting efforts when the BCs were allowed to enter the field of financing exports. The 1980 amendment to the Farm Credit Act allowed Banks for Cooperatives to finance exports by cooperatives and to provide ancillary services as well. These activities have been centralized in the Central Bank for Cooperatives. As a result, the volume of international trade financed for cooperatives rapidly expanded to $1.2 billion in 1984. Equally important has been a four-step risk-reduction program which analyzes country and foreign borrower risk. This program also incorporates CCC loan guarantees and an insurance program. In addition, the central bank helps cooperatives develop export markets by establishing relationships with foreign governments, banks, and purchasing entities.

SUMMARY

Cooperatives employ several unique marketing methods and strategies in dealing with members. Their special owner-patron relationship presents opportunities and problems. Delayed or at-delivery payment and contracting have important ramifications for both members and the cooperatives. Contracting provides farmers with a guaranteed market for their products and an opportunity to coordinate supply with demand. Cooperatives that leave marketing decisions such as time and quantity to the producer operate much like IOFs. Farmers often are attracted to this approach because it gives them maximum freedom and opportunity to use their own marketing skills. The open-market and cash-and-carry approaches are suitable for the marketing of supplies, services, and some commodities (such as wheat).

Group action requires producers to delegate marketing decisions to the cooperative, such as pooling, bargaining, and marketing orders. The first two avoid large financial investments, and the latter one helps correct the free-rider problem. Centralizing marketing decisions of many producers increases the use of professional experts, reduces risk and capital requirements, and enhances market power due to greater control over a larger supply. Marketing contracts are essential for pools and bargaining efforts. From a member's perspective contracts have favorable market and credit features in addition to avoiding the free rider problem. Cooperatives can benefit from reduced solicitation costs, improved bargaining and servicing position with clients, and improved operational efficiencies. Marketing agents in common allow several independent cooperatives to share common marketing resources.

Forward integration into export markets provides farmers with opportunities and poses hazards associated with these extended markets. Opportunities appear to be more promising in nonbulk than bulk commodities such as grain and oilseed, where cooperatives have a comparative disadvantage compared to their competitors. These competitors have major synergistic advantages made possible by their economies of multiple origin, information, and risk-spreading capabilities created by worldwide facilities.

With an understanding of these marketing alternatives, you are now equipped to consider the unique features of cooperative finance.

DISCUSSION QUESTIONS

12-1. Contrast the competitive environment and the likelihood of success for cooperatives that handle nonbulk, value-added products such as nuts and poultry with cooperatives that handle bulk commodities such as wheat, corn, and oilseeds.

12-2. Define a marketing contract and outline the advantage to the cooperative and the member.

12-3. Compare and contrast results of situations in which the producer makes the marketing decisions with situations in which the producer turns the marketing decisions over to the cooperative by pooling or bargaining. Compare freedom, risk, skill, and capital requirements.

12-4. Marketing pools are a distinctive cooperative practice. What are the key practices, or in other words, how do they operate?

12-5. What are the inherent problems of cooperatives when they produce consumer recognizable products? How can they overcome these problems?

12-6. Discuss the advantages of cash at delivery with delayed payment plans.

12-7. In what ways are cooperatives a keystone to the success of marketing orders?

12-8. What is the free-rider problem, and what steps can cooperatives take to minimize this problem?

12-9. What is a pure bargaining cooperative? How effective can bargaining cooperatives be in enhancing producer-member prices? Would the effectiveness vary for bargaining efforts on behalf of more storable commodities (grain) versus rather perishable commodities (fruit or milk)?

REFERENCES

BIGGS, GILBERT, *Status of Bargaining Cooperatives.* Washington, DC: USDA ACS RR 16, Apr. 1982.

BLACK, WILLIAM E., "Marketing Agreements," *Agricultural Cooperatives and the Public Interest,* ed. B. W. Marion, NC 117 Monograph 4, College of Agr., Univ. of Wisconsin-Madison, 1978.

BRANSON, ROBERT E., AND DOUGLAS G. NORVELL, *Introduction to Agricultural Marketing.* New York: Macmillan, 1983.

BUNJE, RALPH B., *Cooperative Farm Bargaining and Price Negotiations.* Washington, DC: USDA ESCS CIR 26, 1980.

BUNKER, ARVIN B., AND TRACEY L. KENNEDY, *American Cooperative Exporters: Fruits,. Vegetables, and Nuts.* Washington, DC: USDA ACS SR 11, July 1984.

BUNKER, ARVIN B., AND TRACEY L. KENNEDY, "Exports by 87 Ag Cooperatives Exceed $3.39 Billion in 1985," *Farmer Cooperatives.* (Feb. 1987):4.

CATEORA, PHILIP R., *International Marketing,* 5th ed. Homewood, IL: Richard D. Irwin, 1983.

DUNN, JOHN R., STANLEY K. THURSTON, AND WILLIAM Farris, *Some Answers to Questions about Commodity Pools.* Dept. of Agr. Econ. EC-509, Purdue Univ., undated.

FARRIS, PAUL L., "Building Bargaining Power--Economic Considerations," *American Cooperative 1963 Supplement.* Washington, DC: American Institute of Cooperation, 1983.

HAMMONDS, T. M., *Cooperative Market Pooling.* Oregon Agr. Exp. Sta. Cir. Info. 657, Nov. 1976.

HARRINGTON, DONNIE E., *The Produce Auction as a Cooperative Venture.* Washington, DC: USDA ACS Staff Report, Mar. 1982.

HAFFERT, WILLIAM, "A Cooperative Look Abroad," *Feed Manage.* (Apr. 1981):32.

MARCUS, GERALD, "Report of Subcommittee on Problems Peculiar to Bargaining Cooperatives," *The Coop. Accountant.* (Summer 1985):67.

MCBRIDE, GLYNN, *Agricultural Cooperatives.* Westport, CN: AVI, 1986.

STAFFORD, THOMAS H., *Financial Performance of Dairy Cooperatives.* Washington, DC: USDA ACS RR 49, June 1985.

Part V
Finance

13

Distribution of Net Income

David W. Cobia,
North Dakota State University

INTRODUCTION

It is impossible to communicate meaningfully or make prudent decisions in a cooperative environment without comprehending essentials of cooperative finance. It is in finance where cooperative principles of user–owner and user–benefits (business-at-cost) are applied.

To fully understand the finance topics discussed, it is necessary to be familiar with the operating statement and balance sheet. It is also helpful to have an understanding of the time value of money and of basic financial management concepts and tools such as financial ratios. Straightforward reviews of these topics are found in the book by Duft (1979), chaps. 4 and 5, and in that of Downey and Erickson, chaps. 6 and 8. Van Horne covers essentials of financial management, and Brigham gives an advanced treatment of it.

There are several ways to handle financial aspects and new ones are still being created. Exploring all of those approaches would make the chapters unwieldy. Therefore, we discuss only the more common ones. A few obscure options will be mentioned, but you should be aware that other variations also exist.

Cooperative finance is complicated by the intertwined relationships among cooperatives. Financial decisions at one cooperative level often have impacts up and down a federated structure because of the direct linkage from patron to local to regional to interregional.

INTRODUCTORY CONCEPTS

Accounting practices and the basic structure and rationale of financial statements are essentially the same for cooperatives as for other firms. Generally accepted accounting principles and practices are applicable for cooperatives as well. However, differences in terminology have been developed to accommodate cooperative transactions which have no parallel in standard accounting practice and to recognize the fundamental distinctions between cooperatives and investor-oriented firms (IOFs). Special accounting practices and terminology used by cooperatives include revaluation of inventories of marketing pools, accounting for investments in other cooperatives, types of member equities, and allocation of gains or losses. Keeping records by patron and product categories so as to conform to the business-at-cost principle is an added burden unique to cooperatives.

There are two sources of data which provide useful information that illustrates the relative importance of components of the balance sheet and operating statement. The Agricultural Cooperative Service (ACS) has periodically conducted a comprehensive financial survey of all farmer cooperatives. The most recent one (Griffin et al.) was for 1976. Although this information is dated, it does show the differences among commodity groups and from one area of the country to another. Geographic differences are given by farm credit districts. The ACS also prepares an annual financial summary of the 100 largest cooperatives in the United States These data are normally published in the September through December issues of *Farmer Cooperatives* (Davidson and Royer).

We will use the simplified balance sheet and operating statements given in Tables 13.1 to 13.3 to illustrate financial concepts. In this section we discuss major differences between the balance sheets of cooperatives and those of other firms to facilitate the discussion on distribution of net income that follows.

A balance sheet is a financial statement that lists the value of the assets a cooperative controls, what it owes to others (liabilities), and the amount owners have invested (net worth or equity) (Table 13.1). The only unique entry on the asset side of this balance sheet is the one represented by investments in other cooperatives. This entry ($2,400 in 1986, Table 13.1) represents equity created as a result of intercooperative business including Banks for Cooperatives (BCs), and perhaps to a lesser extent, of cash investments in those cooperatives. In 1985 the 100 largest cooperatives had an

TABLE 13.1   Simplified Balance Sheet.

Franklin County Cooperative, Inc.
Balance Sheet, Dec. 31, 1986, with comparative figures
for 1985 (thousands).

	1986	1985
	Assets	
Current assets		
Cash	$ 462	$ 392
Marketable securities	680	430
Receivables (net)	1,900	1,870
Inventory	2,400	2,295
Prepaid expenses	78	102
Total Current	5,520	5,089
Fixed assets		
Buildings	5,200	4,980
Equipment	2,300	2,111
Land	200	200
Less accum. depreciation	(3,600)	(3,153)
Total Fixed	4,100	4,138
Other assets		
Equities in other co-ops	2,400	2,300
Total assets	$12,020	$11,527
	Liabilities and member equity	
Current liabilities		
Accounts payable	$ 850	$ 700
Accrued expenses	200	211
Current portion, long-term debt	170	170
Short-term loans	2,532	2,600
Patronage refund payable	224	115
Taxes payable	44	30
Total Current	4,020	3,826
Long-term liabilities	1,700	1,870
Members Equity		
Common stock	2	2
Allocated credits	6,198	5,739
Unallocated reserve	100	90
Total member equity	6,300	5,831
Total liabilities & members' equity	$12,020	$11,527

investment of about $1.2 billion in other cooperatives. These investments accounted for roughly 7% of their assets and represented 20% of their net worth. Equivalent percentages for all U.S. cooperatives in 1976 were about the same. A classification of cooperatives by commodity reveals that farm supply and grain cooperatives had the highest average investment in other cooperatives, representing 12 to 14% of their assets and about 25% of their net worth. Sugar cooperatives had the lowest average (0.1%) of total assets invested in other cooperatives. Nearly one-fifth of all farmer cooperatives had no such investments at all. But some individual local cooperatives, especially farm supply cooperatives that have been long-time federation members or have substantial losses at the local level, often have had a third to a half of their total assets invested in federated cooperatives.

The liabilities section of the cooperative balance sheet has two unique aspects. First, the patronage refunds allocated to members but not yet paid and the proceeds due to patrons from pooling arrangements are included with other accounts payable. Second, a large proportion of short-term (seasonal) and long- term loans are obtained from BCs. A few cooperatives borrow from their members.

Finally and most important, the cooperative balance sheet differs from those of IOFs in its member equity section. Common stock ($2, Table 13.1) is usually purchased voting stock which only qualifying persons can own. It is generally limited to one share per member. Members cannot obtain additional voting rights by purchasing more shares. Further, membership certificates, used to denote voting rights, may appear in place of common stock in nonstock cooperatives.

Allocated credits ($6,198) are retained refunds, per-unit capital retains, or check-offs which are allocated to members on the basis of patronage. A variety of other designations such as *retained refunds, certificates of investments, revolving capital, patronage creditors* and even *surplus* are used to represent this broad category of equity. The term *preferred stock* also appears in the member equity section of some cooperatives. Preferred stock may represent investments by employees and the general public as well as by members. In other instances, retained patronage refunds and per-unit capital retains are classified as preferred stock.

Unallocated reserves ($100) is a category of equity not attributable to a particular member. This equity is generally income from nonmember sources and income on which the cooperative has paid corporate income tax. The relative importance of unallocated reserves, and other major components of the balance sheet for the 100 largest agricultural cooperatives in 1985, are shown in Figure 13.1. With these differences from an IOF's balance sheet in mind, we can now move on to discuss the special considerations cooperatives face when distributing net income.

Allocating net income of a cooperative focuses attention on the key characteristic of cooperatives—business-at-cost. Policies and practices regarding net income allocations are critical because this is how benefits are distributed. Alternative distribution methods discussed are cash and noncash refunds, qualified and nonqualified refunds, dividends, and unallocated reserves. The basis for calculation and rationale for each is

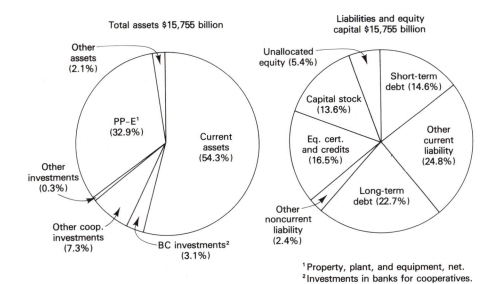

Total assets $15,755 billion

Other assets (2.1%)
PP-E¹ (32.9%)
Current assets (54.3%)
Other investments (0.3%)
Other coop. investments (7.3%)
BC investments² (3.1%)

Liabilities and equity capital $15,755 billion

Unallocated equity (5.4%)
Short-term debt (14.6%)
Capital stock (13.6%)
Eq. cert. and credits (16.5%)
Other current liability (24.8%)
Other noncurrent liability (2.4%)
Long-term debt (22.7%)

¹Property, plant, and equipment, net.
²Investments in banks for cooperatives.

Figure 13.1 Assets, liabilities, and equity of largest cooperatives, 1985. *Source*: AC USDA.

provided. Some cooperatives also experience losses. Alternative methods of handling them are explored.

CALCULATING NET INCOME

Net income is calculated by subtracting costs from income in the standard way (Table 13.2). Total sales less the cost of goods sold equals gross income ($42,750 - $40,000 = $2,750 for Franklin Co. Cooperative). Gross income less operating expenses leaves a balance ($43 for Franklin County Cooperative) which is referred to as operating income. Nonoperating revenue ($1,250) and nonoperating expenses ($396) are listed separately after operating income, to facilitate a separate analysis of the results of operations from method of financing and extraneous business such as rent and interest income.

Net income ($897) is the term for the remaining income after operating expenses ($2,707) and other authorized deductions ($396) are subtracted from a cooperative's total revenue and after other income ($1,250) is added. Net income is the recommended term to represent revenue less expenses; it is preferred to other commonly used terms such as *net profit, net savings, net margins, net proceeds,* or *net earnings.* These terms should be avoided because they are not as generic or they convey misleading concepts and connotations. For example, net savings is appropriate for supply cooperatives but does not apply to marketing cooperatives. Net profit implies that this amount is equivalent to taxable income in computing a cooperative's corporate income tax obligations. Furthermore, patronage refunds should be viewed as adjustments to prices. They are the result of a supply cooperative charging too much for supplies or a marketing cooperative underpaying the patron. Conceptually, then, patronage refunds can be considered as discounts and allowances by a supply cooperatives.

Cooperatives operating on a pooling basis may show neither an income nor a loss. They generally rely on per-unit capital retains as their primary source of equity. In 1976, for example, 108 of the 5,795 cooperatives had no net income or loss. Four of the largest 100 had none in 1985.

TABLE 13.2   Simplified operating statement

Franklin County Cooperative, Inc
Operating Statement, year ending December 31, 1986
with comparative figures for 1985 (thousands).

	1986	1985
Total sales	$42,750	$40,180
Less: cost of goods sold	40,000	37,780
Gross income	2,750	2,400
Expenses		
Payroll	1,200	1,119
Advertising	26	32
Allowance for bad debts	50	68
Depreciation	700	650
Directors' expenses	6	6
Insurance	130	125
Legal & consulting fees	6	5
Utilities	100	96
Vehicle expense	180	175
Taxes	110	100
Repairs & maintenance	145	130
Communications	54	35
Total expenses	2,707	2,541
Operating Income	43	(141)
Other revenue[a]		
Patronage refunds received	250	100
Other	1,000	850
	1,250	950
Other deductions[a]		
Interest expense	396	464
Loss on sale of equipment		
Net income	$ 897	$ 345

[a]Nonoperational income and expense (interest andpatronage refunds) are listed after operating resultsto facilitate analysis of operations.

Marketing cooperatives using a delayed payment plan such as pool-ing modify their income statements (Table 13.3). They frequently deduct advances and final payments to producers ($300) after operating expenses have been deducted from gross income rather than classifying these pay-ments as cost of goods sold. Alternatively, pooling cooperatives may show advances on products delivered by patrons as cost of goods sold in the nor-mal way, particularly if the advances approximate the prevailing market prices.

TABLE 13.3 Example of an operating statement of a marketing cooperative using a marketing pool and per unit capital retains

Marketing Cooperative Inc.
Operating Statement (thousands)
Year ending June 30, 1986.

Sales (products marketed)	$1,250
Ending inventory of finished goods	250
	1,500
Less beg. invent. of fin. goods	500
Gross proceeds	1,000
Operating expenses	650
Net income from operations	350
Credits to growers	
Per-unit capital ret. withheld	20
Cash advances to growers	
Apples	200
Peaches	100
	320
Undistributed proceeds[a]	$30

Source: Adapted from Touche Ross & Co.
[a]May be returned to patrons as net income or as cash advances.

Depreciation and losses on the sale of fixed assets are two entries found on operating statements that can result in serious distortions of the business-at-cost principle and inequitable treatment of members. Using accelerated depreciation methods may penalize patrons in early years be-cause the amount of depreciation subtracted from gross income in the early years can far exceed the actual decline in the value of the asset. This leaves less money available for patronage refunds. Then later, patrons using the more fully depreciated equipment experience lower operating costs and higher patronage refunds. In effect earlier patrons subsidize later patrons. To strictly adhere to the business-at-cost principle, depreciation should be charged over an asset's useful life and equal the decline in the value of the asset from period to period. Straight-line depreciation would probably come closer than accelerated methods, to conforming with the business-at-cost principle. Tax considerations may, however, become relatively more important than strict adherence to the business-at-cost principle.

Inequitable treatment can result from losses on the sale of equipment or buildings. The effect is opposite to that of accelerated depreciation, because patrons during the time of the write-down subsidize earlier patrons. For example, the multimillion-dollar write-down taken by several large regional cooperatives on a Chicago refining venture (Energy Cooperative, Inc.) in 1981 penalized patrons in the following years for a decision made during an earlier time. Some of these distortions of the business-at-cost principle are unavoidable, but others may be planned. For example, spreading risks of new product development over time may be a legitimate objective.

If cooperatives operated on a strictly cost basis, they would never have any net income. But most cooperatives seek to operate and price in such a way that some net income results. Otherwise, it would not be possible for them to grow or to maintain financial integrity.

In this section we explain the unique features of the balance sheet and calculation of net income. Per-unit capital retains or check-offs, used primarily by marketing cooperatives, are discussed under the heading "Equity."

ALTERNATIVES OF NET INCOME DISTRIBUTION

The board of directors has the authority to decide how to distribute net income, keeping, of course, within the guidelines and priorities specified by state and federal law by the cooperative's bylaws. Alternative distribution methods discussed include cash and noncash patronage refunds, qualified and nonqualified patronage refunds, dividends, and unallocated reserves. The basis for calculation and rationale for each is given.

Issuance of patronage refunds ensures conformance to the business-at-cost principle. Other methods erode this principle if they involve distribution of net income from member business.

Distributions made by all cooperatives in 1976 are listed in Table 13.4; those made by the 100 largest cooperatives are given in Figure 13.2 and Table 13.5. A discussion of each allocation follows.

TABLE 13.4 Distribution of net income by U.S. farmer cooperatives, 1976

Allocation	Number of co-ops (%)	Dollars (%)	Tax burden[a]
Patronage refund		85	
Qualified	80	84	M
Cash	80	40	M
Noncash	74	44	M
Nonqualified	2	1	C
Dividends on equity			
Section 521 co-op	11	4	M
Non-section 521 co-op	18	1	C&M
Unallocated reserves	59	7	C
Corporate income tax	50	6	C

Source: Griffin et al. (pp. 80, 83).
[a]M = member, C = cooperative.

TABLE 13.5 How 100 largest cooperatives distributed net income, 1985[a]

Commodity group	Number of co-ops with net income	Total net income (millions)	Patronage refunds	Unallocated equity	Income taxes	Dividends
			Percent share			
Dairy	31	$134.3	76.1	18.1	5.5	0.3
Farm supply	9	76.1	82.0	9.7	7.0	1.3
Diversified[b]	6	63.9	26.6	30.5	32.7	10.2
Fruits & vegetables	9	54.8	56.3	31.4	11.8	0.5
Cotton	4	38.6	100.9	(4.3)	3.3	0.1
Grain	9	22.4	67.1	27.8	5.1	0.0
Rice	4	19.4	0.0	58.3	23.3	18.4
Sugar	4	5.6	27.4	73.5	(0.9)	0.0
Other products[c]	3	6.5	42.8	44.6	12.2	0.4
Total	79[d]	421.6	64.2	21.7	11.3	2.8

Source: ACS USDA.
[a]Before deducting net losses.
[b]Marketing/farm supply cooperatives handling several commodities.
[c]One nut, one poultry, and one livestock marketing association.
[d]Seventeen of 21 other cooperatives incurred net losses and 4 used pool accounting methods with no net income (or loss) activities reported.

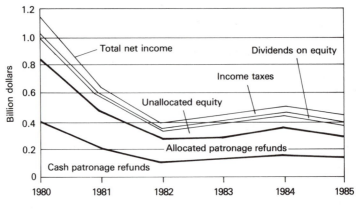

[1]Before deducting net losses.

Figure 13.2 Distribution of net income by 100 largest cooperatives. *Source*: ACS: USDA.

Patronage Refunds

Patronage refunds[1] are distributions of net income (generated only from patron business) returned to patrons in proportion to the value or quantity of their patronage. For example, if a member of Franklin County Cooperative (Table 13.2) accounts for 2% of that cooperative's patronage during the year, the member is credited with 2% (or $17,940) of the cooperative's net income. These allocations may be a straight percentage as above or may be segregated by product or service. They may be distributed in cash, retained by the cooperative, or both. Furthermore, they may be in either qualified or nonqualified form. Federal income tax obligations hinge on this classification. The distinction will be touched on at this point but explained in detail in chap. 16.

The important distinguishing features of *qualified* patronage refunds are that they can be excluded from a cooperative's taxable income, but patrons must agree to include the entire allocations as income when determining taxable income, as though the patron had received them all in cash. Cooperatives can make patronage refunds in cash payments, or they can pay part in cash and retain the rest in the cooperative to increase the patron's equity investment in the cooperative. Most cooperatives allocate net income in this way. In 1976, for example, 90% of all cooperatives reporting net income allocated an average of 84% of it to patrons as qualified refunds.

---

[1]The Internal Revenue Service uses the term *patronage dividends*. This is misleading because the term *dividends* carries with it connotations of a return on investment. Patronage refunds are based on patronage, not investment.

To be qualified and hence exempt from inclusion in the cooperative's taxable income, the refunds must meet the following exacting criteria specified by the Internal Revenue Code. (1) At least 20% of the refund must be paid to patrons in cash. (2) All of the net income must have resulted from patronage-sourced business. (3)There must have been a preexisting obligation by the cooperative to pay the patronage refund. And (4) notification of the refund and cash payment must be made within specified time limits. Patrons must agree to include the entire refund (cash and non-cash) as income when filing their tax. If any of these criteria are not met, the refund is nonqualified. For example, suppose that a patron refuses to acknowledge receipt of the refund by failing to cash the refund check. Such refunds then become nonqualified and the cooperative is then obligated to pay tax on them. The cooperative receives a tax credit when nonqualified patronage refunds are redeemed in cash. Patrons are obligated to report as taxable income all qualified patronage refunds when first issued and non-qualified refunds when redeemed as cash.

Cash Refunds

Putting a high percentage of refunds in the cash form benefits current patrons and may encourage patronage and membership because it can reduce cash flow problems of patrons who are financially strapped. Often such patrons have recently started farming or expanded operations. Members with income in marginal tax brackets higher than 20% are especially interested in a large percentage of cash refunds because they tend to ease the burdens of cash outflows due to tax obligations. Such patrons would have a negative after-tax cash flow on patronage refunds if the minimum cash refund of 20% were given. These patrons might insist on a higher level of patronage refunds in cash to a level that would at least offset the additional taxes from qualified refunds.

Despite their appeal, high cash refunds may compromise a cooperative's financial strength and thus hinder its ability to grow, to borrow funds at attractive interest rates, and to redeem equity of overinvested members. High cash refunds may also delay the accumulation of equity from under invested members.

In 1976 the level of patronage refunds paid in cash varied widely from one category of cooperatives to another. Marketing cooperatives' cash refunds ranged from 27% of total net income paid by grain marketing cooperatives to 87% paid by sugar cooperatives. On the other hand, supply cooperatives paid 36% of their net income as cash refunds. Geographically, distributions ranged from an average of 23% of net income in the Louisville Farm Credit District to 82% in the Sacramento district. The 100 largest cooperatives paid out an average of 58% of their net income in cash in 1985.

Retained Refunds

Retained refunds are allocations of net income made to patrons but retained by the cooperative. They are the noncash portion of qualified refunds or all of the nonqualified refunds. The cooperative's purpose of retaining refunds, rather than returning them to patrons in cash, is to increase patron equity in the organization. Accumulated funds may be used for expansion or to replace previously contributed equity which is scheduled for redemption. Patrons are informed of this action by written notices of allocation.

Nonqualified Refunds

As explained previously nonqualified refunds are allocations of net income made to patrons on which the cooperative assumes income tax responsibility. Members do not report these refunds as taxable income until received as cash. Cooperatives may use nonqualified refunds to allocate income from nonmember business, to reduce or eliminate cash refunds, and to avoid the requirement that members claim refunds as taxable income at least until the refunds are redeemed in cash. A California cooperative, for example, gives each patron the option of taking allocations either as qualified or nonqualified. This practice allows patrons more flexibility in tax planning. Nevertheless, this method is not widely used. In 1976 very few cooperatives allocated net income in nonqualified form. Only 3% of the cooperatives use the method, and they only distributed 0.5% of all cooperatives' net income in this way. Larger cooperatives seem to use nonqualified refunds more, but even then their percentage is low. In 1985 the 100 largest cooperatives allocated merely 8.2% of their net income in nonqualified form. Further, only 10 cooperatives were involved.

Cooperatives that wish to pay high cash refunds or that are already in high corporate income tax brackets may wish to avoid nonqualified refunds (see Chapter 16 for details). Keeping track of nonqualifying refunds may also become a burden for cooperatives and members.

Selecting Bases for Distributing Net Income

As part of determining the way that net income is to be distributed, the board must also determine the unit (physical or monetary) and number of categories that will be used to calculate each member's patronage refund. Cooperatives handling only one product need only one category, and must only decide between physical and monetary units as a basis for distribution.

Physical units should be used when costs and benefits are related to volume. In this case, the board usually selects an industry standard such

TABLE 13.6. Method of distributing patronage refunds by 50 selected cooperative elevators, 1985(%)

Method	Physical	Monetary	Total
Individual grain	18	8	26
Blend	66	8	74
Total	84	16	100

Source: Cobia et al.

as a *packed box* for apples or a *ton* for fertilizer. A monetary measurement should be used when monetary risks such as credit and price changes account for the bulk of the benefits returned to patrons. The relative importance of these factors depends on the situation. Methods used by 50 large train-load grain shipping cooperatives are given in Table 13.6.

Cooperatives that provide a variety of services to a variety of patrons often calculate net income on a department or commodity basis in order to achieve a more accurate reflection of business at cost. Cooperatives that both market commodities and sell supplies usually separate refunds for these two services. Still other cooperatives take the business-at-cost principle one step further by segregating refunds into major categories determined by the commodity marketed or supplies provided: for example, wheat, barley, oats, fertilizer, fuel, and TBA (tires, batteries and accessories). For example, one cooperative returned patronage refunds of 11.3 cents/bushel for wheat and 17.3 for barley. A blended refund would have resulted in the barley patronage subsidizing the wheat patronage.

It is not always practical, however, to calculate income on each transaction just to maintain the business-at-cost principle. In fact, some benefits from cooperative activity result from sharing overhead costs. For this reason, many cooperatives distribute net income without segregating refunds by type of service.

Ideally, boards will for the most part follow business-at-cost policies but make some adjustments to avoid excessive costs associated with maintaining separate records for several products or services.

Dividends on Equity

Some cooperatives distribute part of their net income to equity holders based on their proportion of equity rather than on patronage. Funds allocated in this way are called dividends and are equivalent to dividends paid to stockholders in an IOF except that they are limited to less than 8% of the equities' face value (5 or 6% in some states) or less.[2] Nor-

[2]This limitation was modified in New York. Cooperatives in that state may pay up to 12% as dividends, but they must use the one-member, one-vote rule. No limit exists in a few states.

mally they are paid only on membership stock. Sometimes dividends are paid on allocated retained patronage refunds or per-unit capital retains.

Dividends on equity account for a small proportion of net income. This is not solely a result of legal restrictions, but also because most cooperatives choose not to pay any dividends. In 1976 only one-third of all cooperatives made such distributions. Further, those distributions represented only 2.1% of net income. Geographically, cooperatives in the midwest paid the lowest dividends and those along the eastern seaboard, the highest. Among the 100 largest cooperatives, rice cooperatives paid the highest percentage (18.4%) of dividends, and grain and sugar cooperatives paid none (Table 13.5).

The use of dividends is limited largely because only those cooperatives qualifying under section 521 can deduct equity dividends from the cooperative's taxable income. Otherwise, both the patron and the cooperative must report net income paid as dividends as taxable income. The effect of this tax policy is obvious; in 1976, cooperatives that did not qualify for section 521 status and tax treatment distributed only 1.4% of net income as dividends, while section 521 cooperatives distributed 4%.

There are three arguments supporting the declaration of dividends. First, a case can be made for paying dividends on equity out of net income resulting from nonmember business. Income from nonmember business is analogous to income earned by IOFs because it is generated from investment rather than patronage. Therefore, it only makes sense that income from nonmember business (like the net income of IOFs) should be distributed on the basis of investment rather than patronage. This is what was meant in Chapter 1 regarding cooperatives that have an IOF component— who do business with nonmembers (type 4 patrons in Table 1.3). As a result, they can be viewed as two separate but related businesses: combination of a pure cooperative and a pure IOF.

Second, offering dividend payments is popular among some members. Apparently, these members feel that their investments in the cooperative are, as the saying goes, dead money unless dividends are received. This feeling has some merit if equity is not held in proportion to patronage.

Third, some states require cooperatives to pay authorized dividends before redeeming any equity. Dividends on interest-bearing preferred stock are a priority obligation in these situations.

On the other hand, distributing dividends can also be problematic. Dividends become, in effect, a fixed-interest expense if the board feels obligated to declare dividends because of members' expectations. Furthermore, corporate income tax must be paid on income distributed as dividends unless the cooperative qualifies under section 521. Finally, dividends violate the business-at-cost principle if paid from net income generated by member business.

Unallocated Equity

Sometimes a cooperative will distribute part of its net income as *unallocated equity*. Unallocated equity includes net income retained by the cooperative but not allocated to an individual member or patron. It may come from patronage or nonpatronage income. Rents, interest earned on various funds, and excesses over book value from the sale of property are among the sources of nonpatronage income. (See chap. 14 for a more complete discussion of unallocated equity.)

Income from Nonmember Sources

In this chapter we discuss the possibility of distributing income from nonmember business as dividends. It can be distributed in the same way that income from member business is distributed, including the payment of patronage refunds to nonmembers. Several cooperatives use initial patronage refunds to purchase membership stock for nonmembers qualifying for membership. Any retained patronage refunds in excess of the cost of membership stock can then be credited to the allocated equity of that member in the usual way. Cooperatives may also give a patronage refund to patrons not qualifying for membership and thus avoid corporate income taxes on this income.

ALLOCATING LOSSES[3]

Despite attempts to operate at cost, cooperatives can and do incur operating losses. In 1976, 560 or 10% of all cooperatives reported losses. They totaled $22 million.

As the agricultural economy declined in the early 1980s, cooperative financial performance followed. As a result, the number and size of cooperative losses increased. In previous years cooperatives had become heavily dependent on debt, increasing their leverage to a questionable point, and this practice, together with much higher interest rates, resulted in rapidly rising interest costs and declining net income. Net losses among the 100 largest cooperatives grew from $17 million in 1976 to a peak of $240 million in 1982. Losses dropped to $68 million in 1984 but increased again to $81 million in 1985 (Figure 13.3). Losses among all cooperatives were $174 million in 1984.

Factors Contributing to Losses

A broad array of items have affected agriculture and its cooperatives during the 1980s. In the first place, many cooperatives and farmers were

[3]Revised from material provided by Kenneth Duft (1984).

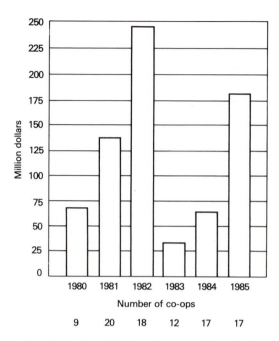

Figure 13.3   Net losses by 100 largest cooperatives, 1985. *Source*: ACS
USDA.

deceived by the prosperity they enjoyed during the 1970s. The economic
vitality of our agricultural economy during this period prompted many
cooperatives to expand through increased use of debt capital. Expensive
technologies were acquired, which also increased debt loads. These debts
with high interest charges attached were disastrous when the economy
later declined.

The depressed farm economy affected cooperatives by decreasing
their volume and margins and also by increasing the number and size of
uncollectible patron accounts. Several local cooperatives and a regional,
FCX Inc., a farm supply cooperative in the Carolinas, were forced into liquidation during the 1980s due at least in part to bad-debt losses.

In addition, the pyramiding of local cooperative equity through layers
of regional cooperatives has long been of concern to cooperative lenders.
In 1976, 25% of the net worth of all cooperatives was represented by investments in other cooperatives. For some local cooperatives the percentage
exceeded one-third. If the situation were allowed to reach some extreme
point, one could easily envision the classic house of cards disaster, wherein
a failure of the federated cooperative causes a collapse of the entire system.
Some lenders, therefore, have elected to discount the value of equities held

in other cooperatives. The wisdom of this act depends on the relative magnitudes of local member-injected capital and of the capital the local cooperative has invested in the regional cooperatives.

Operating losses sustained at the regional level are beyond the direct control of the local cooperatives and can impose a traumatic burden on them. Historically, some local cooperatives have generated little net income from their own operations, relying instead on the patronage earnings distributed by their regional cooperative. When such earnings disappear, local cooperatives are rudely awakened to the operational inefficiencies and income limitations of their own facilities.

In such cases, the status of remaining equity may also be in jeopardy. Shifts in the financial structure of cooperatives with fixed commitments associated with increased debt loads and leased equipment hamper a board's ability to respond to economic misfortune.

Alternatives

To best protect its base of equity capital, a cooperative must have available alternative means of allocating losses should they occur. Alternatives available are (1) to charge losses to unallocated reserves, (2) to reduce a prior year's patronage refund, (3) to charge the losses to current patrons, or (4) to carry them as unallocated losses (Touche Ross & Co). Before deciding which to use, the cooperative should also investigate the legal implications of each of these methods. The IRS, various statutes, and the cooperative's articles and bylaws may restrict use of some of these options.

When mentioned, bylaws most commonly call for the distribution of operating losses in a manner such as this: "Whenever a net loss occurs, said loss shall be borne, insofar as possible, by patrons in the year of said loss, distributed in accordance with their respective patronage with the cooperative during the year of said loss" (Duft, 1984). Quite simply, this means that losses are treated as the inverse of net income. This treatment is intuitively appealing because it is linked to the business-at-cost principle.

Reducing unallocated reserves is the most popular method of allocating losses (Table 13.7). Of the cooperatives sustaining losses in 1976, 77% offset losses by reducing unallocated reserves. Cooperatives like the method because it sidesteps the otherwise necessary action of allocating losses to patrons during the year of the loss. The burden for the loss depends on how the unallocated reserves are generated. If a cooperative chooses to reduce unallocated reserves, it can retain the tax-loss benefit rather than pass it on to the member. Cooperatives can also carry losses as unallocated losses if unallocated reserves are insufficient.

TABLE 13.7 Distribution of operating losses by 560 farmer cooperatives with net losses, fiscal year 1976

Distribution of losses	Number of cooperatives	Net losses distributed(%)
Unallocated reserves	429	53.5
Patronage		
Allocated equity[a]	138	29.0
Cash[b]	38	16.7
Dividends[c]	86	(0.7)
Income tax	n/a	1.5
Total[d]	560	100.0

Source: Griffin, et al.
[a]Deducted from previously retained refunds.
[b]Deducted from marketing proceeds due members, members were billed, charged to members' accounts.
[c]Payment of dividends on equity by cooperatives with losses.
[d]Does not add to 560 because some cooperatives allocated losses more than one way.

Allocating losses to current patrons according to patronage, the second most popular method, is also advantageous because it is clearly in harmony with cooperative principles and with IRS regulations. Those who patronize the cooperative during the year of the loss share in that loss on a prorata basis. However, problems of fairness arise in using this approach. (Those problems are discussed later in the chapter.)

A convenient approach to allocating losses to patrons is to reduce the value of previously allocated patron equity. Nearly 30% of all cooperative losses were allocated this way in 1976. Cooperatives can allocate losses to all equity holders on the basis of allocated equity instead of current patronage. In this case, the cooperative spreads losses over its entire membership in the year of the loss. This may be the most equitable approach if the loss is related to a major decision of an earlier year. The disadvantage is that inactive patrons who may not be responsible for the loss share in it simply because they have equity, whereas newer patrons who as yet have little or no equity get a free ride.

Another approach is to allocate losses and collect them in cash from patrons. This is usually handled in one of three ways. Marketing cooperatives often deduct the amount from marketing proceeds. Charging losses to members' individual accounts or billing them directly are options open to any cooperative. Because of their directness, however, these last two methods may be more difficult to carry out because of possible negative reactions from patrons.

An alternative is to carry losses forward under the net operating loss provisions of section 172 of the IRS code. Thus all present and future patrons share in the loss while maintaining net worth. Again, however, cooperatives should research the legal implications of their actions, because there are differences in the options and limitations of section 521 and non-section 521 cooperatives (Kothmann).

Another option for cooperatives is to combine the loss of one operation with the net income of another operation. (See the "Netting Losses" section in Chapter 16 for the rationale and legislation covering this practice.)

Tax Issues

Sometimes the method a cooperative chooses to allocate losses will depend on the effects of certain tax laws. For example, if a qualified retained patronage refund is reduced subsequently by having a loss charged against it, patrons can include this loss in the year's operations, just as if they had lost actual cash. A loss allocation has a positive value to each patron equal to the amount of the loss multiplied by the patron's marginal tax rate. However, if cooperatives charge the loss against unallocated equity, there are no current income tax benefits for members.

Unfortunately, tax courts, IRS regulations, and private letter rulings have not resolved all issues, and inconsistencies still exist. The following paragraphs summarize the current status of existing tax laws and rulings regarding cooperative losses.

The specific method of allocating losses among past, present, and future members is a judgment that is subject to certain restrictions. Netting of a loss between major functions is allowable for any cooperative. IOFs are allowed to carry losses forward under section 172 of the Internal Revenue Code, but the IRS discourages cooperatives from doing so. The agency feels that cooperatives should charge their losses to patrons in proportion to patronage (the reverse of patronage refund) during the year of the loss. Tax courts have not supported this contention. The IRS recommendation that losses should be allocated on an annual basis ignores to some extent the problems discussed in the following section.

Problems

Allocating losses to patrons in the year of the loss has intuitive appeal. But strict adherence to this principle may prove to be inequitable. Some extraordinary losses may have unusual circumstances associated with them, making it inequitable to allocate the loss entirely to current patrons.

Why, for example, must current-year patrons be penalized for a current-year loss when it can clearly be shown that the loss resulted from a major decision made by the cooperative two or more years earlier? For example, overly optimistic expansion in the mid-1970s left cooperatives with costly capacity in excess of needs in the 1980s. Resulting losses would, under the foregoing provision, be borne by current-year patrons, while the allocated equity of those patronizing the cooperative during the period of growth decisions is untouched. A parallel situation is a loss incurred by a regional cooperative two years before it is experienced at the local level.

Similar questions of equity can arise if cooperatives elect to offset their losses against future years' earnings. Is it fair and equitable to burden future patrons with the errors of their predecessors?

A related question regards netting. Is it fair for a cooperative to offset losses in one of its divisions with income generated by another division, assuming that each division, to some degree, also has different patrons?

A problem plaguing northwest cooperatives is created by so-called floating tonnage, characterized by growers who choose to take their produce from one cooperative to another in successive years. As these growers withdraw their volume from one cooperative, their unused capacity may contribute to an operating loss for the cooperative. Yet the burden of the loss, if allocated to current patrons, is borne by those who remained loyal to the cooperative, not by the growers who took their business elsewhere. When cooperatives are confronted with a high proportion of such floating tonnage, real chaos results and serious questions of what is fair and equitable arise.

Benefits to cooperatives derived by allocating losses to patrons are not always positive. The allocation of losses directly to the member-patron may provide some personal income tax relief for well established and economically secure patrons. If it does, those members, at least, will probably accept it. But for other patrons who are largely undercapitalized and who are experiencing losses, the direct allocation of a loss is less well-received.

Cooperative bylaws usually determine the options available for handling losses. However, they often pose additional problems. Too often, bylaws are ignored or forgotten. And even when they are examined, these documents often contain little, if any, mention of procedures for allocating losses. Further, provisions that do exist are often nebulous, sometimes even mandating actions or policies that reduce the cooperative's ability to protect its equity.

Seemingly, the cooperative should be able to allocate its losses by using methods that least diminish its equity base and that are fair to its membership. Yet the IRS has informally discouraged bylaw provisions allowing cooperative boards of directors total discretion regarding the allocation of losses. To make matters more confusing, the U.S. Tax Court, when challenged, has rejected the IRS's arguments, especially when the IRS has forbidden cooperatives to carry over net operating losses pursuant to section 172. Clearly, the cooperative seeking to protect its equity capital base when confronting an operational loss may find itself in a dilemma.

SUMMARY

Benefits of owning and patronizing cooperatives are distributed as patronage refunds and dividends. Some income, especially from nonmembers, is often retained by the cooperative as tax-paid unallocated reserves. Patronage refunds payable to individual patrons may be calculated on the basis of monetary value or physical units, as a blend or by departments. Patronage refunds may be made to the patron in cash or retained by the cooperative to build up patron investment and to provide necessary equity. The tax obligation is assumed by the patron if the retained portion is qualified and by the cooperative if it is nonqualified. Dividends paid on owner equity are particularly appropriate for income generated from business on which no patronage refunds are given. Allocations to unallocated reserves typically come from the same sources, and cooperatives have corporate income tax liabilities on allocations to this account.

Cooperatives may deduct losses from unallocated reserves and allocated equity. They may also charge losses directly by reducing marketing proceeds due members or by charging losses to their accounts.

DISCUSSION QUESTIONS

13-1. Trace the possible allocations of net income, and discuss the pros and cons of each.

13-2. What are the alternative units or categories for distributing net income to patrons, and under what conditions is one preferred over another?

13-3. How do arguments for returning net income as patronage refunds versus paying dividends on equity differ when one considers member versus nonmember sourced income?

13-4. Not many cooperative leaders seem to be familiar with nonqualified refunds. How are they different from qualified refunds, and under what conditions would they be preferable?

13-5. Many successful cooperatives do not pay dividends on equity. Why not?

13-6. How may accelerated depreciation and gains or losses from the sale of equipment distort the business-at-cost principle?

13-7. How might pooling avoid problems associated with allocating losses?

13-8. What are the pros and cons of alternative methods of allocating losses?

13-9. Ms. J. J. Oviatt received a patronage refund check for $35 on her business with the Tippecanoe County Co-op. Assume that the cooperative issued qualified refunds and that the minimum cash refund was paid. The cooperative's total costs were $625,000 on $640,000 of total revenue. What was Ms. Oviatt's total refund and total patronage with the cooperative? What is the purpose of the minimum cash refund and how was Ms. Oviatt probably credited with the balance of her refund?

13-10. Why do many cooperatives give cash refunds greater than 20%?

REFERENCES

BRIGHAM, EUGENE F., AND LOUIS C. GAPENSKI, *Financial Management Theory and Practice,* 4th ed. Hinsdale, IL: Dryden Press, 1985.

COBIA, DAVID W., et al., *Pricing Systems of Trainloading Country Elevator Cooperatives.* Agr. Econ. Report 214, North Dakota Agr. Exp. Sta., Dec. 1986.

DAVIDSON, DONALD R., AND JEFFREY S. ROYER, *Top 100 Cooperatives, Financial Profile.* Washington DC: USDA ACS Jan. 1987.

DOWNEY, W. DAVID, AND STEVEN P. ERICKSON, *Agribusiness Management,* 2nd ed. New York: McGraw-Hill, 1987.

DUFT, KENNETH D., *Principles of Management in Agribusiness.* Reston, VA: Reston, 1979.

DUFT, KENNETH D., "Equity Ramifications to the Allocation of Cooperative Operating Losses," *Agribusiness Management.* Washington State Univ. Cooperative Ext. Serv., Oct. 1984.

GRIFFIN, NELDA, et al., *The Changing Financial Structure of Farmer Cooperatives.* Washington, DC: USDA ESCS FCRR 17, Mar. 1980.

KOTHMANN, S. S., "Cooperative Losses and Netting," *Coop. Accountant.* (Winter 1983):61.

TOUCHE ROSS & CO., *Accounting and Taxation for Cooperatives,* 4th ed. San Francisco: 1978.

VAN HORNE, JAMES C., *Fundamentals of Financial Management,* 6th ed. Englewood Cliffs, NJ: Prentice-Hall, 1986.

14

Equity
and Debt

David W. Cobia,
North Dakota State University

Thomas A. Brewer,
Pennsylvania State University

INTRODUCTION

The cooperative's need for funds, as portrayed on the asset section of the balance sheet, is much the same as that of any other business. Funds are needed to purchase land, buildings, equipment, inventory, or other assets and to pay operating expenses or meet unforeseen financial contingencies. Sources of funds to meet these needs are as widely varied as those of any other business, but some methods of raising equity capital are utilized only by cooperatives. Cooperatives typically acquire and handle equity differently than do investor-oriented firms (IOFs). The unique user-owner linkage of cooperatives makes it both possible and necessary for a cooperative to secure owner-equity capital by unique methods. (See Introductory Concepts of Chapter 13 for a brief description of a cooperative's balance sheet.)

EQUITY

Equity is the investment that member-patrons make in the assets of their cooperative. A key financial responsibility of members is to provide equity in proportion to benefits received in the past or anticipated in the future. An adequate equity base is essential if a cooperative is to provide services desired by members, survive adversity, and obtain credit.

Equity is risk capital; it exists to serve as a buffer during periods of economic misfortune. Any losses experienced by the cooperative are subtracted from the cooperative's equity pool until it is exhausted. Thus, a strong equity base provides security for lenders and makes it possible for borrowers to receive more favorable interest rates. Unfortunately, some cooperatives learn this the hard way. For example, after experiencing several years of operating losses, some northeast milk-marketing cooperatives were forced to require members to furnish more equity capital, because their lenders would not renew or make new loans until member-equity levels were at least partially restored.

Fortunately, problems like this can usually be avoided, if not in a crisis, because the cooperative's board of directors governs the status of member equity and through their action the pool of equity can be maintained. Typically, the board decides each year how much to assess, how much to retain, and how much to redeem. The board is therefore able to respond to good times and bad by supplementing, sustaining, or reducing the pool of equity as conditions warrant.

Equity is used not only to provide funds for operations, but also as a measure of member interest. Members should compare the marginal rate of return associated with additional investments in the cooperative with those of other investment alternatives available to them. To maximize returns to invested capital, members should continue to invest in cooperatives as long as marginal returns (including benefits of refunds, favorable prices, and dividends) are as great as or greater than those from other investments. A difficult part of this task is evaluating the intangible and indirect benefits of investment in a cooperative. Returns from such benefits as access to markets and a reliable source of supply are difficult to quantify. On the other hand, when members invest in a cooperative, they must also understand that their equity capital, although it may have special benefits, is risk capital and is therefore subject to loss.

Unique Features of Cooperative Equity

Because of their unique ownership structure, cooperatives differ a great deal from other businesses in the way they gain, allocate, and return owner-equity. First, only qualifying persons, generally agricultural producers, can become members and own common stock or obtain membership certificates in a cooperative. The right to vote is tied to such ownership. Further, control of the cooperative is generally democratic, allowing one vote per member, rather than having the number of votes determined by the shares owned.

Second, net income is generally returned to patrons in the form of patronage refunds in proportion to the business done through the cooperative—it is not based on the shares of stock owned. Similarly, patrons provide equity in anticipation of benefits arising from patronage rather than in expectation of capital appreciation or dividends.

Third, equity is redeemed by the cooperative at book or par value, whichever is less. Therefore, the value of most cooperative equity does not change the way the value of IOF equity does. (Exceptions to this generalization are discussed later.) However, some cooperatives whose assets have market values in excess of their book value have considered redeeming equities at appreciated levels. This generally occurs when assets (usually land) have appreciated or when income from nonmember business has accumulated as unallocated reserves. Generally accepted financial policies and accounting procedures to handle these situations are not well established. In addition, only rarely can holders of cooperative equity sell it for cash. Equity holders of many other businesses can usually sell their equity anytime and to anyone.

Fourth, cooperatives often raise equity indirectly through the use of retained patronage refunds or per-unit capital retains. Other businesses can raise equity by retaining earnings but cannot assess their customers to contribute equity.

Finally, unlike equity of most corporations, a substantial portion of cooperative equity is temporary because cooperatives have an implied obligation to redeem it. (Only preferred stock purchased for investment and unallocated reserves can be considered permanent.) However, the board can decide when to redeem equities and can thus protect the financial integrity of the cooperative. Mandatory obligations to redeem do not now exist.

Similarities between cooperative and IOF equity are fewer but noteworthy as well. Cooperatives, for example, have the same requirements for capital to invest in plant and equipment and to provide operating capital. Also, since equity is risk capital, losses associated with business reverses are absorbed by equity just as they are in other firms. All liabilities must be satisfied in case of liquidation before any cash can be returned to equity holders. Finally, like the equity in IOFs, equity in cooperatives has no definite dividend, interest rate, or maturity date when it must be redeemed in cash. Any financial instrument that has a mataurity date must be classified as a liability.

Raising Equity Capital

Cooperatives often find it difficult to raise equity capital because the normal incentives for investment, as discussed above, are not present. As a result, members sometimes resist supplying their fair share of equity.

They, in effect, become free riders unfairly burdening those who provide more than their share of needed equity. Fortunately, however, some factors unique to cooperatives tend to offset these difficulties.

For example, cooperatives often acquire equity (per-unit capital retains or retained refunds) as a result of patronage. Although a back-door approach, this does relieve member-investors of the obligation to provide cash up front. However, traditional lenders often are willing to lend money to patrons for investment in their cooperative, especially if it is apparent that the investment will enhance the earning capacity and loan-repayment ability of the patron. This situation can occur, for example, if a cooperative provides access to a market that would not otherwise be available.

Another factor to be considered is the impact of federal and state income taxes on accumulation of equity. It is sometimes argued that cooperatives can build net worth faster than IOFs by retaining earnings because cooperatives do not have to pay corporate income taxes. This argument has merit only for those cooperatives that have net incomes over $50,000 and who pay the minimum proportion of qualified patronage refunds in cash. Those cooperatives that issue nonqualified refunds pay federal corporate income taxes at the same rate as any other corporation. Cooperatives issuing qualified refunds must pay patrons at least 20% of the refund in cash; this 20% more than offsets the cash drain associated with the marginal corporate tax rate of 15% on net income of less than $50,000.

It is true that cooperatives with net incomes over $50,000 can retain a larger share of equity than an IOF. With net income over $75,000 the cooperative's advantage over other businesses is 14% (20% cash refunds to patrons versus 34% marginal tax rate on profits over $75,000). But even at these higher net income levels, this advantage is a bit theoretical since many of the larger cooperatives refund more than the required 20% minimum in cash. Furthermore, many IOFs have found ways of sheltering income which results in low effective tax rates for them. Certainly, the assumed advantage is not accurate for the top 100 cooperatives, because in 1985, they, as a group, paid members cash refunds that on average accounted for 53% of their net income. Of course, current tax laws may change and could shift any advantage in any direction.

Cooperatives also face member pressures to redeem retained equity— a problem IOFs seldom face. (However, IOFs may have pressure to pay stock dividends.) Granted, section 521 cooperatives do not have to pay corporate income tax on equity-based dividends, but this advantage is relatively unimportant since these dividends are generally limited to less than 8% and a declining number of cooperatives are finding it worthwhile to maintain section 521 status.

Because incentives for investing lie with patrons since only they can benefit, and are absent for others, cooperatives must rely on member-patrons to provide equity capital. This limits the sources from which a cooperative can attract risk capital. Cooperatives are further limited because potential contributors frequently find it difficult to obtain funds for such uses.

Types and Sources of Equity

Cooperatives can build up a pool of member equity by utilizing one or more of three methods. These are direct investment, retained patronage refunds, and per-unit capital retains. Farmer cooperative members had allocated equity in their cooperatives worth $6.6 billion in 1976. The relative importance of these three methods of acquiring equity are summarized in Figure 14.1. In 1985, the largest 100 cooperatives reported a combined net worth of $5.6 billion. Unfortunately, a breakdown by equity accumulation method is not available for the 100 largest cooperatives.

Direct investment

Direct investments include cash purchases of common or preferred stock, membership certificates, or other forms of equity. The initial funds for starting a cooperative are generally raised in this way and are collected

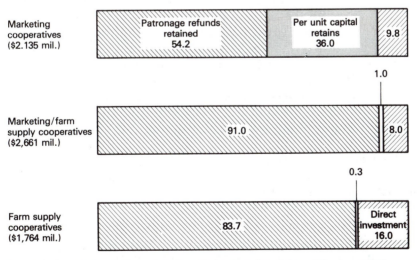

Based on percent of total allocated equity capital outstanding at close of fiscal year 1976.

Figure 14.1 Methods of acquiring allocated equity by farmer cooperatives, 1976. *Source:* ESCS, USDA.

directly from member-patrons. Existing cooperatives can also use the direct investment method to accumulate equity.

Direct investment accounts for the smallest share of total cooperative member-equity. For example, in 1976, 55% of all U.S. farmer cooperatives obtained some equity through direct purchases, but these purchases represented only 10.7% of total allocated equity. Allocated equity from direct investment by members ranged from a low of 4% among cooperatives classified as handling "other products" to an average of 16% in farm supply cooperatives and 36% in sugar marketing cooperatives. Direct investment is relied on as a means of raising equity more heavily in the east than in the rest of the country. In fact, cooperatives in the Wichita district acquired only 1% of their equity through direct investment by members compared to 28% in the east. Membership certificates are not relied upon to supply much equity either. They were used by only 393 of 5,795 cooperatives in 1976, and purchased membership certificates accounted for only 0.4% of all equity in U.S. agricultural cooperatives (Griffin et al.).

Advantages There are two major advantages to the direct investment method of obtaining equity. First, it provides the initial funds required to start a cooperative, and second, it can be used as a critical test of member interest. Often members campaign vigorously for a cooperative to make a major investment but fail to support it once it is made. If, on the other hand, these members decide to make a direct investment, they are likely to be genuinely committed. Direct investments seem to be most successful when they are linked directly to use of the cooperative. For example, potential members of American Crystal (a sugar marketing cooperative) invested $105 per acre for the right to deliver production. Members of Pro Fac (a fruit, vegetable, and specialty products marketer) also contributed capital based on deliverable acreages or tonnages of specific commodities such as snap beans and chipping potatoes. But, despite these benefits, the use of the direct investment method as a means of raising equity is limited. It is usually difficult and sometimes unsuccessful.

Disadvantages Direct investment has limited appeal among members because its ownership provides for limited or no return linked directly to that investment. Furthermore ownership provides little opportunity for capital appreciation, and the transferability of shares is limited. These restrictions are especially restrictive for farmers hard-pressed for cash. Moreover, it is much more difficult for cooperatives to collect direct investments from members than it is to retain refunds or per-unit capital retains. The latter two can be collected automatically, but a special effort is required every time that direct investments are used to collect equity. Because of these limitations, management must generally engage in an ambitious and costly educational effort each time it attempts to obtain equity through direct investment by members.

Retained patronage refunds

Retained patronage refunds are portions of net income allocated to members but retained by the cooperative. They are, in fact, new investments made in the cooperative by those who are patronizing it.

Retained refunds represent the bulk (77%) of cooperative equity. This method is used, more than any other, as a means of raising equity (Figure 14.1). In 1976 only 11% of all agricultural cooperatives did not use this method. The proportion of equity accumulated through retention of patronage refunds ranged from an average of 28% among sugar cooperatives to 94% for grain cooperatives. Geographically, the differences are even greater. In the Sacramento Farm Credit district cooperatives accumulated 26% of their equity by retaining patronage refunds. In the Wichita district it was 99%.

Advantages Retaining patronage refunds is a popular practice because once established, it is an easy and systematic method of generating equity funds. Each contribution, for example, is directly related to patronage in the year just passed. Further, the method is particularly well suited for supply and service cooperatives, where per-unit capital retains (the other major method) do not work well. Retained refunds can also be used to encourage nonmember patrons to become members, because they can use retained refunds to purchase membership stock. This does not require members to make a direct cash outlay.

Disadvantages A problem associated with relying on retained patronage refunds as a major source of equity accumulation is that the quantity of retained refunds is dependent on net income, which fluctuates with the fortunes of the cooperative. For example, the 100 largest cooperatives as a group lost equity in 1982 due to operating losses. Also, equity accumulated by retaining refunds can lead to misunderstandings among members. For example, they may tend to think of retained refunds as a debt owed them by the cooperative rather than as an investment. If they view it as a debt, they expect it to be repaid whereas if they understand that it is risk capital, they should realize they may get it, receive benefits in other ways, or could lose it. Retained refunds may appear deceptively easy and too reliable a source of equity and thereby lead to unwise expansion or expenditures. For these reasons, some marketing cooperatives balance the use of retained patronage refunds with the per-unit capital retain method of accumulating equity.

Per-unit capital retains

Per-unit capital retains are patron investments in the cooperative that are based on the value or number of units handled for each patron. Marketing cooperatives are the major users of this method of accumulating equity capital. These patron investments are deducted from the proceeds of products marketed.

For example, a milk-marketing cooperative might decide to retain 10 cents/cwt from each member's marketing proceeds for the purpose of building equity. Each month the cooperative calculates the amount due each member for milk delivered by multiplying the price to be paid by the number of hundredweights delivered. From this amount, the cooperative deducts 10 cents for each hundredweight marketed and retains it as member equity. The member receives a check for the gross value of the milk minus the member-equity capital withheld. At the end of the year the cooperative provides each member with a statement indicating the amount of capital invested through this deduction procedure. A member who delivers 1,000 cwt. of milk to the cooperative would have contributed $100 of equity capital during the year (1,000 cwt x $0.10). Per-unit capital retains may be allocated to patrons in either qualified or nonqualified form. They are taxed in a similar way to patronage refunds, except that cooperatives are not required to refund 20% of these retains in cash.

The proportion of equity raised through per-unit capital retains averaged 12% in 1976. Just over 5% of all cooperatives used this method, and 58% of those using it were California or Florida fruit and vegetable cooperatives operating on a pooling basis. Large dairy marketing cooperatives commonly use this method of acquiring equity.

Advantages Cooperatives that use per-unit capital retains favor them because they are not affected by the level of net income of the cooperative and therefore provide a more stable source of equity accumulation than do retained refunds. Per-unit capital retains have most of the advantages of retained patronage refunds.

Disadvantages Cooperatives may avoid per-unit capital retains, however, because cash that would otherwise be available to the farmer is reduced by the amount of the retain. Use of the per-unit capital retain method does not work well in supply or service cooperatives. Retains give the appearance of a price increase rather than an investment. In marketing coopertives, per unit capital retains give the appearance of a price decrease; thus, patrons tend to believe the cooperative is not price competitive. Per-unit capital retains also have the same disadvantages as retained refunds. They are often considered as a cooperative debt by members and may seem, to cooperative's board and management, to be too easy a source of equity.

Size of retains If a cooperative uses per-unit capital retains, the board must decide what rate of retain to use. In making this decision, the board should consider (1) the amount of capital required, (2) the speed with which capital needs to be accumulated, (3) the planned rate of equity redemption, (4) member understanding of the process including redemption, and (5) competitive environment of the cooperative.

Unallocated equity

Equity can also be accumulated by building up funds that are not allocated to any member, patron, or other individual account by any form of certificate or book credit. Instead, this equity would show up as member equity on the balance sheet but in an unallocated account. It may come from sources such as nonoperating income (interest, rent, etc.), acquisitions of businesses whose purchase prices are less than the book value of the assets, or from net income (from nonmembers or even members) which was not allocated or refunded. It can also result from the sale of assets in cases where market values are greater than book values.

About 15% of the 100 largest agricultural cooperatives' equity was held in unallocated form in 1985. This was also the average reported by all agricultural cooperatives in 1976. Most cooperatives at that time had at least some unallocated equity, but 17% did not. Several cooperative leaders have observed cooperatives with more than 90% unallocated equity. The relative importance of unallocated reserves among product groups ranged from an average of 0.7% for rice to 38% for livestock and wool cooperatives. Geographically, this percentage ranged from an average of 0.1% of equity in Texas to 52% in the Springfield Bank for Cooperatives (BC) district.

Reasons for using unallocated equity Cooperatives in some states are legally obligated to build unallocated equity up to a specified level. Other states are neutral. Still others specifically permit unallocated reserves for such purposes as covering possible losses. Building unallocated equity is also helpful in relationships with creditors. Creditors have suggested that allocated equity should be supplemented with a higher proportion of truly permanent capital—that cooperatives should adopt more stable financial structures less dependent on actual or implied redemption obligations. Unallocated equity reassures these lenders, because it reduces the pressures placed on the board for equity redemption.

During periods of inflation, some cooperatives build unallocated reserves to reduce cash outflows. Net income may, in effect, be overstated at such times, because replacement costs for inventory and capital assets are understated.

In addition, cooperatives that do not have section 521 status often retain net income derived from nonmember business as unallocated reserves to avoid double taxation. Otherwise, if these cooperatives were to allocate the money to members, both the members and the cooperative would have to report it as taxable income.

Returning current income to members may be a nuisance to both members and cooperatives, and the amounts may be so small as to have little economic value to members of cooperatives such as bargaining or livestock shipping associations. These types of cooperatives have low equity requirements. They accumulate equity from operations, credit it to unallocated reserves, and return any excess income above capital requirements to patrons as cash refunds. This system is simple and avoids uncertainty about equity redemption. It avoids complex financing methods in which boards must decide whether to use qualified or nonqualified, cash or noncash patronage refunds, and how and when to redeems members equity.

Problems with unallocated reserves For all its ease and simplicity, use of unallocated equity as a method of accumulating capital has serious inherent problems. First, when cooperatives use net income from member business to build unallocated reserves, they violate the principle of service at cost. Furthermore, individual members' ownership of unallocated reserves is less apparent, which in turn may lead to three additional problems. The ties between the cooperative and its members are eroded. Members hold little identifiable equity in the cooperative. The net result is that they lose a sense of ownership. Second, a high level of unallocated reserves may keep a cooperative from satisfying the requirement that active members must own 50% of a cooperative's equity for it to qualify and obtain commodity loans in the name of its members under the government price support programs. Third, equitable distribution of unallocated reserves in case of liquidation may prove awkward and perhaps impossible. State laws may require any remaining funds in unallocated reserves to be distributed to patrons according to past patronage; however, it may be difficult to make such calculations because of lost records, and it may be impossible to locate some recipients. Cooperatives may also have bylaws that provide current patrons with a claim on residual funds.

There is another concern about high levels of unallocated reserves. As the proportion of unallocated equity increases, management may become more independent and less subject to member control. It is argued that management feels less responsible to members and that members with very little allocated equity will not be as concerned about the cooperative and will leave hired managers in control. Substantial unallocated reserves also provide an incentive for some patrons to press for liquidation so as to acquire the cash value of those reserves or to place restrictions on acceptance of new members to avoid dilution of them.

Equity from Nonmember Sources

One way of avoiding some of the problems discussed above is to aquire equity from nonmembers. Retained refunds are often allocated to nonmenber patrons. This results in nonmembers holding equity in the cooperative. Also, peferred stock can be sold to the public if the bylaws of the cooperative permit. This can provide cooperatives with another rather permanent source of capital. Preferred stock seldom carries voting or membership privileges but has a preference in the distribution of net income in that preferred stockholders must generally be paid a stipulated amount in dividends before net income can be allocated to patrons. Preferred stockholders have priority over common stockholders if the cooperative is liquidated.

Exchange of Equity

A few cooperatives offer opportunities for or at least do not prevent the purchase and sale of equities or the transfer of allocated equities among members. Most do not allow these transactions. Where they are allowed, the price is privately negotiated, and as a result may be sold at face value or a discount from or premium to it. Transactions must usually be approved by the board and recorded by the cooperative.

More limitations are placed on the transfer of equities such as common stock and membership certificates which carry voting rights than on equities without voting rights. Discussion in this section is confined to equity which is restricted to qualifying members. Equities sold as direct investments to the general public can be exchanged but are not a part of this discussion.

In practice, only a small amount of cooperative equity capital is exchanged. Very few cooperatives have an active market for their allocated equity. Equity exchanges most often occur in cooperatives using the base-capital plan, particularly if patronage rights are linked to equity. They are also used by cooperatives employing revolving funds. Tri-Valley Growers in San Francisco, California has received some publicity for its equity exchange (Cobia et al.). Other examples include American Crystal Sugar and National Grape. These cooperatives, however, are a minority.

The provision in many bylaws that capital stock may be redeemed only by the cooperative at book or par value, whichever is lower, deters trading even if it were permitted. Members of a cooperative using a revolving fund can evaluate equities and determine transfer prices if the cooperative has a consistent revolving policy. But without such a program, it would be difficult to establish a value mutually agreeable to prospective buyers and sellers. Even then the buyer is assuming a risk because circumstances may force the board to change the revolving policy.

Those members who buy equity do so either because they can acquire it at a discount and later redeem it at face value or because it is tied to needed marketing rights. Those selling equity may desire a discounted cash value now rather than its face value later, or they may no longer need marketing rights associated with the equity.

Cooperatives may become interested in the exchange of equity among their members to help shift it from overinvested to underinvested members. To motivate the latter group to furnish its share of equity, the board may require underinvested members to pay interest on their shortfall. A market for equities can also be desirable since it increases the liquidity of allocated equity and thus enhances its value as collateral. In addition, equity exchange can facilitate equity redemption and adjustments in patronage rights among members; these exchanges are necessary when equity is linked to patronage rights.

However, cooperatives may not wish to encourage the exchange of equities, because their values may fluctuate too much and thus encourage speculation, which many have tried to avoid. Moreover, the base for establishing a transaction price may be too limited. For example, members of cooperatives using an unpredictable equity redemption program would have difficulty determining appropriate sales prices for equity. Some cooperatives also have had trouble keeping equity in the hands of producers when a speculative interest is present.

Legal problems may be associated with equity exchange as well. Cooperatives considering the adoption of an equity exchange program need to evaluate carefully legal requirements and financial commitments (Cobia et al.) before taking any actions.

DEBT

Even though all cooperatives accumulate equity, most of them still have to borrow money. With the exception of the BCs and member patrons, the sources that cooperatives borrow from are the same ones used by other firms (Figure 14.2 and Table 14.1).

Because cooperatives are not isolated from the economic environment or the economic health of their patrons, the relative importance of debt has fluctuated, to some extent, with the times as well as with changes in management philosophy. Until 1962, for example, farmer cooperatives as a group had very little debt. At the close of 1962, only 59% of all U.S. farmer cooperatives carried borrowed funds on their balance sheets. By 1976, this percentage had climbed to 79. Debt was used to finance 58% of total assets in 1976 and again in 1984.

Royer uses the interest-coverage ratio to indicate how this development might hamper a cooperative's ability to survive. The ratio is

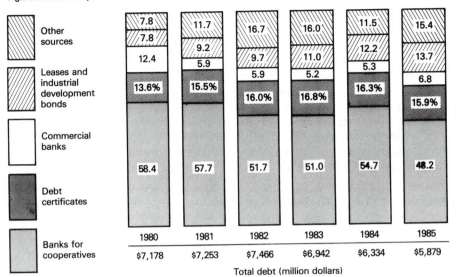

Figures in bars are percent of total debt.

		Other sources
		Leases and industrial development bonds
		Commercial banks
		Debt certificates
		Banks for cooperatives

	1980	1981	1982	1983	1984	1985
Other sources	7.8	11.7	16.7	16.0	11.5	15.4
	7.8	9.2	9.7	11.0	12.2	13.7
Leases/IDB	12.4	5.9	5.9	5.2	5.3	6.8
Commercial banks	13.6%	15.5%	16.0%	16.8%	16.3%	15.9%
Banks for cooperatives	58.4	57.7	51.7	51.0	54.7	48.2
Total debt (million dollars)	$7,178	$7,253	$7,466	$6,942	$6,334	$5,879

Figure 14.2 Sources of debt by 100 largest cooperatives, 1980–1985.
Source: ACS, USDA

TABLE 14.1 Sources of debt by 100 largest cooperatives, 1985

Source	Co-ops using each source		Debt outstanding from each source (millions)			Percent of total
	Short term	Long- term	Short term	Long term	Total	
Banks for cooperatives	50	79	$ 916	$1,918	$2,834	48
Bonds, notes, & certificates issued by cooperatives	18	35	214	720	934	16
Industrial development (revenue) bonds	—	47	—	408	408	7
Capitalized leases	—	36	—	401	401	7
Commercial banks	15	17	162	236	398	7
Insurance companies	—	16	—	317	317	5
Other nonfinancial entities (co-ops & IOFs)	7	27	78	133	211	4
Commodity Credit Corp. & other government sources	6	3	147	3	150	2
Commercial paper	4	—	127	—	127	2
Other sources	5	50	9	90	99	2
Current portion of long-term debt	93	—	642	(642)	—	—
Total			2,295	3,584	5,879	100

Source: ACS USDA.

Equity and Debt   255

computed by dividing earnings before interest and income taxes by the annual interest expense. It measures the firm's ability to meet current interest payments. It also provides a method of evaluating the cooperative's ability to take on additional debt by taking into account not only the increase in the amount of debt, but also the level of earnings and interest rate associated with it. Royer reports that

> between 1962 and 1982, the estimated average interest coverage ratio of farm supply cooperatives among the 100 largest cooperatives fell from 11.94 to .94. The average large farm supply cooperative could not cover its interest expenses in 1982 although it could have covered them almost 12 times in 1962. Cooperatives appear to be more heavily leveraged than competing firms in the same industries (Royer, p. 83).

Clearly, too high a level of debt is very risky particularly when net income and interest rates both fluctuate severely.

Relative Ease of Aquiring Debt

Very little difference exists between cooperatives and IOFs in their ability to obtain standard short-term trade credit for such items as supplies, utilities, wages, and even raw products. Debts for these items simply show up as accounts payable in the current liabilities section of the balance sheet. Accounts payable may also include the cash portion of patronage refunds to be paid, dividends payable, or the amount of a declared equity redemption.

Creditors extending short-term credit to cooperatives evaluate them in much the same light as they would the cooperative's IOF competitors. Since short-term debt is typically backed by liquid assets such as inventory, it is considered quite secure by creditors.

However, cooperatives do have *disadvantages* when borrowing long-term capital. Most lending agencies do not compete for this segment of the market, nor are they particularly receptive when approached because they do not understand cooperatives or how they operate. Lending agencies evaluating cooperatives as possible borrowers become leery of them. They do not understand such cooperative financial transactions as patronage refunds or the use of per- unit capital retains to accumulate equity. Nor do they understand equity redemption plans that are used to keep equity in the hands of current patrons.

The cooperative's implied, if not explicit, obligation to redeem equity supplied by past members is particularly bothersome. Lenders feel uncertain when evaluating the creditworthiness of an organization if a high proportion of its equity is to be distributed to those who contributed it, even if it is to be replaced by member's contributing new risk capital. The dynamic nature of cooperative equity accounts does not reassure lenders looking for permanent risk capital to secure loans and protect them from adverse business developments.

On the other hand, some lenders may have the mistaken idea that cooperatives have a considerable advantage in their ability to meet loan obligations. Cooperatives do not have to pay corporate income tax on net income distributed as qualified patronage refunds. Lenders may overlook a cooperatives' obligation to refund part of its net income in cash. A realistic analysis of the cooperatives' cash flow, however, may show its earning potential combined with tax and cash refund obligations, rarely give it any real advantage over other business when it comes to its ability to pay debts.

The greatest real advantage cooperatives have in the acquisition of debt capital is that they can borrow from their own special lenders, the Bank for Cooperatives (BCs). These federated cooperatives (see Chapter 3) serve cooperatives, which are their clients. Therefore, they understand cooperative finance. In fact, they are cooperatives themselves.

In addition, a few large commercial banks have developed a level of expertise by working with cooperatives and understanding their unique features. Because many of the cooperative loans made by these banks are seasonal and secured by current assets, they may not be as concerned about the nonpermanent nature of cooperative equity. For these banks, the ability to generate funds to liquidate seasonal loans is more important than the ability to generate net income or to maintain a strong balance sheet. Their most important consideration is whether they can get their money back.

One banker expressed a common attitude by saying: "We don't discriminate either for or against cooperatives. In general, we don't care who or where a dealer or cooperative is, as long as it meets our [performance] criteria . . . . We shift our business in a given area depending on results" (Hanrahan, p. 9).

Alternative Sources

The variety and relative importance of each source of borrowed funds among the 100 largest cooperatives is given in Table 14.1 and depicted in Figure 14.2. BCs are the major source, but they are declining in importance. Various forms of certificates of indebtedness are second in importance. Industrial revenue bonds, capitalized leases, and commercial banks each accounted for about 6.8% of total borrowed capital. Industrial revenue bonds issued by city or county governments were used by 42 of the 100 largest cooperatives. Long-term leases (reported by 36 of the 100 cooperatives) were capitalized to avoid misrepresenting future obligations. Cooperatives with long-term leases restate their balance sheets by increasing assets up to the current value of the leased asset and by increasing their liabilities the amount of the present value of future lease payments.

In 1976 more than 60% of the borrowed capital of all farmer cooperatives' came from BCs. About 20% had loans from commercial banks, and these loans represented about 10% of their borrowed funds. The remainder came from debt securities (18.9%), from leases and industrial revenue bonds (3.6%), from other cooperatives (1.9%), and from other sources (3.9%).

Banks for Cooperatives

BCs are a unique source of credit for cooperatives. It is important to remember that the BCs are owned by their borrowing cooperatives and that those cooperatives must therefore take the necessary measures to keep the BCs operating securely. For example, when a cooperative obtains its first loan from a BC, the borrowing cooperative must purchase a share of class C stock ($100 par value). Later, if larger loans are approved, the cooperative must make an additional investment in class C stock, the amount of which is in proportion to the size of the loan. The borrowing cooperative then acquires equity in the BC in the traditional way, via retained allocated patronage refunds. The relationship between the BCs and their member cooperatives is equivalent to that between other cooperatives and their members. For this reason, BCs also have some of the same challenges in their relationships with borrowers that cooperatives have with their member patrons. For example, loan officers of BC banks monitor financial progress of borrowing members but are reluctant to make uninvited suggestions unless the quality of the loan appears to be threatened.

However, BCs do have strict policies that regulate borrower eligibility for their loans. To qualify for loans, cooperatives must meet specific organizational requirements (see Chapter 3).

Cooperatives borrowing from BCs must also meet the financial requirements established by the bank. These requirements vary among BCs, over time, and according to prospects for individual cooperatives. BCs evaluate applications in much the same way as do other lending agencies. Such factors as repayment ability, collateral, balance sheet, and income statement changes or trends, quality of management, and member support enter into approval of a loan.

BCs make seasonal and long-term loans. The former are used to finance short-term seasonal needs such as inventories, which, in agribusiness, may represent a relatively large portion of total liabilities at certain times during the year. Long-term loans are used to finance long-lived assets such as land, buildings, and equipment.

BC redemption programs and interest-rate policies vary among the banks. However, certain generalizations about these policies provide a contrast to those of other banks. BCs historically obtained the bulk of their funds from the agency[1] money market through the Farm Credit System. Therefore, their cost of funds have been relatively low. Traditionally, interest rates charged BC borrowers have been based on the average interest charges being paid by the BC on its borrowed funds. Commercial banks, on the other hand, charge a marginal rate (the rate they must pay for new funds). A cooperative, then, to minimize its interest expense, might borrow from a BC when interest rates are rising (and then the BC's average rate would be lower than the current market rate) and from a commercial bank when they are falling. This may partly explain the decline in the relative importance of BC loans and the financial difficulty of many BCs during the 1982–1986 period. Interest rates reached their peak late in 1981 and were in a general decline until 1986.

BCs have tended to charge an average interest rate to borrowers regardless of risk, whereas commercial banks have used a differential rate which depended on the perceived risk associated with each loan. BCs have recently been moving in the direction of marginal-cost pricing, incorporating both their current cost of funds and the cost of servicing loans with different levels of risks.

As mentioned in Chapter 3, the organizational structure and operating practices of BCs are in a state of flux. Substantial changes in the Farm Credit System, including the BCs, are likely. Their role in providing credit to cooperatives for export markets is described in Chapter 12.

Other Cooperatives

To obtain trade credit, local cooperatives may choose to borrow from federated cooperatives with which they are affiliated. However, this can be a dangerous practice because a local cooperative's credit standing can deteriorate, forcing the federated organization to assume control of the local cooperative, thus reversing the preferred flow of automony. If the local cooperative becomes insolvent and the federated cooperative assumes ownership, a centralized system is created inadvertently.

Commodity Credit Corporation

The Commodity Credit Corporation (CCC) can sometimes be used as a source of short-term capital by cooperatives. They may obtain a loan from the CCC on agricultural commodities they have in inventory and for which a loan rate exists. Generally, the interest rate is competitive with other short-term rates and the level is established by formula based on Treasury bill rates. If the loan is forfeited, meaning that the loan provides

---

[1]The official title of the agency market is the Federal Farm Credit Banks Funding Corporation.

the best available price, no interest is charged on the loan. The commodity is turned over to the CCC.

Cooperatives must meet several eligibility requirements to participate in this program (Dylla). Member-patrons must own and control the cooperative, and they must supply 80% of the inventory for which the loan is made.

Members[2]

Members often provide various forms of debt. They may appreciate an opportunity to leave their money in the cooperative if they receive interest and are assured of a payback date. A range of financial instruments may be used. They vary from individual notes to bond issues, which must be registered with the Securities and Exchange Commission (SEC).

Demand deposits held by members are a new but questionable source of debt capital. Patron demand deposits provide patrons with investment opportunities and liquidity requirements and at the same time help to finance the cooperative. Numerous uncertainties remain regarding the legal and securities regulations that may affect this practice. Any cooperative considering such a program is advised to review the proposal with legal counsel and to pay particular attention to all related local, state, and federal statutes.

In its simplified form, it involves a cooperative accepting deposits from members and paying interest on these deposits, typically at a rate below that paid on loans from the BC, but above the rate commercially available to most members. Members can withdraw these deposits at their convenience; consequently, grower demand deposit accounts listed as current liabilities make no contribution to the equity position of the cooperative. Such deposit accounts merely supplant more traditional forms of short-term debt capital.

Funds received as patron deposits and not required to support cooperative operations can be invested by the cooperative in short-term securities. Since the rate of return on these deposits generally exceeds the rates associated with alternative commercial opportunities (CDs, IRAs, etc.), nonmembers, including employees, often express interest in them. As a result, some cooperatives accept deposits from the general public, while others may team up with a cooperative member who simply acted in their behalf.

This cooperative practice began simply and was initially of very modest proportions. To begin with, a fruit-packing cooperative established the practice purely as a matter of grower convenience. Final pool settlements to growers and grower payments to their cooperative for

---

[2]Most of this section was revised from material provided by Kenneth Duft, Washington State University.

production supplies often occurred simultaneously in the early spring. To simplify accounting procedures, cooperatives began to retain final pool payments from the previous harvest as a means of offsetting the cost of the current year's production supplies.

Since the time interval between these transactions was short and the sums of money were small, the practice was almost inconsequential to cooperatives and growers alike. However, to compensate the grower for funds retained, the cooperative agreed to pay interest on each account balance. The rate was generally set at a level exceeding that paid by commercial banks for CDs, short-term investments, IRAs, and savings accounts, but below that rate of interest paid by the cooperative for its seasonal borrowing. The practice, therefore, was justified on the basis of its financial advantages to both the cooperative and its depositor-patrons.

Advantages Factors favoring this practice appear to be convenience and interest income for members. Growers can make immediate deposits from their pool payments and purchase supplies against their account balance and/or withdraw previously made deposits according to their needs. The rate of interest paid on patron deposits can also reach respectable, if not attractive, levels. Members also like the method because they are always free to withdraw their deposits to seek higher returns or for other purposes.

Several advantages also accrue to the cooperative that accepts demand deposits. For example, as account balances grow, the cooperative can reduce its seasonal borrowing. And insofar as the cooperative pays grower-depositors at interest rates lower than those it pays for other debt capital, its operating costs are reduced. Furthermore, the cooperative itself maintains direct control over such costs, since its board can simply adjust the rate of interest paid on demand deposits, constrained only by the patrons' perceptions of other investment options. The board can also change the level of demand deposits in the same way. As demand deposit accounts reach significant levels, funds needed to support seasonal product inventories can be provided by producer patrons. Further, as demand deposit funds replace borrowed capital, liens on inventories and other lender-imposed restraints on cooperative operations are removed.

Unanswered Questions The overall credibility and impact of grower demand deposit accounts remain untested. Numerous questions remain unanswered.

For example, cooperatives that consulted an attorney about initiating such a practice appear to have uncovered a legal and regulatory void. Cooperative legislation, articles of incorporation, and bylaws are silent on this issue. State statutes provide neither explicit restrictions nor an enabling environment for such a cooperative practice. However, in 1986 an Arkansas court held that such instruments "were securities under the Arkansas securities law and should have been registered or properly exempted" (Baarda, p. 18).

Cooperatives also need to deal with the danger of exposing investors to great risk. Most cooperatives have not taken action to secure the positions of depositors. While commercial banks and other financial institutions provide Federal Deposit Insurance Corporation protection for demand deposits, no counterpart protection exists for cooperative entities. Individuals participating in a demand deposit program may discover that they are *residual claimants*. It is questionable just how many depositors are knowledgeable of this exposure. Certainly, cooperatives should either inform depositors of this risk or protect them from it.

Similarly, cooperatives need to consider their own exposure to risk as well. Cooperatives have little protection from the financial adversities resulting from a large and unexpected withdrawal of deposits. Those firms that acknowledge this concern have established a line of credit with a lender to backstop a large and unanticipated withdrawal. However, such a credit line rarely covers more than 30% of the total account balance.

In addition, demand deposit practices raise several security issues. The SEC restraints are probably not applicable to this practice as long as member-patrons are the sole depositors. However, when a cooperative accepts deposits from the general public, especially from across state lines, major regulatory issues arise. Cooperatives do not typically prepare prospectuses for potential investors, nor do they regularly file financial reports with the SEC (unless, of course, they have coincidentally established the sale of public stocks or bonds). A policy restricting deposits to members may therefore be a prudent one.

Finally, cooperatives must guard against finance problems associated with demand deposits. These deposits represent a true debt of the worst possible kind; that is, they are payable on demand. The timing and magnitude of demand withdrawals is largely unpredictable, making it very difficult for cooperative management to control either its cash flows or its working capital position, and cooperative lenders are reluctant to establish large lines of credit to backstop unpredictable withdrawals. Therefore, when reviewing a cooperative's financial position and performance, most cooperative lenders have established the procedure of removing demand deposits from their consideration of the cooperative's liquidity, solvency, and profitability. To overcome this problem, cooperatives might be wise to create a shallow safety net by instituting a policy requiring 30 days notice before withdrawal. They could also balance their reliance on this source of funds by more fully utilizing traditional sources which have proved to be safe.

Traditional Sources

Traditional sources of debt used by cooperatives include trade credit, industrial revenue bonds, insurance companies, commercial banks, and debenture bonds. These types of debt are not discussed at length because they are common to other types of businesses.

Cooperatives sometimes acquire funds by such standard methods as selling a division or other assets, or by reducing inventories and accounts receivable. In addition, they can reduce the need for funds by leasing rather than owning plant and equipment. Other alternatives, such as mergers or joint ventures, may provide other ways for cooperatives to gain access to the capital required for some of their objectives.

CREDIT POLICIES

Although extension of credit is a use of capital rather than a source, it has unique features. Management must keep in mind that the member-patron ownership structure of cooperatives contributes directly to a few unique and serious problems with accounts receivable. Management should also follow the guidelines and principles given for handling accounts receivable (see Van Horne, for example).

Often, members feel justified in building up high levels of accounts payable (accounts receivable on the cooperative's balance sheet) with their cooperative because they own it. They say: "It is just like owing money to myself." This attitude leads to several special concerns. For example, cooperatives in this situation need to evaluate more carefully than usual whether or not they are charging an interest rate on unpaid balances which truly reflects the costs of credit provided. Credit often costs the cooperative more than is charged on overdue accounts receivable, particularly if they are small accounts. This is not the way it should be; the full costs of administration, collection, and bad-debt loss should be borne by those who are responsible.

Some members request that their accounts payable be offset by redemption of their equity. This is a dangerous and threatening practice because the cooperative's equity can be quickly eroded. Furthermore, it is also very unfair and inequitable to remaining loyal members.

Bad-debt loss has been so severe that some cooperatives have had to liquidate. In some cases, one large account has been enough to put the cooperative under. Such incidences raise many issues of fairness. The board and management are responsible for developing and administering policies and procedures that will maintain the financial integrity of the cooperative and protect it from such abuses.

To avoid problems with overdue accounts, some local supply cooperatives have established a strict cash or advance payment policy. The Frenchman Valley Farmer Cooperative in Nebraska has gone to the other extreme by aggressively seeking to extend loans to members. They feel that they are in a position to know more about their members' ability to repay than any other agency and thus serve as a major source of capital for their members. A few federated cooperatives extend credit to local cooperatives that pass it on to farmer patrons. The Omaha Bank for Cooperatives has had a program of this type. However, cooperatives other than those in the Farm Credit System are not banks and to serve in that capacity may lead to unexpected problems.

SUMMARY

Cooperatives have opportunities to and methods of accumulating capital which are not available to other businesses. Most notable of these are retained patronage refunds and per-unit capital retains. These funds are automatically accumulated in the course of business—the former, as a percentage of net income not distributed in cash, and the latter, as a deduction from the price received by or in addition to the price paid by patrons for their products or supplies. Direct investments in capital stock are also used as a method of accumulating member equity. Typically, they are made to purchase voting stock. They are not very important as a source of equity because most benefits are distributed according to patronage rather than on levels of investment. Net income from nonmember business is a major source of unalloctted reserves.

Banks for Cooperatives are the major providers of debt capital. They are owned by the cooperatives to which they make loans. Cooperatives also obtain debt from other nontraditional sources such as other cooperatives, the Commodity Credit Corporation, and members.

Cooperatives must pay careful attention to credit policies and patron accounts receivable. They must avoid the problems associated with owners being in debt to their own organization. These problems include bankruptcy due to extensive bad-debt losses, the subsidizing of members using credit by those who pay cash, and the dangerous belief of a few members that they can offset their overdue accounts by relinquishing their equity in the cooperative.

Adequate capitalization plans include choosing from among capital accumulation methods as well as establishment of policies for its rotation. Cooperatives must be concerned with paying off debts and redeeming equity of inactive and overinvested members. Redeeming equity is the topic of the next chapter.

DISCUSSION QUESTIONS

14-1. How do the incentives for making equity investments differ between IOFs and farmer cooperatives? What do the investors expect to gain in each case, and how do they expect to obtain those gains?

14-2. Under what conditions might a marketing cooperative adhering to the business-at-cost principle require an increase in per-unit capital retains?

14-3. What is meant by the transient nature of cooperative equity capital, and why should that be of concern to lenders making long-term loans to a farmer cooperative?

14-4. Cooperatives' member-equity claims are generally redeemed at face value. Upon dissolution of the cooperative, they might be redeemed at higher or lower values. Explain how this could happen.

14-5. If a cooperative were heavily leveraged (i.e., if it had much more debt or liabilities than members' equity on its balance sheet), what dangers would it face, and how would you expect creditors to react?

14-6. The debt/equity ratio for two cooperatives is 0.85. I.J. Pool cooperative has its membership under contract and operates under a marketing pool with very little competition. The other, Matison Co. Elevator Cooperative, operates on a cash-and-delivery basis and is faced with considerable competition. Compare and discuss the debt/equity ratio for both cooperatives.

14-7. If a cooperative decides that it should operate without any debt, what would such action imply, and how might it affect members and/or patrons?

14-8. How should a cooperative member evaluate the returns on equity capital supplied to the cooperative? What are the sources of these returns, and how might they be determined?

14-9. Per-unit capital retain plans seem to work well for marketing cooperatives but not for farm supply cooperatives. Can you develop logical reasons as to why this is so?

14-10. In establishing a credit policy at a farm supply cooperative, what would be wrong with letting accounts receivable from each member rise until they reach the book value of the members' share of funds allocated?

14-11. Cooperatives and IOFs alike can raise equity capital by selling stock, but the rights conveyed to the owners of such stocks are different. What are the differences? Can you think of at least three?

14-12. Why is it more difficult for a cooperative to raise money from the investing public than it is for an IOF?

REFERENCES

BAARDA, JAMES, "Legal Corner" *Farm. Coop.* (Jan. 1987):18.

COBIA, DAVID W., ET AL., *Equity Redemption: Issues and Alternatives for Farmer Cooperatives.* Washington, DC: USDA ACS RR 23, Oct. 1982.

DYLLA, LARRY, "Financing Alternatives for Cooperatives," *Coop. Accountant.* (Spring 1987):25.

GRIFFIN, NELDA, ET AL., *The Changing Financial Structure of Farmer Cooperatives.* Washington, DC: ESCS FCS RR 17, Mar. 1980.

HANRAHAN, MICHAEL S. *Attitudes Toward Cooperatives by Bank Trust and Professional Farm Managers.* Washington, DC: USDA ACS SR 1, Dec. 1980.

ROYER, JEFFREY S., "Strategies for Capitalizing Farmer Cooperatives" *Farmer Cooperatives for the Future,* ed. Lee F. Schrader and William D. Dobson, NCR-140, et al., workshop, St. Louis, MO: Dept. of Agr. Econ., Purdue Univ., Aug. 1985.

VAN HORNE, JAMES C. *Fundamentals of Financial Management,* 6th ed. Englewood Cliffs, NJ: Prentice Hall, 1986.

15

Equity Redemption

David W. Cobia,
North Dakota State University

Jeffrey S. Royer,
Agricultural Cooperative Service, USDA

Gene Ingalsbe,
Agricultural Cooperative Service, USDA

WHY EQUITY REDEMPTION?

Equity redemption is returning equity in cash to member-patrons who have previously invested it. Over the years, patrons build up equity from direct investments, per-unit capital retains, and retained patronage refunds. But as individual patronage declines or ceases, cooperatives need to redeem equity to avoid overinvestment by some patrons. Equity redemption plans provide a means of returning these funds to the member-patrons who invested them.

Overinvested patrons are those who have invested more than their proportionate share of equity, based on patronage, in a cooperative. When there are overinvested patrons, the cooperative is not financed according to use. Some patrons are providing more than their share according to their current level of patronage. Others are providing less than their share. Finally, patrons often have been told that they will eventually receive their equity investments in cash.

Are Members Financing Cooperatives According to Use?

A serious flaw in the performance of cooperatives is the failure to redeem equity of overinvested members and to secure more funds from those not providing their share. Many cooperatives have adequate plans for equity redemption, but most do not. A long history of ignoring investor rights has created investor inequity problems.

In a 1974 survey of 857 randomly selected cooperatives, 29% had no planned program to redeem equity, and 37% redeemed it only upon largely unpredictable events such as death. The remaining cooperatives had systematic programs of equity redemption. However, even these programs were deficient. From 10 to 42% of the equity in these cooperatives belonged to members who were no longer active patrons (Brown and Volkin). These data understate the problem because they do not reflect the over- or under-investment of active patrons.

Although the problem is widespread, its seriousness varies around the country and among commodity groups. For example, more grain and supply cooperatives (85%) had redemption plans than any other commodity group. Very few livestock and wool cooperatives had redemption plans (less than 20%), but most of these cooperatives have small amounts of assets. Many are financed solely by unallocated reserves. Among other commodity groups, 46% of tobacco cooperatives, 68% of cotton cooperatives, and 71% of dairy cooperatives had redemption plans (Brown and Volkin, p. 6). Dairy cooperatives without redemption plans reported that 78% of their members were inactive. These inactive members held 61% of their cooperatives' equity. Fruit and vegetable cooperatives in this same category had 33% inactive members who accounted for 6% of the equity.

Similar differences surfaced in a survey made by researchers at Purdue University. They found that the oldest equities in 59 midwestern grain

TABLE 15.1 Comparison of oldest equities of cheese and grain marketing and supply cooperatives, 1979

	Type of cooperative	
Item	Wisconsin cheese	Midwest grain & farm supply
Average age of oldest equity (years)	10.6	34.2
Range of oldest equity in each cooperative (years)	5–17	3–90[a]
Cooperatives with equities over 19 years (number)	0	33
Sample size (number)	28	59

Source: Cobia et al.
[a]Frequency at the extreme was four co-ops at 50 years, three from 60 to 65, two at 78, and one at 90 years.

marketing and farm supply cooperatives averaged 34 years compared with 11 years for 28 Wisconsin cheese cooperatives (Table 15.1). A classic example of deficient equity redemption involves Andrew J. Volstead, coauthor of the Capper-Volstead Act. He had equity in a cooperative in 1979, although he had died 32 years earlier (*Western Horizons*).

What is the Result?

This failure to redeem equity is a serious injustice to member-patrons. Members have a reasonable right to expect redemption of allocated equities unless they are lost because of normal business risk. If a cooperative does not expect to redeem equities, it should question whether it should allocate them, especially in qualified form. Members should not have to pay income taxes on allocations if there is little likelihood that they will ever be redeemed in cash.

Failure to redeem systematically the equity of inactive and overinvested members means that these members bear the burden of financing a cooperative for services they do not use, a violation of cooperative principles. In addition, funds held by overinvested members are not available for their use in most cases, not even as collateral. Furthermore, often there is ownership without control because inactive members lose voting rights in many states.

These conditions have led to political pressure for legislation mandating redemption. Members and competitors of cooperatives have argued, with some justification, that without equity redemption, the present value of any funds invested in a cooperative is zero. (A few large-volume patrons have reacted to this situation by leaving the cooperative and building their own facilities.) Finally, a negative image has been generated of cooperatives, their directors, and management on the part of general public, farmers, and even some members.

Some cooperatives know that it is bad business not to redeem equity. One manager's spontaneous response to a question on this topic was: "It is the kiss of death if co-ops don't redeem." Unfortunately, however, many requests for redemption made directly to a cooperative often have been ineffective—usually because of the cooperative's weak financial position resulting from a situation that may have persisted far too long. Board members are torn between requests for redemption by inactive members and the need for equity to support services required by current patrons. Some managers argue that "inactive members don't need the cash as much as the cooperative does." Although this may be true, the cooperative nevertheless is committing an injustice against those who cannot reclaim their invested capital and use it as they see fit.

Some members have become so frustrated with cooperative management's reluctance to redeem overinvested funds that they have sometimes taken cooperatives to court only to find that the discretionary and judiciary powers of the board have priority. Because these members often do not have a vote in the cooperative and because many courts will not substitute judicial judgment for that of the board, some members have sought and often found sympathy from their elected officials. This has produced some attempts to pass legislation to force redemption—none of them have yet been successful. However, the governmental watchdog agency, the U.S. General Accounting Office (GAO), has recommended that if cooperatives do not voluntarily redeem equity, legislation mandating redemption and/or dividend payments should be enacted.

Advantages of Redemption

Members, the board, and management must weigh the trade-offs associated with adopting or changing an equity redemption program. The relative importance of each issue depends on the cooperative's individual situation, but one thing is certain—some kind of program is essential.

Cooperatives and their members gain several benefits when they follow carefully tailored equity redemption plans that take into account the needs of all members. In the first place, a policy of consistent equity redemption would reduce conflict between active and inactive members. In addition, cooperatives need to have a systematic redemption plan to encourage active patronage. Ownership tends to become proportional to use in cooperatives with a systematic plan. This condition by itself, however, is meaningless unless ownership results in direct benefits to members. More than 65% of the farmers surveyed for the GAO report seemed to recognize this, indicating that they would start or increase patronage in a cooperative if they did not have to wait so long for redemption of equities. Equity redemption helps members more easily see the linkage between investment and benefits. They see their equity working for them because of benefits from patronage (as part of their retirement and insurance programs) and therefore have an enhanced understanding of and an appreciation for cooperative principles.

Equity redemption would also negate the effectiveness of arguments highlighting negative cash flow associated with some patronage refunds and low present value of allocated equities. This, in turn, would reduce pressure for mandatory redemption legislation. Further, cooperative equity might be more valuable as collateral for loans. Lenders are hesitant to lend members money based on equity in a cooperative if their equities do not have a ready market or if the cooperative has a poor record of redemption. If members could get loans more easily because the cooperative redeemed equity consistently, there would be an additional incentive for patrons to ascribe value to equity in their cooperative rather than to view it as so-called dead money.

Finally, if cooperatives followed policies on timely redemption of equity, their boards and managements would be forced to develop and implement effective financial plans. This could be the most important benefit of a strong redemption plan because good financial planning is a prerequisite to better financial performance.

Disadvantage of Aggressive Redemption

Despite its benefits, an aggressive redemption plan has some drawbacks that cooperatives should consider as well. Obviously, a cooperative may have to compromise its financial position by using funds to redeem equity rather than to enhance its financial position. However, a redemption plan could make it easier, not more difficult, to service equity, because more patrons would be willing to make new investments in the cooperative. Income from new investments is desirable as long as they are financed by those who benefit.

The cooperative redeeming equity might have to reduce its cash patronage refunds and/or dividends, thus creating conflicts among some members. A cooperative should not abandon a sound financing plan just to aid in resolving the financial stress of current members. Regardless of the redemption plan used, underinvested members probably will have to supply more funds. This may discourage patronage, especially by financially strapped farmers. Nonetheless, members can usually find a way to provide equity, if patronage is profitable for them. Unfortunately, a financially sound plan that provides flexibility and equitable financing may be difficult for members to understand.

However, cooperatives can overcome or minimize most problems created by redemption, particularly if they deal with all aspects of the situation in a timely manner. Obviously, much work is involved, not only in developing a new program, but in educating members and, most important, in facing financial realities.

When Not to Expect Redemption

Members must recognize that redemption may be delayed by situations other than poor performance. For example, cooperatives that consistently offer favorable prices give patrons immediate benefits. The resulting narrow margins do not generate sufficient funds to meet both cash flow and redemption requirements. Also, members in some cooperatives have benefited from equity provided by their predecessors. Therefore, when these members become inactive they should expect to leave their equity in the cooperative until new members accumulate their share.

ALTERNATIVES

A common misconception in cooperative literature is that a particular redemption plan is linked to a particular source of equity. In fact, any source of equity can be used with any redemption plan—they are independent.

The four plans used to redeem equity are (1) the revolving fund plan, (2) the base capital plan, (3) the percentage-of-all-equities plan, and (4) the special situation plan. Combinations of various features of these plans also exist. Because an equity redemption plan should be tailored to each situation, one program cannot be universally recommended.

Special situation plans (redemption is initiated by a special situation of a member such as death or moving away) are by far the most popular (Table 15.2). Fifty-nine percent of all cooperatives redeem equities in this way. However, 20% combine special situation plans with other plans, leaving 39% using special situation plans exclusively. The revolving fund plan is second in popularity. Fewer cooperatives use the base capital and percentage-of-all-equities plans.

Revolving Fund Plan

Under a revolving fund plan, a cooperative pays off or retires in cash the oldest equities on a first-in, first-out basis, or, in other words, in the same chronological order in which they were allocated. In the example used to illustrate this plan (Table 15.3), the cooperative's objective is to accumulate allocated equity of $1,500. Retention begins with $500 in year 1

TABLE 15.2 Equity redemption plans used by 857 randomly selected U.S. farmer cooperatives, 1974

Type of plan	Percent
Revolving fund	29
Base capital	1
Percentage-of-all-equities	2
Special situation only[a]	39
No plan	29
Total	100

Source: Brown and Volkin (pp. 5, 8).
[a]An additional 20% of the other cooperatives having plans also used special situation.

and reaches the designated level by the end of year 3. Then, as the cooperative retains new equities ($500 in year 4), it redeems old equities ($500 in year 1).

In year 5, the cooperative retains $1,000, perhaps because it has realized better than expected net income or because it has deliberately decided to shorten the revolving period. For whatever reason, by retaining $1,000 in year 5, it is now able to redeem equities of $500 each from years 2 and 3.

Member A contributes varying amounts of equity each year according to use of the cooperative. Beginning in year 4, A's equity is $300. In year 4, when the cooperative decides to redeem equities from year 1, A gets $50. In year 5, when the cooperative decides to redeem equities from years 2 and 3, A gets $250.

Length of revolving period The length of a revolving period is a compromise between the time necessary to accumulate equity and the time necessary to redeem it. Revolving periods range from 18 months to more than 30 years. Surveys have found the average revolving period to be about 10 years.

It has been recommended that most revolving periods should be less than seven years to keep investment more in line with patronage and reduce or eliminate the need for redeeming equities out of sequence (Cobia et al.). Thus most cooperatives using the revolving fund need to shorten the length of the cycle.

TABLE 15.3 Illustration of revolving fund operation

Year	Beginning equity	Patronage allocations retained	Equity amount redeemed	Equity years redeemed
		Cooperative level		
1	$ 0	$ 500	$ 0	—
2	500	500	0	—
3	1,000	500	0	—
4	1,500	500	500	1
5	1,500	1,000	1,000	2,3
6	1,500	500	500	4
		Member A		
1	0	50	0	—
2	50	100	0	—
3	150	150	0	—
4	300	200	50	1
5	450	200	250	2,3
6	400	200	200	4

Source: Royer & Ingalsbe.

Special adaptations Some cooperatives have created complex revolving plans by incorporating features of other programs or by setting up several revolving funds in the same cooperative. Cooperatives commonly incorporate features of the special situation plan into a revolving fund, and they generally give high priority to redemption of equity in estates. Adding such features lengthens the revolving period. The shorter the revolving period, the less need for special adaptations to accommodate hardship cases.

Many cooperatives have segregated revolving funds by sources of investment and major products. At one time Dairymen Inc. had five revolving funds based on the source of investment (per-unit capital retains and net income in two divisions and income on nonmember business), in addition to equities acquired from mergers.

Cooperatives separating revolving funds by products, such as feed and fertilizer or soybeans and corn, return benefits according to each major product's income. Local cooperatives sometimes establish separate revolving funds to distinguish income from their own operations from that received from other cooperatives. This structure highlights the origin of equity and provides a means for comparing performance—including equity redemption.

Advantages and disadvantages of the revolving fund The revolving fund is popular because it is easily understood and administered; equities are approximately proportional to use if revolving periods are short; and cooperatives can absorb a bad year or increase equity by extending the length of the revolving period.

Disadvantages associated with the revolving fund include an indefinitely extended revolving period created by increased needs for funds or by poor operating results. Also, disparities between benefits received and capital invested emerge when margins vary substantially over time, and among different products or services covered by the same fund. Finally, members may develop unrealistic expectations by assuming that redemption will occur on a fixed schedule regardless of their cooperative's financial condition.

Base Capital Plan

Description of a typical plan Cooperatives that use a base capital plan determine a member's equity obligation annually, based on the cooperative's need for capital and on the member's use of the cooperative. Underinvested members continue to invest, using the methods previously outlined. They may be required to pay an interest charge on the amount of their underinvestment. Overinvested members generally begin to receive at least partial, if not full, redemption of their excess investment.

The six-member cooperative in the illustration (Table 15.4) operates on a five-year base period and has $18,250 member equity. The board has determined that the cooperative needs an additional $250 of equity capital. The board then allocates each member's equity obligation based on patronage over the past five years. The member's percentage of the cooperative's total business determines the percentage of the member's new equity obligation. Comparing the new obligation with the member's equity account, the board then determines what adjustment, if any, is necessary in the member's equity obligation.

In Table 15.4, member A needs to contribute $350 more than already invested. The board determines this amount by dividing A's patronage ($120,208) by the cooperative's total volume ($1,092,796) to determine A's percent (11%), which determines A's equity obligation of $2,035 ($18,500 x 0.11). Because member A has only $1,685 in equity, $350 more is needed. In member C's circumstance, a negative amount of $215 is determined. This amount could be redeemed. The burden of redeeming equity of over-invested members is shared by all members, according to their use of the cooperative during the base period.

Sometimes cooperatives use a variable cash patronage refund plan along with the base capital plan. The percentage of the refund that is in cash varies according to the member's obligation for equity. An overinvested member receives a greater percentage of the patronage refund in cash than the underinvested member receives. Unless underinvested members are required to meet their capital obligations immediately, it is impossible to retire all the overinvested equities of other members. Several options for handling this problem are explained in greater detail in Cobia et al.

Two important adaptations of the base capital plan make the linkage between equity and patronage direct and unambiguous. One places the responsibility for equity immediately on the member, according to anticipated use. A common approach is to tie a patron's equity requirement to market access or to

TABLE 15.4 Illustration of five-year base capital plan

Member	Beginning equity	5-year patronage total	Share of co-op's business(%)	adjusted equity obligation	Over or under invested
A	$ 1,685	$ 120,208	11	$ 2,035	-$350
B	3,345	207,631	19	3,515	-170
C	2,805	152,991	14	2,590	+215
D	5,515	327,839	30	5,550	-35
E	4,550	284,127	26	4,810	-260
F[a]	350	—	—	—	+350
Total	18,250	1,092,796	100	18,500	-250

Source: Royer and Ingalsbe·
[a]Inactive member.

a service provided by the cooperative, such as sugar beet refining, sugarcane crushing, vegetable processing, fruit packing, or grain drying and storing. American Crystal Sugar Co. required prospective members to sign an acreage contract that required an investment of $105 per acre.

Northern Pacific Grain Growers (now a part of Harvest States Cooperative) extended this approach by requiring member cooperatives to invest according to anticipated volume, arguing that if the local cooperatives were to have access to their regional cooperative in bad as well as good times, they had to provide equity accordingly.

Another approach created by CF Industries in Long Grove, Illinois, is appropriate for cooperatives that require more assets for manufacturing one product than they require for manufacturing another. This cooperative calculates member-equity obligation based on assets employed in the manufacture of each type of fertilizer.

Advantages and disadvantages of the base capital plan The base capital plan is the most equitable plan because it links investment to use rather than to returns or earnings retained from members. It also enables management to alter equity requirements to meet changing needs of the cooperative. Finally, the base capital plan is the only plan that provides a logical framework in which cooperatives can require underinvested members to pay an interest fee to compensate overinvested members.

This plan can present problems because underinvested members, especially new farmers, may be unable to provide their equity share immediately. In addition, some boards hesitate to increase equity requirements to meet increasing capital needs. They find this action more difficult than others, such as lengthening the revolving period. Because of its complexity, the plan may be difficult to understand and administer. It especially may not work well in cooperatives with a constant turnover of members.

Percentage-of-All-Equities Plan

When a cooperative uses the percentage-of-all-equities approach, it retires a percentage of all outstanding equity, regardless of issue dates. In other words, the cooperative reduces the equity of all members by the same percentage. Only 2% of all farmer cooperatives report using this method, although at one time it was used by CENEX, a federated farm supply cooperative at St. Paul, Minnesota, and adopted by a few of CENEX's members.

In Table 15.5 we have illustrated the operation of this plan for a cooperative with $2,000 in allocated equity. After the business year is over and net margins are allocated according to established policy, the new total is $2,500. But the board, assessing the coming year, determines that only $2,300 is needed. Consequently, the board can redeem to members $200 in allocated equity (10% of the cooperative's $2,000 in allocated equity before

TABLE 15.5 Illustration of percentage-of-all-equities plan at cooperative level

Item	Dollars
Allocated equity at beginning of year	2,000
Patronage allocations retained	+ 500
Equity available at end of year	2,500
Equity required	2,300
Equity redeemable (2,500 minus 2,300)[a]	200

Source: Royer and Ingalsbe
[a]10% of allocated equity at beginning of year.

the most recent year's allocation). The results of a 10% redemption to five members are illustrated in Tables 15.5 and 15.6. Note that the cooperative redeems the same percentage of all equities, regardless of the amounts of the equities or the length of time the equities has been held.

Advantages and disadvantages of percentage-of-all-equities plan Cooperatives use the percentage-of-all-equities plan because it enables them to award a new patron by prompt equity redemption. It also allows them to charge full margins to generate needed cash flow, because members participate promptly in the return of generous margins. Moreover, the policy is easy to understand and administer, and it can readily be adjusted to different operating results. It works especially well for cooperatives with stable membership and patronage.

When cooperatives use this approach, they can run into difficulties as well, because the transfer of ownership from overinvested members to current patrons is extended and, in fact, is never completed without an additional provision for closing out equity accounts of former members. The plan appears to be least adapted to local farm supply cooperatives, particularly if membership changes and patronage is relatively erratic.

TABLE 15.6 Illustration of percentage-of-all-equities plan applied to members

Member	Beginning equity	Percentage of equity redeemable	Amount to be redeemed
A	$ 750	10	$ 75
B	250	10	25
C	250	10	25
D	500	10	50
E	250	10	25
Total	2,000	10	200

Source: Royer and Ingalsbe.

Special Situation Plan

A special situation plan is one by which a change (such as death) in the situation of a member qualifies that member's equity for redemption. It is by far the most common plan. It is not known to what extent this high level of use reflects planned redemption or is merely a reaction to extenuating circumstances.

In a special situation plan, a cooperative accumulates and retains new equity until the prescribed situation or event occurs. Upon verification of the situation and after the other administrative work is completed, the cooperative redeems the entire amount of equity at one time or over a period of years subject to board approval.

Several events and conditions that trigger equity redemption are, in order of use, as follows: (1) death; (2) termination of farming; (3) retirement or qualification because of age; (4) patron call or on-call or on-demand request by the patron; (5) hardship, including bankruptcy; (6) relocation away from the trade area; (7) resigned membership or ceased patronage of the cooperative; and (8) application of equity to uncollected accounts receivable (Table 15.7). This last option is not recommended because it es tablishes a dangerous precedent for the cooperative and is unfair to mem-

TABLE 15.7 Frequency of situation used to trigger redemption of equities by 857 randomly selected farmer cooperatives, 1974

Event	Role of equity redemption in event[a]	Type of cooperative[b]			
		Marketing	Farm supply	Service	Total
Estates	1	57	51	50	54
No longer farming	1	24	13	30	19
Age	1	10	10	—	10
On call	2	5	13	10	9
Hardship	1	2	5	—	3
Other		2	8	10	5
Moved from trade area	1				
No longer patronizing co-op	2				
Member resigned	2				
Apply to accounts receivable	2				
Case by case	c				
Total		100	100	100	100

[a]Role that short-run benefits from redemption may play in prompting the event: (1) events over which the member has no control or benefits from redemption are unlikely to be a major factor in decision; (2) events that the member can control & short-run benefits from redemption may be a major factor in decision.
[b]Brown and Volkin.
[c]Depends on circumstances.

bers (see "Credit Policies" in Chapter 14). Another potentially dangerous practice reported is to consider requests for redemption on a case-by-case basis rather than to specify the priorities in advance.

The combinations of specific events used, time elapsed between event and redemption, and method and limits on payment vary widely. Practices range from redeeming equity only in one situation, such as death, to redeeming equity in many situations. For example, one cooperative used five of the situations listed in Table 15.7 to initiate redemption. A cooperative may establish an upper limit on total annual redemption and then redeem equities according to a predetermined set of priorities.

Planning for cash flow requirements Special situations plans generally require an unpredictable amount of cash. In some cases cooperatives do not know who will qualify for redemption based on age, because members' birth dates usually are not in the cooperatives' records. This introduces another obstacle to financial planning. However, cooperatives can minimize this obstacle by compiling records of the probability of the occurrence of each situation and of the corresponding equities involved. Using these data as a basis, the cooperative can then make plans to provide the necessary cash flow. Some events may be cyclical and/or related to the local agricultural economy.

Special adaptations Death and age qualifications for redemption discriminate against incorporated businesses and institutions with perpetual lives, such as schools and government agencies, as well as trusts, incorporated farms, and other cooperatives. Only liquidation would prompt redemption to these patrons, and in time these institutions would hold more than their share of equity.

This problem can be overcome in two ways. Lake-to-Lake Cooperative in Wisconsin redeemed the equity held by a corporation corresponding to the proportion of the equity held in the corporation by persons meeting redemption criteria. Assume that a person died who owned 30% of a corporation that held equity in a cooperative. The corporation would receive 30% of its equity in cash, which normally would be paid to the deceased person's estate on a pass-through basis.

A second approach that has more general application is used by Agland Industries of Eaton, Colorado. This cooperative redeems equity of incorporated members at the same rate as it redeems the equity of natural patrons. For example, if 4% of the equity of natural patrons was redeemed because of death or age, 4% of the equity of corporate members would be redeemed in that year.

Advantages and disadvantages of the special situation plan A special situation plan can be a good option for the following reasons: Its financial burden on the cooperative is generally light, it is easily understood, and it is often popular with members. It provides a safety valve for high-priority cases when another plan is not functioning as intended, and it is especially well adapted to cooperatives that pass most of their monetary benefits to patrons immediately in the form of favorable prices.

On the other hand, some cooperatives do not use the plan because it fails to meet the financing-according-to-use test. Members may provide capital based on use, but equity is not always redeemed as patronage declines. As mentioned before, financial planning is also more difficult for cooperatives that use this plan because of the unpredictable nature of situations that affect redemption.

Perhaps the strongest objection to the plan is that waiting to redeem equity until members die prevents them from personally benefiting from the overinvested equity. Finally, redemption initiated by events controlled by the member can place a financial burden on the cooperative and be unfair to other members. Redemption in these cases (second column, Table 15.7) probably should not be granted, or at least should be given low priority.

EVALUATION

The four plans can be combined in several ways. Separate redemption programs can be created for each source of equity. For example, cooperatives can place per-unit capital retains in a base capital plan, retained net income in a revolving fund plan, original paid-in capital in a special situation plan, and net income from another cooperative in a percentage-of-all-equities plan. The justifications for these linkages are that the base capital plan is well adapted to per-unit capital retains, that original paid-in equity is generally voting stock and often not a significant part of equity, and that patronage with a federated cooperative is likely to be stable.

Redemption plans should be evaluated in light of their potential impact on the members and the cooperative. Principal factors to consider are as follows:

1. Facilitate capital acquisition to provide necessary cash flow. A cooperative must have adequate capital.

2. Ensure that members supply equity in proportion to their current patronage.

3. Provide flexibility to accommodate a wide range of financial operating results and members' characteristics and needs.

4. Recognize that the board of directors must control the redemption policy on behalf of all the members.

5. Recognize tax and other statutes and existing contracts with creditors and others.

6. A plan should be easily understood by members and employees in contact with members. Members do not enthusiastically support programs which they find difficult to understand or in which they cannot see benefits.

7. Another desirable characteristic is ease of administration, including low operating costs.

Changing from one equity redemption plan to another does not correct a problem of inadequate margins or retains. If a low or negative cash flow is the primary cause of dissatisfaction with a redemption program, the cooperative must face this problem first. A positive cash flow is necessary to make any equitable capital program work.

FACILITATING PLANS

Members needing cash, moving from the area, or retiring from farming may ask for early redemption of their allocated equity. It is sometimes difficult for cooperatives to grant these requests, because retiring equities out of sequence is unfair to remaining members and may create a financial burden for the cooperative. Some cooperatives, however, have developed procedures that minimize the otherwise adverse financial impact of early redemption on the cooperative and provide flexibility for members. Examples of these plans include exchange of equity among members, redemption at a discount, and conversion of equity to debt securities.

Equity Exchange

A properly conceived equity exchange program could help cooperatives maintain the necessary funds and could alleviate problems associated with redeeming equities out of order. Overinvested members, for example, could sell their equities to potential and underinvested members without the cooperative's being directly involved. Cooperatives can thus augment and facilitate their equity redemption programs by developing incentives for and mechanics of equity exchange programs (see "Exchange of Equity" in Chapter 14).

Redemption at a Discount

Cooperatives can alleviate equity redemption problems by redeeming some equity at a discounted rate. For example, when some members choose not to provide their share of equity for the same length of time as other members, a levy against the members' short-fall investment is justified for the cooperative to achieve equitable treatment. The cooperative can thus protect its financial position and avoid discriminating against its remaining members.

One California cooperative has developed an approach for redeeming at a discount and for setting the discount rate. This cooperative replaces equity redeemed early by a loan, the entire principal of which is to be repaid by the cooperative in the year when redemption would otherwise occur. A fixed interest rate on the loan and tax consequences to the cooperative are used in setting the discount rate. Participating members thus satisfy their responsibility to finance the cooperative, and the cooperative treats gains and costs from discounted redemption as nonmember business. Interest expense on the loan and the income taxes associated with the nonmember income offset the gain to the cooperative and ensure a neutral effect on the other members.

A discount program places early retirements on a more equal basis and reinforces the concept of members' responsibility to finance their cooperative, while recognizing extenuating circumstances of individual members. In addition, it explicitly recognizes the present-value concept.

However, problems are associated with discounting equity out of the regular redemption sequence. Such a program could have an adverse effect on the balance sheet of some cooperatives when debt replaces equity. Furthermore, this program is not well adapted to equity redemption programs in which anticipated redemption dates cannot be estimated.

Establishing an equitable and fair discount rate is often difficult, and income tax consequences are not yet well established. Also, some members, who do not understand or appreciate the present value concept, may feel that all redemptions should be at book value. Therefore, if a cooperative adopts a program for early redemption of equities at discounted values, it should carefully specify and explain the procedures in its bylaws or articles of incorporation.

Unallocated Equity

Cooperatives sometimes create a relatively large share of unallocated equity to reduce pressures for redemption or speed up the redemption of allocated equities. The cooperatives have less equity to redeem. Therefore, a given cash flow can support a faster equity turnover of the remaining allocated equity. However, because less equity is allocated, total benefits to patrons may still be reduced.

Conversion to Debt or Preferred Stock

Another alternative to redeeming equity in cash is to convert it into interest-bearing debt. This relieves a cooperative's temporary cash flow problem and gives members an additional option.

Cooperatives must use this method with caution. Many have had serious financial difficulties because they converted equity to debt without making offsetting reductions in previously existing debts to maintain a reasonable debt/equity ratio. Western Farmers Association, for example, went into bankruptcy partly because of this approach. A safer way to exchange equity for debt is to offer members the option as an alternative to cash and then to substitute the conversions for existing debt. This increases the options open to members without seriously eroding the debt/equity ratio. Normally, the interest rate given members on this debt is one or two percentage points below a bank loan rate. The difference is used to cover the costs of managing the program. Members like this arrangement. Their participation increases, they receive a relatively high rate of return, and the cooperative's interest expense may decrease while its image improves. But cooperatives must beware of the serious consequences of eroding liquidity if previously existing debt is not correspondingly reduced.

EXTERNAL INFLUENCES

Federated Cooperatives

Federated cooperatives are at least one step removed from farmer-members, but they affect redemption of farmers' equity through their influence on their members' equity redemption practices. This occurs because of the federated cooperatives' financing methods, performance, equity redemption policy, and member services.

Financing methods and performance Nearly all of the equity in some local cooperatives, particularly supply and grain marketing cooperatives, is tied up in investments in federated cooperatives. It is difficult for these local cooperatives to redeem equity while supporting their federated cooperatives, unless the federated also redeems the local's equity. Often the net income reported by these local cooperatives comes almost entirely from business with their federated cooperatives. If the federated cooperative does not provide the local with cash flow through cash patronage refunds or equity redemption, the local is limited in what it can do.

Most federated cooperatives have long-term and stable associations with their member cooperatives. Therefore, redemption of inactive-member equity is less of a concern for federated cooperatives. Even so, disparities exist. Fair- share investments from each member do not, of course, reduce the total requirement; they only allow it to be distributed equitably among the entire membership.

Methods of adjusting members' equity Although some federated cooperatives have no systematic program for adjusting member equity, others use a range of methods, including those discussed earlier. When a member cooperative is financially unable to redeem the equities of its inactive members, federated cooperatives may assist by purchasing preferred stock in this local cooperative or by making other financial arrangements on a case-by-case basis to help the cooperative with estate redemptions.

Increasingly, federated cooperatives have adopted programs in which they participate systematically in their members' equity redemption. Some cover only redemption of estates, while others include redemption of the equity of retired farmers, usually those reaching a specific age such as 65 or 70. One large federated cooperative with a redemption program that participates in both estate and age redemptions is CENEX of St. Paul, Minnesota.

Several federated cooperatives consistently pay a high proportion of their patronage refunds in cash. Cooperatives with this policy generally redeem little or no equity. Only the equity of member cooperatives that no longer qualify for membership is redeemed.

Leadership of federated cooperatives Federated cooperatives are in a unique position to provide assistance and leadership. They can, for example, establish an equity redemption plan that involves participation from member cooperatives on a voluntary basis. For example, Indiana Farm Bureau Cooperative Association (IFBCA) has tailored a base capital plan that sets up a flexible, equitable redemption and cash refund program for its members. IFBCA developed bylaw provisions, reviewed the plan with the Internal Revenue Service, established planning and accounting procedures, and provided information booklets and personnel to help local cooperatives establish the plan.

Even though the plan was flexible, local directors still had to establish capital budgets and decide on the amount of equity to be redeemed. IFBCA could give local cooperative boards pro forma financial statements for different levels of redemption and cash refunds, as well as prepare checks and associated statements for members.

Lending Agencies

Creditors often play a significant role in setting equity redemption limits. Retirement programs reduce the cash available for loan repayment and therefore may hinder a cooperative's debt-servicing capacity or future borrowing ability. Creditors therefore have a vested interest in equity redemption plans and policies. Some Banks for Cooperatives require bank approval for any redemption.

Cooperatives must develop their redemption programs in light of their creditors' policies. BCs pursue a variety of policies regarding their clients' equity redemption practices. They all, however, are more restrictive toward cooperatives in relatively weak financial positions.

SUMMARY

Most cooperatives have an obligation to redeem equity in cash to inactive and overinvested members. Four types of plans are used. The revolving fund is a first-in, first-out plan. It is popular, easy to understand, and flexible. The base capital plan links member investment requirements directly to patronage during a base period. A percentage of all equities regardless of age are redeemed in the percentage-of-all-equities plan. The most frequently used and generally the most inequitable plan is the special situation plan. In this case redemption is initiated by a special situation or event such as death or retirement.

Cooperatives use exchange of equity, unallocated reserves, conversion equity to debt, and redemption at a discount to facilitate and reduce the burden of equity redemption. Credit agencies and federated cooperatives influence redemptive practices.

DISCUSSION QUESTIONS

15-1. Why is equity redemption so important, and what are the consequences of not doing it?

15-2. Describe the operation and major advantages and disadvantages of each major equity redemption plan (revolving fund, base capital plan, percentage-of-all-equities, and special situation plans).

15-3. How does the special situation plan discriminate against incorporated members? What methods can be used to solve this problem?

15-4. How can the following programs or practices facilitate or reduce the burden of equity redemption?

   a. Redemption at a discount

   b. Equity exchange

   c. High proportion of unallocated reserves

   d. Conversion of equity to debt

15-5. What warnings should be given a cooperative contemplating the above programs listed in question 15-4?

15-6. How do federated cooperatives influence the redemption performance of their members?

15-7. Under what conditions would equity redemption be ill-advised?

16

Taxation

Jeffrey S. Royer,
Agricultural Cooperative Service, USDA

INTRODUCTION

Net income of farmer cooperatives is generally taxed according to the *single-tax principle*. This principle, recognized by federal income tax law, ensures that cooperative net income is usually taxed at either the cooperative level or the patron level, but not at both. The principle is based on the concept that cooperatives are nonprofit extensions of the business enterprises of the patrons who own them. Consequently, the net income of cooperatives is generally taxed only once. This taxation of farmer cooperatives, as well as of other corporations operating on a cooperative basis, is defined in subchapter T of the Internal Revenue Code.

SUBCHAPTER T

Subchapter T, which consists of sections 1381-1388 of the Internal Revenue Code, applies to "any corporation operating on a cooperative basis" except mutual savings banks, mutual insurance companies, and cooperatives engaged in furnishing electric energy or telephone service to rural areas.

Subchapter T provides that in addition to making deductions allowed other businesses, cooperatives can exclude from their taxable income certain distributions of net income or allocations paid patrons. Subchapter T also specifies additional distributions that may be deducted by certain cooperatives, commonly called exempt cooperatives or section 521 cooperatives and defined under section 521 of the code. All farmer cooperatives, whether or not they qualify for section 521 treatment, file annual federal income tax returns on Form 990-C.

Patronage Refunds[1]

Subchapter T provides that in determining their taxable income, cooperatives may exclude certain patronage refunds and per-unit capital retains they pay patrons. *Patronage refunds* are amounts paid patrons from the net income of a cooperative on the basis of quantity or value of business done with or for the patrons under a preexisting legal obligation. Patronage refunds do not include amounts paid to patrons based on earnings from business not done with or for patrons. They also do not include amounts paid members based on earnings from business with nonmember patrons to whom smaller amounts are paid for substantially identical transactions.

In determining taxable income, a cooperative may deduct from its income any patronage refunds that are paid in cash, qualified written notices of allocation, or other property with respect to patronage occurring during the tax year. The cooperative must pay a patronage refund during the payment period for the tax year to make it eligible for deduction. The *payment period* begins the first day of the tax year and ends on the fifteenth day of the ninth month after the close of the tax year, which is the date by which the cooperative must also file its Form 990-C return. Allocations made to patrons after the 20 1/2-month payment period do not qualify as patronage refunds and must be included in the cooperative's taxable income. Distributions paid in cash to patrons after the payment period must also be included in the patrons' taxable income.

A *written notice of allocation* is any capital stock, revolving fund certificate, retain certificate, certificate of indebtedness, letter of advice, or other written notice that discloses to the recipient the amount allocated to the patron and the portion of the allocation that is a patronage refund. A written notice of allocation that qualifies for deduction from a cooperative's taxable income is called a *qualified written notice of allocation* or *qualified allocation.*

To qualify a written notice of allocation for deduction, a cooperative must pay at least 20% of the patronage refund in cash or by qualified check.

---

[1]Although the Internal Revenue Code uses the term *patronage dividends,* this book uses *patronage refunds,* to avoid confusion with dividends paid on capital stock.

In addition, the patron must either have the opportunity to obtain the total refund in cash within 90 days after the allocation is made or consent in one of three ways to have the noncash portion treated as if it had been distributed in cash and reinvested by the patron in the cooperative.

By consenting to have the retained portion of the refund treated as if it had been paid in cash, the patron agrees to include the stated dollar or face amount of the total refund as ordinary income earned during the year in which it was received. The patron may do this by (1) agreeing in writing, (2) joining or continuing as a member of the cooperative (as long as the cooperative has a bylaw adopted after October 16, 1962, providing that membership constitutes such consent and members have received written notification and a copy of this bylaw), or (3) endorsing and cashing a qualified check.

A *qualified check* is a check or other instrument that is redeemable in cash and paid as part of a patronage refund. Imprinted on it is a statement that endorsing and cashing the instrument constitutes patron consent to include in taxable income, as provided in federal income tax laws, the stated dollar amount of the written notice of allocation that is also part of the patronage refund.

A cooperative cannot deduct a patronage refund allocation from its taxable income unless all requirements for qualified status are met. If the cooperative does not receive patron consent or if it does not pay at least 20% of a patronage refund in cash, the allocation is not considered a qualified allocation. However, if the allocation is made before the end of the cooperative's payment period and otherwise meets the definition of a patronage refund, it is considered a *nonqualified written notice of allocation* or *nonqualified allocation*. Nonqualified allocations are included in the cooperative's taxable income and therefore are not immediately taxable income for patrons.

There is no requirement that 20% of nonqualified patronage refund distributions be paid in cash. If a distribution includes both cash and a nonqualified allocation, the cash portion is included in the patron's taxable income and is deductible from the cooperative's income.

Although a cooperative cannot deduct nonqualified allocations from current income in the year the allocations are distributed, it can deduct redemptions of nonqualified written notices. The cooperative pays tax on nonqualified notices when they are issued. When they are redeemed in cash or other property, the cooperative can deduct the payments to patrons from its income, and a patron who receives a redemption includes the amount of the payment received in taxable income.

The cooperative's tax in the year the allocation is redeemed is the lesser of either (1) the tax for the current year after deducting the redemption from current income or (2) the tax for the current year without the deduction less the reduction in tax that would have occurred in prior years if the allocation had originally been issued as qualified. If the reduction in prior years' tax is greater than the current year's tax without the deduction, the cooperative receives a refund. Determination of the reduction in prior years' tax can be complex, particularly if it involves losses or redemptions of allocations issued in more than one year.

Relatively few cooperatives use nonqualified written notices of allocation. In 1986, 13 of the 100 largest U.S. farmer cooperatives reported issuing nonqualified allocations. These allocations represented 11.9% of the total patronage distributions made by the 100 cooperatives. Some nonqualified allocations result from the failure of cooperatives to receive nonmember consent. Because the bylaws regarding consent apply to members only, cooperatives that wish to pay patronage refunds to nonmembers must obtain consent through a marketing contract or some other means.

Numerical comparison The differences between qualified and nonqualified written notices of allocation are demonstrated in Table 16.1. The results in this table depend on two important assumptions: First, there is a decrease in the marginal tax rate of the patrons between the dates when the allocations are issued and redeemed. This could occur if a substantial number of patrons retire before redemption or if there is a reduction in federal income tax rates. Also, the cooperative has $25,000 in other taxable income in the year of redemption. Equally valid comparisons under different assumptions could yield much different results.

In this example, the cooperative earns $100,000 in net income the year the allocations are made. If the cooperative chooses to distribute its net income in qualified form, it must pay at least 20% in cash. The entire net income is deductible from federal taxable income, and the cooperative pays no tax. If the cooperative pays 20% cash patronage refunds, its net cash flow is thus $80,000.

On the other hand, patrons who receive qualified distributions include the entire amount in their taxable income. If they are in the 28% marginal tax bracket, they collectively pay $28,000 in income tax on the distributions. This exceeds the $20,000 they receive in cash patronage refunds. Thus they incur a negative cash flow of $8,000 due to the qualified written notices of allocation.

In the year the cooperative redeems the qualified written notices of allocation, it incurs a cash drain of $80,000, and the patrons receive an $80,000 cash flow. There are no tax consequences to either the cooperative or the patrons.

If the cooperative chooses to allocate its net income as nonqualified written notices of allocation, it does not pay cash patronage refunds. However, it includes the notices in its taxable income and pays $22,250 in corporate income tax. The cooperative's cash drain is $2,250 greater than

TABLE 16.1 Comparison of qualified and nonqualified written notices of allocation

Item	Written notices of allocation	
	Qualified	Nonqualified
Year of Allocation		
Cooperative		
Net income	$ 100,000	$ 100,000
Cash patronage refunds (20%)	(20,000)	0
Federal income tax (22.25%)	0	(22,250)
Cash flow	$ 80,000	$ 77,750
Patrons		
Cash patronage refunds (20%)	$ 20,000	$ 0
Federal income tax (28%)	(28,000)	0
Cash flow	($ 8,000)	$ 0
Year of Redemption		
Cooperative		
Equity redemption	($ 80,000)	($ 100,000)
Income tax refund (27%)	0	27,000
Cash flow	($ 80,000)	($ 73,000)
Patrons		
Equity redemption	$ 80,000	$ 100,000
Federal income tax (15%)	0	15,000)
Cash flow	$ 80,000	$ 85,000
Net Cash Flow, Both Years		
Cooperative	$ 0	$ 4,750
Patrons	72,000	85,000
Total	$ 72,000	$ 89,750

it would have been if the cooperative had distributed its net income in qualified form and paid 20% in cash.

Patrons who receive the nonqualified written notices do not include the allocations in their taxable income. They pay no federal income tax on the allocations and do not incur the negative cash flow from qualified allocations. Because the patrons' marginal tax rate is higher than the cooperative's, nonqualified written notices of allocation result in a lower total cash drain to the cooperative and its patrons than do qualified notices.

When the cooperative redeems the nonqualified written notices of allocation, it earns an income tax deduction. The cooperative would save $22,500 in tax by recomputing its tax for the year in which the allocations were made as if it originally had issued them in qualified form. The cooperative currently has $125,000 in taxable income and ordinarily would pay $30,750 in income tax. If it deducts the redemption of nonqualified written notices from its current taxable income, it would reduce its tax by

$27,000. This method results in the greatest tax savings to the cooperative. Therefore, it is the method used, and cash drain from the redemption is $73,000.

Patrons who receive the redemptions must include them in their taxable income. Because the patrons are now in the 15% tax bracket, their income tax due to the allocation is only about half what it would have been in the year of allocation. Thus the patrons' net cash flow is $85,000.

In this example, the total cooperative and patron cash flow for the two years is greatest for nonqualified written notices of allocation. This is due to the patrons' ability to defer the tax until they are in a lower tax bracket and to the cooperative's ability to shelter income that otherwise would be taxed at a higher rate. Not all situations, however, produce this result.

Choosing allocation method The effective tax rates of both cooperatives and patrons may vary from year to year, depending on business success and investment decisions, among other factors. To minimize taxes, both cooperatives and patrons should attempt to recognize income in years in which their effective tax rates are lowest. Patrons with low tax rates may prefer qualified allocations because they are paid partly in cash.

Patrons with high tax rates may wish to delay receiving income and therefore would prefer nonqualified allocations. Cooperatives can also use nonqualified allocations to avoid negative patron cash flows due to tax on qualified allocations.

Cooperatives have some flexibility in using nonqualified allocations to manage their taxes. Allocation of nonqualified paper is more attractive to cooperatives during years in which other taxable income is low. On the other hand, the deduction for redeeming nonqualified paper is most useful in conserving cash flow during years when taxable income is high. Once a cooperative begins redeeming nonqualified allocations issued earlier, this deduction can be used to shelter the cooperative from taxes based on the current year's allocation of nonqualified paper.

Per-Unit Capital Retains

Per-unit capital retains are investments in a cooperative made by patrons based on the dollar value or physical volume of products marketed through the cooperative. Cooperatives withhold per-unit capital retains according to a bylaw provision or membership agreement that authorizes the cooperative to make a specified deduction for capital purposes from proceeds due members or from cash advances. These retains should be distinguished from deductions authorized to cover operating expenses.

Per-unit capital retains are allocated to patrons and taxed in a manner similar to patronage refund allocations, except they do not depend on cooperative net income. Cooperatives notify individual patrons of per-unit capital retain allocations by giving them *per-unit retain certificates*. A per-unit retain certificate is any written notice that discloses to the recipient the dollar amount of a per-unit retain allocation made by the cooperative.

Recipients of per-unit retain certificates may consent to include the amount of the retains in their taxable income by agreeing in writing or by joining or retaining membership in a cooperative with a bylaw agreement. A cooperative must issue a certificate before 8 1/2 months after the close of the tax year to deduct the certificate from its taxable income.

If the recipient agrees to include a per-unit retain certificate in taxable income, the certificate is a *qualified per-unit retain certificate* and the cooperative deducts the amount of the certificate from its income in determining taxable income. The principal difference between tax treatment of qualified per-unit retain certificates and written notices of allocation (patronage refunds) is that the Internal Revenue Code recognizes per-unit capital retains as fundamentally different in concept and therefore does not require that 20% be paid back to patrons in cash.

If the patron does not agree to take a per-unit retain certificate into account, it is a *nonqualified per-unit retain certificate,* and the cooperative cannot deduct it in determining taxable income. However, cooperatives redeeming nonqualified per-unit capital retains can deduct the amount of the redemptions.

Nonpatronage Income and Unallocated Equity

Patronage refunds do not include *nonpatronage-source income,* which is incidental income that is not directly related to the marketing, purchasing, or service activities of a cooperative and that merely enhances the cooperative's overall profitability. This income can include rents received, investment revenues, gains on the sale or exchange of depreciable property and capital assets, and amounts from business done with the federal government. It can also include income from business done with or for nonmembers but not distributed to them.

Because nonpatronage-source income cannot be distributed to patrons as part of a deductible patronage refund, it is generally included in cooperative taxable income. However, a cooperative holding section 521 tax status can exclude from its taxable income nonpatronage income distributed to patrons on a patronage basis. On the other hand, cooperatives without section 521 tax status must pay income tax on nonpatronage-source income regardless of whether they distribute it to patrons.

Many cooperatives retain nonpatronage-source income remaining after income tax as tax-paid unallocated equity. A cooperative may also retain some net income from member business as unallocated equity. Because this net income is not distributed as patronage refunds, it is included in cooperative taxable income.

SECTION 521

Subchapter T provides that some cooperatives may deduct from their taxable income certain additional distributions made during the payment period. These cooperatives, commonly called *exempt cooperatives* or *section 521 cooperatives*, are defined under section 521 of the Internal Revenue Code. The term *exempt* is misleading because tax treatment under section 521 is not the same as section 501, which exempts some organizations from paying federal income tax. On the contrary, section 521 cooperatives often can have taxable income.

Under subchapter T, only a section 521 cooperative can deduct from taxable income nonpatronage income distributed to patrons on a patronage basis and dividends on capital stock, in addition to qualified patronage refunds and per-unit capital retains. Capital stock includes voting and nonvoting common stock, preferred stock, or any other form of capital represented by capital retain certificates, revolving fund certificates, letters of advice, or other evidence of proprietary interest in a cooperative. In filing their individual income tax returns, patrons must include as ordinary income any dividends on capital stock or nonpatronage income distributions received from a section 521 cooperative.

To qualify under section 521, a cooperative must meet the following requirements:

1. It must be a farmer, fruit grower, or similar association organized and operated on a cooperative basis to (a) market farm products or (b) provide farm supplies and equipment.

2. If organized on a capital stock basis, substantially all the cooperative's voting stock must be owned by agricultural producers who market farm products or purchase farm supplies through the cooperative. [As interpreted by the Internal Revenue Service (IRS), this condition is met if at least 85% of producers holding voting stock market or purchase through the cooperative during the year.]

3. Dividends on capital stock are limited to 8% per annum or the legal rate in the state in which the cooperative is incorporated, if greater.

4. Financial reserves of the cooperative must not exceed those that are necessary or required by state law.

5. The cooperative must conduct no more than 50% of its marketing business and no more than 50% of its purchasing business with nonmembers. Purchases for those who are neither members nor producers must not exceed 15% of the cooperative's total purchases. Business done for the federal government is not taken into consideration in determining eligibility for section 521 status.

6. Nonmembers must be treated in the same manner as members with respect to business transactions such as pricing, pooling, payment of sales proceeds, or allocation of patronage refunds.

7. The cooperative must maintain permanent records of the patronage and equity interests of all members and nonmembers.

Cooperatives seeking to qualify for section 521 tax status must request and receive a "letter of exemption" from the district director of the IRS. This is done by filing Form 1028, which provides basic information on how the cooperative is organized and operated. Once a letter of exemption is obtained, a cooperative does not refile Form 1028 unless it substantially changes its organization or methods of operation. Section 521 status is subject to examination by the IRS at any time. If the IRS finds that the original application misstated facts about the cooperative's organization or operation or if these conditions have changed, the IRS may revoke section 521 status and require the cooperative to recalculate its income tax.

Selecting Income Tax Status

Changing economic conditions affecting the operation of cooperatives have decreased the number of cooperatives holding section 521 status. In 1970, 62 of the 100 largest U.S. farmer cooperatives held section 521 status. In 1986, only 21 of the 100 largest cooperatives held section 521 status.

Selection of the best status for a particular cooperative depends on several considerations. A cooperative considering a change in status should examine its situation with respect to these factors:

1. *Dividends on capital stock.* Cooperatives with section 521 tax status are permitted to deduct dividends on capital stock in determining taxable income. Cooperatives without section 521 tax status include these dividends. Thus a cooperative that pays substantial dividends on capital stock may prefer to hold section 521 tax status.

2. *Nonmember and nonproducer business.* Cooperatives with section 521 status are required to treat members and nonmembers equally. Thus, if these cooperatives pay patronage refunds, they must pay them to both members and nonmembers. Cooperatives without section 521 status are permitted to pay patronage refunds only to members. They can also retain earnings from nonmember business remaining after income tax as tax-paid unallocated equity. Thus cooperatives with substantial nonmember income may choose to increase permanent capital by operating without section 521 status.

Section 521 also limits nonmember business to 50% and nonproducer purchasing business to 15%. Some cooperatives, particularly farm supply cooperatives operating in suburban communities and providing lawn and garden supplies or petroleum products to nonmembers, may have difficulty complying with these requirements. These cooperatives must limit nonmember or nonproducer business or choose to operate without section 521 tax status. In some cases, nonmember or nonproducer business may be helpful in achieving economies of size necessary for efficiently providing services to members.

3. *Nonpatronage income.* In determining taxable income, cooperatives with section 521 tax status are allowed to deduct nonpatronage income distributed to patrons on a patronage basis. Cooperatives without section 521 tax status must include nonpatronage income in their taxable income. After paying tax on nonpatronage income, these cooperatives may choose to retain what remains as unallocated equity or distribute it to patrons. Distributions of nonpatronage income are not patronage refunds, and patrons must pay tax on them regardless of their cooperative's tax status.

4. *Business activities.* Cooperatives with section 521 status are restricted to marketing farm products, purchasing farm supplies and equipment, and providing related services. Nonfarm business activities are prohibited. On the other hand, cooperatives without section 521 status may engage in other business activities. Thus, in considering tax status, a cooperative should weigh the benefits of engaging in other business activities as an advantage of not holding section 521 status.

5. *Effect on other cooperatives.* A cooperative considering adopting or dropping section 521 tax status should consider the effect it may have on other cooperatives either through taxation of dividends it issues ,or on the tax status of the other cooperatives through the look-through principle explained later in this chapter.

6. *Securities laws.* Cooperatives with section 521 tax status are exempt from the registration and prospectus requirements of federal securities laws. Cooperatives without section 521 status must register interstate securities if otherwise required by securities laws. Exemption from registration does not relieve cooperatives from liability for violating the antifraud provisions of the laws.

7. *Other considerations.* A cooperative contemplating a change in tax status should take its income tax bracket into consideration. The deductions due to section 521 status are more valuable the greater the cooperative's tax rate. Because cooperatives with section 521 tax status must keep permanent patronage and equity records of both members and nonmembers, cooperatives considering adopting or dropping section 521 status also should consider the increased costs or savings in bookkeeping expenses that would result.

Tax Alternatives Available to Cooperatives

Table 16.2 presents three tax alternatives available to cooperatives. The cooperative has the choice of serving both member and nonmember patrons and qualifying for section 521 tax treatment (A.1), serving both member and nonmember patrons and not qualifying for section 521 treatment (A.2), or serving only member patrons and qualifying for section 521 treatment (B). At first, assume that the cooperative does 85% of its patronage business with members and 15% with nonmembers. Eight percent of the cooperative's income initially is from nonpatronage sources.

TABLE 16.2 Tax alternatives available to cooperatives

Item	Member business (85%)	Nonmember business (15%)	Total
A. Serving both member & nonmember patrons			
1. Paying patronage refunds to both members & nonmembers (qualifying for section 521 tax treatment)			
Net income:			
From patronage sources	$153,000	$27,000	$180,000
From nonpatronage sources	17,000	3,000	20,000
Total net income	$170,000	$30,000	$200,000
Less dividends on capital stock	(8,500)	(1,500)	(10,000)
Patronage distributions	$161,500	$28,500	$190,000
2. Paying patronage refunds only to members (not qualifying for section 521 tax treatment)			
Net income from patronage sources	$153,000		$153,000
Less patronage-source dividends on capital stock	(7,650)		(7,650)
Patronage refunds	$145,350		$145,350
Nonmember income			$ 27,000
Other nonpatronage income			20,000
Total nonpatronage income			$ 47,000
Plus patronage-source dividends on capital stock			7,650
Taxable income			$ 54,650
Less federal income tax			(13,663)
Less dividends on capital stock			(10,000)
Available for unallocated reserves or member patronage distributions			$ 30,987
B. Serving only member patrons (qualifying for section 521 tax treatment)			
Net income:			
From patronage sources	$153,000		$153,000
From nonpatronage sources	20,000		20,000
Total net income	$173,000		$173,000
Less dividends on capital stock	(10,000)		(10,000)
Patronage distributions	$163,000		$163,000

Source: Adapted from Abrahamsen (p. 235).

Only member patrons hold capital stock, and for the sake of simplicity, state income tax is ignored.

To qualify for section 521 tax treatment while serving both members and nonmember patrons (A.1, Table 16.2), the cooperative must treat members and nonmembers alike. Thus net income from both patronage and nonpatronage sources is distributed identically across members and nonmembers. Dividends on capital stock are taken in the same proportions from member and nonmember net income, even though members own all capital stock.

Because the cooperative operates under section 521, it can deduct from taxable income dividends on capital stock and nonpatronage income distributed to patrons on a patronage basis, in addition to patronage refunds (from patronage sources). Thus if the cooperative distributes net income in the manner shown in A.1, it pays no income tax. Members receive $161,500 in patronage distributions (patronage refunds plus nonpatronage income) and $10,000 in dividends on capital stock. Nonmembers receive $28,500 in patronage distributions.

If the cooperative does not pay patronage refunds to nonmembers, it does not qualify for section 521 tax treatment (A.2). Dividends on capital stock and nonpatronage-source income, whether distributed to members on a patronage basis or not, are included in the cooperative's taxable income. Income from nonmember business that is not distributed to nonmembers is considered nonpatronage income. The cooperative deducts dividends on capital stock proportionately from patronage (member) and nonpatronage income.

In A.2 of Table 16.2, the cooperative distributes to members as patronage dividends only the net income from member business remaining after deductions of the proportionate share of dividends on capital stock. The cooperative includes dividends on capital stock from patronage income (member business), as well as all nonpatronage income (nonmember business and other sources), in its taxable income.

After deducting federal income tax and dividends on capital stock from taxable income, the cooperative has $30,987. It can add this amount to unallocated reserves or make additional patronage distributions to members. If the cooperative distributes this amount, members receive a total of $176,337 in patronage distributions in addition to $10,000 in dividends in capital stock. Because these distributions are not considered patronage refunds, members also pay income tax on them.

If the cooperative serves only members (B), it may qualify for tax treatment under section 521. In such a case, the cooperative can exclude from its taxable income all patronage refunds, dividends on capital stock, and distributions of nonpatronage income. The cooperative pays no federal income tax, and members receive the entire net income of the cooperative—$163,000 in patronage distributions and $10,000 in dividends on capital stock.

INVESTMENT TAX CREDIT

Congress has revised tax laws from time to time to encourage investment in certain types of depreciable or amortizable property. As an incentive, firms have been allowed to apply a percentage of such investments as credit against their federal income tax. Although current tax laws do not include a provision for investment tax credit, they do include special rules for the use of investment credit by cooperatives.

Under these rules, which will become effective if investment tax credit incentives ever are reinstituted, a cooperative cannot carry unused investment credit backward or forward to other tax years as other business firms do. Instead, the cooperative must allocate credit not used in the year earned to patrons on a patronage basis, similar to the way it allocates patronage refunds.

Many cooperatives have used investment tax credit in the past to build tax-paid unallocated equity reserves against losses. In fact, cooperatives have retained patronage-source income as unallocated equity to generate taxable income and absorb investment credit. Some cooperatives may have absorbed investment credit to avoid the administrative costs that would result from allocating the credit to individual patrons, as well as potential problems from recapture.

TAXATION OF FEDERATED COOPERATIVES

Cash and noncash qualified patronage refunds distributed to a cooperative by another cooperative are considered patronage-source income of the recipient cooperative for the tax year in which the distribution is received. If the recipient cooperative in turn distributes this income in qualified form to its patrons within 8 1/2 months of the close of its tax year, it can exclude this income from its taxable income. Otherwise, the cooperative must pay tax on the income.

The treatment of dividends on capital stock is more complex and depends on the tax status of both cooperatives. Dividends from other cooperatives are considered nonpatronage income and generally are included in the taxable income of recipient cooperatives without section 521 status. A cooperative without section 521 status is allowed an 80% deduction on dividends received from other cooperatives without section 521 status. The aggregate amount of the deduction is limited to 80% of the taxable income of the recipient cooperative.

A cooperative operating under section 521 is permitted to deduct from its taxable income all dividends received and subsequently distributed to its patrons, regardless of the tax status of the cooperatives that issue the dividends. If a section 521 cooperative does not distribute dividend income to its patrons, tax treatment of the income is identical to that of cooperatives without section 521 status.

The *look-through principle* is used to determine if a federated cooperative satisfies the nonmember and nonproducer business limitations imposed on cooperatives eligible for section 521 tax status. Under the look-through principle, a federated cooperative determines if it meets the requirements by looking through member cooperatives to the membership and producer status of its ultimate patrons.

Member cooperatives with section 521 status qualify as producers because of their income tax status. Member cooperatives without section 521 status do not qualify automatically as producers because they are not required to restrict business with nonmembers and nonproducers. Thus a federated cooperative must look through these cooperatives to their patrons to determine whether they qualify as producers. It may be difficult for a federated cooperative to satisfy the nonmember and nonpatron limitations if it has member affiliates that do not operate under section 521.

OTHER TAXES

Cooperatives usually are subject to other taxes on the same basis as other businesses. Cooperatives pay sales, payroll, license, gasoline, property, and excise taxes. However, state corporate franchise taxes, which are annual levies on the net worth or capital of a corporation, usually exempt farmer cooperatives or tax them at lower rates.

States with income taxes generally follow the basic provisions of federal tax laws and regulations in taxing cooperatives. The cooperative usually deducts patronage refunds from income as long as there is a preexisting obligation on the part of the cooperative to distribute these refunds to patrons. There is some variation, however, in how states treat dividends on capital stock and nonpatronage-source income.

TAXATION OF OTHER COOPERATIVES

Section 501

The tax rules already described in this chapter apply to most farmer cooperatives. However, other cooperative businesses may be taxed differently. For example, section 501 of the Internal Revenue Code exempts some organizations from paying federal income taxes. These organizations include religious, charitable, and civic institutions and nonstock, nonprofit corporations owned and operated for the benefit of members. Many of these organizations, including some credit unions, mutual insurance associations, and rural electric and telephone companies, are organized and operated as cooperatives.

Farm Credit System

Federal Land Banks, Federal Land Bank Associations, and Federal Intermediate Credit Banks are also exempt from federal income tax under section 501 of the Internal Revenue Code. However, Banks for Cooperatives and Production Credit Associations currently are subject to federal income tax. Because both types of organizations allocate their earnings to borrowers on a patronage basis, they are taxed under subchapter T.

Consumer Cooperatives

Consumer cooperatives that sell retail goods are taxed in substantially the same way as farmer cooperatives, except that consumer cooperatives are not eligible for section 521 treatment. They pay income tax on dividends on capital stock and on additions to unallocated reserves, but they deduct patronage refunds from taxable income. Patrons do not include these patronage refunds in their taxable income, because the refunds are reductions in the costs of goods and services provided by the cooperatives, rather than additions to income.

A BRIEF HISTORY OF COOPERATIVE TAXATION

To understand the reasoning behind current cooperative tax rules, it is helpful to learn how they evolved. The relationship between federal income tax laws and farmer cooperatives began in 1909 with passage of the Corporation Tax Statute. This law placed a tax on the net income of corporations and joint-stock associations but provided for exemption of agricultural and horticultural associations operating on a mutual basis. Since then, taxation of cooperatives has evolved with frequent legislative review and a steady stream of IRS rulings interpreting the legislation.

The Revenue Act of 1913 was the first tax measure adopted after the Sixteenth Amendment to the Constitution legalized income tax. The act did not mention cooperatives specifically but exempted from paying income tax "certain types of non-profit concerns, including agricultural and horticultural associations." The IRS first interpreted this to include only nonstock dairy cooperatives making patronage refunds. However, this ruling soon was modified to include dairy cooperatives with or without capital stock. The ruling also provided that only amounts actually paid to members could be excluded from taxable income. Funds retained at the end of the year were taxable.

In following years, Congress extended exempt status to other cooperatives. The Revenue Act of 1916 gave federal income tax exemptions to agricultural marketing cooperatives dealing only with members. The Revenue Act of 1921 gave exemptions to purchasing cooperatives that acquired supplies and equipment for members at actual costs and that operated with capital stock. In 1922, IRS regulations under this act also authorized the accumulation of reasonable reserves.

Congress also made the requirements of qualifying for tax exemption less stringent. Regulations issued under the Revenue Act of 1924 used the word "producer" instead of "member." They also provided that cooperatives could do a limited amount of business with nonmembers. The Revenue Act of 1926 specified that purchasing cooperatives could deal with nonmember nonproducers as long as this business did not exceed 15% of total business. And the Revenue Act of 1934 provided that business done with the federal government should not be included in determining a cooperative's taxable income.

By 1934, the concept that farmers' marketing and purchasing cooperatives were not required to pay income tax on net income distributed as patronage refunds was well established. However, no legal consideration had been given to the income tax responsibilities of patrons for patronage refunds. Nevertheless, no significant change in cooperative tax legislation occurred between 1934 and 1951.

The Revenue Act of 1951 attempted to collect a single tax at the patron level on patronage refunds and on dividends on capital stock. Although the Internal Revenue Code was revised completely in 1954, no changes were made regarding farmers' marketing and supply cooperatives. However, section 101 (12)(A) and (B) of the code became sections 521 and 522.

As it does today, section 521 specified conditions necessary for a cooperative to exclude from taxable income dividends on capital stock and nonpatronage income distributed to patrons. Section 522 subjected cooperatives to income tax, but allowed them to deduct patronage refunds distributed within 8 1/2 months of the close of their tax year, as long as they had a preexisting obligation to do so.

The effectiveness of the Revenue Act of 1951, as incorporated in the Internal Revenue Code of 1954, was compromised severely by two court rulings during the 1950s. In *Commissioner v. B. A. Carpenter* (1955), the court held that noncash patronage refunds were reportable by members at "fair market value." In *Long Poultry Farms, Inc. v. Commissioner* (1957), the court ruled that patrons reporting income tax on an accrual basis did not have to report noncash patronage refunds as income in the year the allocation was made.

As a result of these decisions and subsequent IRS rulings, the Revenue Act of 1951 no longer achieved its congressional intent. Therefore, Congress took corrective action with the Revenue Act of 1962. This act created subchapter T of the Internal Revenue Code and repealed section 522. Most of the significant portions of section 522 were included in section 1382(c) of subchapter T.

CURRENT ISSUES CONCERNING COOPERATIVE TAXATION

Elimination of Single-Tax Treatment

Some opponents of the federal tax treatment of cooperatives contend that section 521 and subchapter T of the Internal Revenue Code should be eliminated. They point out that net income of investor-oriented corporations are taxed twice—once at the corporate level as corporate income and once when distributed to shareholders as dividends. These opponents argue that because corporations must pay tax on dividends on capital stock and retained earnings, section 521 and subchapter T deductions give cooperatives an unfair advantage in raising capital and competing against other firms.

Proponents of current cooperative tax laws suggest that single-tax treatment is not unique to cooperatives but is applied to other business forms, including partnerships and subchapter S corporations. They also point out that subchapter T treatment is not restricted to agricultural cooperatives. Almost any business that chooses to distribute income to patrons on the basis of patronage and according to a preexisting obligation can exclude this income from its taxable income.

Backers of cooperative tax treatment also argue that opponents overlook the tax benefits other corporations receive. Special deductions and credits allow many corporations to reduce their effective tax rates below what cooperatives pay (Baarda and Ingalsbe).

Some cooperative advocates themselves are in favor of eliminating section 521. They reason that the benefits of section 521 are not that valuable, as evidenced by the number of cooperatives that choose not to qualify for it. They suggest that the existence of section 521 costs all cooperatives because of bad public relations and community ill will.

The debate over how cooperatives should be taxed is long-standing. For recent views, see the articles by Baarda and Ingalsbe, Sexton and Sexton, and Schrader. For complex legal arguments against current cooperative tax treatment by a former IRS commissioner and National Tax Equality Association[2] spokesman, see the Caplin article.

---

[2]An objective of the National Tax Equality Association, an association of investor-oriented corporations, is the repeal of the single-tax treatment of cooperative patronage refunds and per-unit capital retains.

Operating on a Cooperative Basis

There has also been controversy about the requirements necessary for organizations to qualify for cooperative tax treatment. Subchapter T of the Internal Revenue Code applies to section 521 cooperatives and other corporations "operating on a cooperative basis." The meaning of this phrase has been subject to much debate and litigation between cooperatives and the IRS. The IRS has interpreted the phrase to exclude some practices used by cooperatives, and it restricts the applicability of subchapter T to only those cooperatives that refrain from these practices.

For example, the IRS maintains that to be eligible for subchapter T, a cooperative must limit nonmember business to 50% or less (Rev. Rul. 72-602, 1972-2 CB 510). However, in *Conway County Farmers Assn. v. United States*, 588 F. 2d 592 (8th Cir. 1978), this revenue ruling was held to be unduly restrictive and contrary to congressional intent. The cooperative involved had over 60% nonmember business.

Netting Losses

Cooperatives' use of *netting* (combining a loss in one area of operation with net income in another for income tax purposes) has also caused disagreement about the meaning of operating on a cooperative basis. For years, cooperatives and the IRS have disagreed about how much discretion cooperatives have in handling losses. The IRS has maintained that "operating on a cooperative basis" for the purposes of subchapter T means that all patrons must be treated on an equitable basis. To the IRS, this has meant that a loss in one operation must be borne entirely by the patrons of that unit and cannot be shared with patrons of other units except under special circumstances.

Cooperatives have argued that the Internal Revenue Code places no special restrictions on handling losses and that cooperatives should be able to use any procedure available to businesses in general. Cooperatives contend that netting is a standard business practice and that cooperation requires some sharing of benefits and risks among all patrons. This includes netting losses and gains among operations—whether among farm supply and marketing functions or along product or commodity lines.

Several federal courts have ruled in favor of cooperatives when the IRS challenged the way that cooperatives handled losses. The courts consistently have held that fairness decisions should be made by members acting through their boards of directors and not by the IRS. Nevertheless, in January 1985 the IRS released an administrative ruling disallowing patronage refund deductions from the income of a cooperative that netted.

After the release of this ruling, cooperatives, with the support of the Secretary of Agriculture, asked Congress to revise the code to clarify to the IRS that netting had been permissible since subchapter T became law in

1962. Congress responded by including in the Budget Reconciliation Act of 1985 an amendment to subchapter T stating that a cooperative with a loss in one area of operation may net this loss with income in another area. The law also requires cooperatives that net to notify patrons of the netting methods used.

There is general agreement that this legislation will not resolve all disputes between cooperatives and the IRS over netting. For example, the legislation does not directly address the argument in the January 1985 IRS ruling that extensive board discretion over netting, exercisable after the close of the tax year, negates the preexisting legal obligation necessary for claiming a patronage refund deduction. The extent and timing of board discretion appears to remain an issue. Consequently, the IRS probably will continue to test the limits of its authority.

Other Issues

The IRS continually rules on issues of importance to cooperatives. Usually, the rulings apply to the eligibility of an individual cooperative for section 521 or subchapter T treatment or to the acceptability of a practice that affects the tax paid by the cooperative. These issues include "operating on a cooperative basis," the "substantially all" capital stock condition for section 521 status, and definitions of terms such as producer and patronage-source income. Although applied to individual cooperatives, these rulings often produce important consequences for other cooperatives.

SUMMARY

Net income of farmer cooperatives is generally taxed according to the single-tax principle. This principle, recognized by subchapter T of the Internal Revenue Code, ensures that cooperative net income is usually taxed at either the cooperative or patron level, but not both. Subchapter T provides that, in addition to deductions allowed other businesses, certain distributions of cooperative net income or allocations paid patrons should be excluded by cooperatives in determining taxable income.

Cooperatives may deduct qualified allocations of patronage refunds and per-unit retain certificates from taxable income. To deduct patronage refunds or per-unit capital retains from its taxable income, a cooperative must obtain patron consent to include both the cash and noncash portions in their income. In addition, at least 20% of qualified patronage refund distributions must be paid to patrons in cash.

Patronage refunds and per-unit capital retains that patrons do not agree to include in their taxable income are called nonqualified. A cooperative must include nonqualified allocations in its taxable income, but it can deduct cash redemptions of nonqualified allocations. Patrons include redemptions of nonqualified allocations in their taxable income. Both qualified and nonqualified distributions must be made within 8 1/2 months after the close of the cooperative's tax year.

Subchapter T specifies additional distributions that may be deducted by cooperatives meeting conditions specified under section 521 of the code. These cooperatives, commonly called exempt or section 521 cooperatives, deduct nonpatronage income distributed to patrons on a patronage basis and dividends on capital stock from taxable income.

Choice of tax status and taxation of cooperatives within federated systems are complex. Current cooperative taxation has evolved over many years. Today there are a number of issues concerning cooperative taxation, including elimination of single-tax treatment. Many of the issues arise from the steady stream of IRS rulings interpreting cooperative tax laws.

DISCUSSION QUESTIONS

16-1. What requirements must a cooperative meet to qualify a written notice of allocation for deduction from its taxable income?

16-2. What are the three ways by which patrons can consent to include patronage refunds in their taxable income?

16-3. What are the differences in how qualified and nonqualified written notices of allocation are taxed at the cooperative level? At the patron level?

16-4. Why might a patron prefer to receive a patronage refund in the form of a nonqualified written notice of allocation?

16-5. Are cooperatives that receive section 521 tax treatment exempt from federal income tax? Explain.

16-6. What distributions of income can a section 521 cooperative deduct from its taxable income that other cooperatives cannot?

16-7. Must a business be a farmer cooperative to be eligible for subchapter T or section 521 tax treatment? Explain.

16-8. What considerations should a cooperative weigh when deciding whether to qualify for section 521 tax status?

16-9. Must a cooperative that desires to deduct patronage refunds from taxable income apply for section 521 tax status?

16-10. What are some of the arguments for preserving the single-tax treatment of cooperatives? What are some of the arguments for eliminating it? Which do you favor?

REFERENCES

ABRAHAMSEN, MARTIN A., *Cooperative Business Enterprise*. New York: McGraw-Hill, 1976.

BAARDA, JAMES R., AND GENE INGALSBE, "Tax Treatment of Cooperatives: Easy to Explain, Fair, Not Unique," *Farm. Coop*. (June 1982):20.

CAPLIN, MORTIMER, M., "Taxing the Net Margins of Cooperatives," *Georgetown Law Rev*. 58(1969):6.

FREDERICK, DONALD A., *Tax Treatment of Cooperatives*. Washington, DC: USDA ACS CIR 23, Sept. 1984.

HOLLIS, C. DAVID, AND CHARLES H. INGRAHAM, *Farmer Cooperatives and Federal Income Taxes: Is Exempt Status More Beneficial?* Research Bull. 1039, Ohio Research and Development Center, Wooster, Sept. 1970.

ROYER, JEFFERY S., AND ROGER A. WISSMAN, "Nonqualified Allocations Offer Alternative Method of Retaining Funds," *American Cooperation 1985*. Washington, D.C: American Institute of Cooperation, 1985.

SCHRADER, LEE F., "Cooperatives' Tax Treatment Fundamentally Sound, Fair," *Farm. Coop*. (Feb. 1978):14.

SCHRADER, LEE F., AND RAY A. GOLDBERG, *Farmers' Cooperatives and Federal Income Taxes*. Cambridge, MA: Ballinger, 1975.

SEXTON, RICHARD J., AND TERRI ERICKSON SEXTON, "Taxing Co-ops," *Choices* (second quarter 1986):21.

U.S. DEPT. OF AGRICULTURE, Farmer Cooperative Service, *Legal Phases of Farmer Cooperatives*. Washington, DC: USDA FCS Info. 100, May 1976.

Part VI
Management

17

Managerial Skills,
Functions,
and Participants

Roger G. Ginder,
Iowa State University

Ron E. Deiter,
Iowa State University

DEFINITIONS AND MANAGEMENT SKILLS

Cooperative management is the process of pursuing cooperative objectives by utilizing the resources available to the organization, including people, capital, and facilities. Those who manage these resources must have the necessary professional skills to work with people and to make sound decisions. Whether a cooperative is a success or a failure often depends on whether there is management or mismanagement.

It is not the purpose of this chapter to duplicate business-school management textbooks by presenting complicated models or terminology. Our purpose, instead, is to explain a few of the unique facets of cooperative management and to relate them to more traditional management concepts.

Successful business organizations are frequently said to have "good management." But what is it that really distinguishes good management from mediocre management or poor management? Among other things, management skills play an important role. Just as secretaries, accountants, engineers, attorneys, computer programmers, and others must possess skills to do their jobs effectively, managers need certain skills to do their jobs effectively as well.

Management skills and decision-making skills are often treated as synonymous, yet modern managers must possess other skills in order to be effective, including interpersonal relations skills and goal-setting skills (Table 17.1). In fact, these may be more important in the management of cooperatives than in the management of investor-oriented firms (IOFs).

Cooperatives often face problems, uncertainties, and alternatives that are similar to those of their IOF competitor firms. However, because cooperatives are owned by their users, they differ from IOFs in the way they approach and define problems. Cooperative decisions are usually expected to result in improved service to patrons, enhanced member profitability and feasibility, as well as competitive financial returns on invested capital.

Cooperative management also requires broader interpersonal relation skills than does IOF management, because cooperative patrons are also cooperative stockholders. This creates a number of special communication problems not present in IOFs, whose communication with the stockholders can be completely separate and distinct from any communication with customers. A cooperative board of directors must understand and guide members' expectations for the cooperative. The general manager must not only understand patron needs (as would be the case in any IOF) but must communicate with patrons on a personal basis more frequently and maintain a higher degree of visibility.

Top IOF management may maintain very close communication with their boards of directors, who usually represent the interests of a majority of the stockholders. But with few exceptions, the chief executive officer (CEO) of an IOF is rarely visible to the customers of the firm. Exceptions

TABLE 17.1 Basic management skills

Decision making	Interpersonal relationships	Goal setting
Defining problems	Communication	Setting personal & organizational goals
Choosing among alternatives	Understanding individuals & group	
		Motivation
Delegating decisions	Leadership	
		Reward systems
Making decisions under uncertainty	Policies & authority	
		Managing stress
	Management values	
Creative decision making	Management ethics	Managing conflict and change

Source: Anderson (p. 17).

include Lee Iacocca of Chrysler, Frank Perdue of AW Perdue and Sons, and a few other CEOs who have elected to become active in advertising campaigns. But even when CEOs are widely known and recognized by patrons, the one-member, one-vote relationship between patrons and the board is not present, and patrons rarely have direct access to the general manager. Also, several members of the boards of large IOF corporations are often appointed by management rather than vice versa. For cooperatives, however, this rarely happens.

The democratic control by cooperative owner-patrons places a higher premium on political skills as well as communication skills in cooperative management. Managers cannot afford to be isolated, and in some situations they must exercise finely honed political skills in resolving conflicts among different groups in the cooperative membership regarding patron-centered issues.

Cooperative managers also have a different code of ethics and values than those of IOF managers. This occurs because cooperative principles prescribe equitable treatment of members, democratic control, and distribution of net income based on patronage. Indeed, it was the ability to affect these business ethics and trade practices which has led to the formation and continued farmer support of many cooperatives. Consequently, a patron-first policy is difficult to avoid. For example, cooperatives must be extremely cautious when approaching ethical decisions that involve trade practices which disadvantage patrons but increase return on assets. In particular, practices that prey on members' lack of understanding or information are not likely to be acceptable.

Goal setting is a third skill area that is critical to good management. Business organizations, whether cooperative or IOF, cannot continue to function efficiently in the long run without clearly defined goals and objectives. Goal setting must be done for the short term, the intermediate term, and the long term. In the following section we explain more fully goal setting in cooperatives and who is involved.

MANAGEMENT FUNCTIONS

Management skills are useful only when applied to actual managerial functions that contribute to the accomplishment of the organization's mission. Most management experts identify four or five basic management functions (Anderson; Beierlein et al.; Downey and Erickson; Drucker; Duft; Greene; and Roy). The differences between functions as presented by these experts are usually in the classification of activities rather than in the nature of the activities themselves. For our purposes we will categorize the functions as *planning, organizing, directing, staffing,* and *controlling.*

Planning

Planning (deciding future direction and goals to be pursued) is probably the most important of the management functions. Without knowledge of goals and direction, there is no point of reference for organizing, staffing, motivating, or controlling. These other functions must be conducted in pursuit of the goals established in the planning process.

The role of planning in the cooperative is identical to the role of planning in the IOF. It provides direction and purpose to the deployment of material and human resources. It aids in communication of direction and purpose throughout the firm. Done correctly, it serves as a cohesive force, binding together very different activities at all levels in the firm and working toward a common goal for time periods as short as a day or as long as many years.

Boards and managers sometimes confuse planning with the plan itself. Planning is a continuous process rather than an activity to be completed when a document called a plan has been produced. As conditions change, plans need to be revised. Even long-range plans should be updated annually and, in some cases, more frequently. The job of planning is never completed as long as the business continues to operate. It is the process whereby the cooperative continuously evaluates itself and the outside forces that affect it and then charts future activities.

It is customary to divide planning activities into two categories—operational and strategic (Table 17.2). Long-range or strategic planning commits the company to a particular, relatively irreversible direction requiring large amounts of resources. Planning that deals with shorter-range problems, that does not affect the overall direction of the firm, that invol-

TABLE 17.2   Operational versus strategic planning in cooperatives

Decision aspect	Type of planning	
	Operational	Strategic
Frequency of change	Frequent	Infrequent
Time span of plan	Short run (up to 1 yr.)	Long run (1-20 yrs.)
Effects on asset structure	Little or none	Extensive
Commitment of resources	Small	Large
affect on general direction of co-op	No change	Substantial
Degree of risk	Small or none	Large
Reversibility of decisions	Highly reversible	Difficult to reverse
Orientation	Employee	Customer/investor

ves smaller amounts of resources, and that is more easily reversed is called short-run, tactical, or operational planning. Consequently, we frequently hear the terms "long-range plan" and "short-range plan" as if they were two separate plans. However, it is not practical for a firm to have two plans. Both strategic aspects and operational aspects of the plan should be melded together so that they are consistent and only one plan results.

For example, if a cooperative specializing in petroleum and fertilizer decides to enter the feed manufacturing business, the decision is strategic because of the time required to carry out the decision, the resources required for acquiring plant equipment and personnel, the change in the nature of the business, and the degree of difficulty in reversing the decision. Closely related and subordinate to the strategic plan is a plan for marketing the feed products produced. However, this is an operational decision as it involves smaller quantities of resources and is more easily reversed. The same principles would apply to exiting the business.

Strategic plans would be involved when downsizing, liquidating fixed assets, or changing the nature of the business. Operational plans would be developed to reduce and reorganize production personnel, change promotional and marketing activities, and perhaps make changes in logistical activities. Again, plans to enter and exit business activities may have both strategic and operational components.

Organizing

The second management function is organizing the human and capital resources of the cooperative. It involves matching those resources with the work necessary to accomplish the firm's objectives. Two processes that take place in organizing are specialization and coordination (Anderson).

Specialization is the process of gaining efficiency by dividing work into separate, easily understandable components that can be carried out effectively by an individual employee. This permits the employee to gain experience and special knowledge about a narrow portion of the work and to become very efficient in performing it. The ultimate example is the assembly line in which workers perform a very narrowly defined set of operations on a large number of units with little knowledge of operations performed either before or after the unit reaches them. Generally, in cooperatives the jobs are not as narrowly defined; nonetheless, such jobs as fertilizer and chemical application, feed formulation, grain drying, operation of a cotton gin, and milk processing are specialized.

Since much work performed in agricultural cooperatives is seasonal in nature, the cooperative manager faces additional problems in specializing employees not encountered in less seasonal industries. Both cooperatives and IOFs involved in agriculture need to train employees in more than one specialty, and employees clearly need to understand the priorities of the manager.

When work is divided into specialized areas, cooperative managers also need to coordinate various specialized jobs by transferring materials and information between persons and departments. Coordination includes all the steps required to ensure that specialized jobs get performed smoothly toward the accomplishment of strategic goals and objectives.

Directing

Beyond dividing work into and coordinating specialized areas, managers must constantly direct or, in other words, focus the efforts of employees on the strategic objectives of the organization. There is an amusing but accurate expression that describes the necessity for this kind of directing: "When you're up to your neck in alligators, it's difficult to remember that your objective is to drain the swamp." Directing can help employees keep end objectives in mind, even though employees often face a myriad of problems in conducting the day-to-day activities related directly to their jobs. Similarly, directing can assist employees in solving the technical problems of operations and in overcoming conflicts among units or persons that interfere with the achievement of strategic objectives.

Directing extends into higher management levels. For example, the CEO or general manager can get caught up in management problems and occasionally may lose sight of overall company objectives. At such times, it therefore may be necessary for the board to direct the general manager.

The following list summarizes the activities that are part of the directing function (Anderson).

1. Communicate the objectives of the organization to subordinates and work with subordinates to design plans and procedures for achieving those objectives.

2. Motivate subordinates to achieve their objectives or to implement their part of the plan by rewarding learning and achievement with pay, recognition, and peer support.

3. Help subordinates solve technical problems related to reaching goals by encouraging team effort or group management.

4. Overcome conflicts among subordinates and introduce change. At various times subordinates will disagree with cooperative goals and the means of achieving them. It is the manager's role to resolve these conflicts and to ensure cooperation among subordinates. In other instances, organizational change is necessary for increased effectiveness, but organizational change must be introduced and managed correctly if it is to succeed.

Staffing

The need for staffing, the fourth function of management, also arises from the cooperative's plan. The staffing function is frequently divided into four activities: selection, training, development, and appraisal. Once the cooperative has identified the work, the material, and the human resources required to achieve the plan, the cooperative must then select and recruit persons with appropriate skills for jobs within the organization. However, even a person with the appropriate skills is rarely capable of moving into a specific job without some training or, in other words, without special instruction about the job's unique characteristics in the particular firm. For example, an accountant has the proper skills to do accounting work but must be trained to fill a position in the accounting department in a particular firm. Staffing also includes developing skills among those already employed and providing feedback on performance.

Development is similar to training but is not as closely related to a specific job. It includes enhancing personal management skills and capabilities that make the individual eligible for promotion to higher positions in the organization. Appraisal involves regular discussion of the employee's performance, compared to the expected level of performance or performance standards for the job.

Controlling

A fifth function of management is controlling, which means making periodic measurements to determine whether or not various performance standards, goals, and objectives are being met as expected and to reconcile divergences. There are actually four steps in the controlling function.

The first step is to establish acceptable levels of performance, or performance standards. These should not be confused with goals, since an acceptable level of performance in most cases is much different (lower) than the level that might be established as a goal.

The second step is to measure performance of either the individual, the department, or the business itself according to performance standards and cooperative objectives. If these standards and objectives are clear, specific, and quantifiable, then measurement is much simpler.

The third step is to compare the actual measured results with standards or expected results. Actual results will rarely be exactly the same as the expected results specified by the standard; they may be slightly less than expected in some cases and slightly better in others. Consequently, managers in most cases allow for a small deviation around the expected or standard performance. However, when the actual results deviate too far from the expected results, the fourth step in the process—corrective action—becomes necessary.

At this point, effective managers will act according to a principle called "management by exception." This principle states that positive management action should be focused on those areas where deviations from the standard or expected results occur and where problems exist rather than on those areas where there are no apparent problems.

Before taking corrective action, the management must identify the cause for the deviation. Sometimes changes occurring outside the firm make past goals, objectives, or standards obsolete. When this happens, management may need to alter standards in order to account more accurately for actual conditions. On the other hand, if there seem to be no problems with cooperative goals, objectives, or performance standards, managers will probably choose to increase efforts to motivate departments or individuals to improve their performance.

In summary, control is a process of establishing standards, measuring performance, comparing performance with standards, and taking corrective action if necessary. It is not, as many believe, a process of controlling people. It fosters continual evaluation of business performance and appropriate adjustments to changes inside and outside the cooperative, thus helping to assure the survival and health of the cooperative.

THE PLANNING PROCESS

The cooperative planning process itself does not differ from that used by IOFs. However, the selection of the cooperative's overall objectives and the determination of an acceptable balance between financial returns and other goals usually involve more board input than is necessary for an IOF. Furthermore, a cooperative's major objective is that of improving member-patron profitability, which is an objective frequently absent or is of much lower priority in an IOF.

Planning is perhaps the most important of the five management functions. A number of alternative planning processes are available for use by cooperative managers. The one that we will discuss here involves six steps: (1) definition of the firm's mission, (2) evaluation of the firm's present position (strengths/weaknesses and opportunities/threats), (3) development of objectives, (4) selection of strategies, (5) preparation of an action plan, and (6) implementation of the action plan.

Defining the cooperative's mission means specifying the purpose for the organization or, in other words, clearly stating the reasons for the cooperative's existence. A mission statement should reflect the intent of the cooperative membership and should be consistent with the articles of incorporation and bylaws. Cooperative missions do change over time, but a mission statement should not change frequently.

For example, the mission of a large number of midwestern elevator-supply cooperatives originally included the procurement of coal to be used as fuel by patrons. This has long since ceased to be a major function in these cooperatives. Therefore, they have had to revise their mission statements in some cases to reflect a long-term change of purpose. To avoid the need for frequent revision, cooperative boards and managers need to make mission statements general enough to allow the cooperative flexibility in operating in a dynamic industry. On the other hand, the statements must be specific enough to confine the organization to the activities that stockholders have mandated—otherwise, the mission statement is meaningless.

After a cooperative has defined its mission, it must start planning with a clear understanding of the cooperative's present position. Plans are not likely to be achievable unless they are based on a realistic assessment of the existing state of the cooperative and the business environment.

To assess its current position, the cooperative needs to consider (1) those factors within the cooperative and under the control of the board and management and (2) those factors outside the cooperative and beyond the control of the board and management. Some planners differentiate between the two as (1) *strengths* and *weaknesses* and (2) *opportunities* and *threats,* using the acronym SWOTs.

An assessment of strengths and weaknesses might include an evaluation of the cooperative's management and personnel, fixed assets, production capabilities, location and trade area, financial position, rolling stock, and other internal factors. The process usually includes an individual comparison of each factor with its competitors. In this way, the board can identify specific areas of strength and weakness. This process also helps managers to set goals and objectives for the future.

To assess opportunities and threats, the cooperative identifies those legal, business, and social factors outside the cooperative's control that may have profound effects on the firm's performance. Among these factors are government regulations, state and national economic conditions, world trade and economic conditions, industry trends, and demographic trends. All of these can either provide business opportunities for the cooperative or threaten its survival.

Since the environment outside the cooperative is continually changing, assessment of the cooperative's position is never really completed, even though management may commit a plan to paper. Should the cooperative fail to stay in touch with market forces by ignoring internal weaknesses or external threats, it may fail to survive. Similarly, should the cooperative fail to capitalize on internal strengths or external opportunities in the business environment, it will fail to reach its potential as a business organization.

The third step in the planning process is setting goals or establishing objectives that will guide the firm's efforts. Objectives state what is to be accomplished by using the cooperative's human and capital resources. They also provide a cohesive force within the organization, binding the members, the board, the professional management, and the employees together and communicating in a clear and concise way to all what the cooperative intends to achieve.

Strategic objectives set the overall direction for the cooperative. Good strategic objectives will contain concise information on (1) what is to be achieved, (2) when it is to be achieved, and (3) what quantitative measures will indicate whether the objective has been achieved.

As indicated earlier, identification of strengths, weaknesses, opportunities, and threats helps management to determine both appropriate objectives and possible strategies for achieving those objectives. Generally, strategies that depend heavily upon cooperative strengths are better than those that require performance in the cooperative's weak areas. In addition, strategies that minimize exposure to external threats are preferred.

Cooperative strategy selection, the fourth step in the planning process, usually involves more board involvement than those of their IOF counterparts. Thus, some observers believe that the professional management in competing IOFs have greater flexibility in selecting and implementing strategies. For example, a board may insist on continuing a department that is completely disadvantaged.

Strategy selection frequently is closely related to the development of action plans, the fifth step in the planning process. In fact, in many cases strategy selection may not occur as a distinct step but rather, as part of establishing objectives or developing an action plan. Strategy selection need not be a separate step in the process, as long as management consciously evaluates their alternatives at some point.

The final step in the planning process is to implement action plans that prescribe specific actions to be taken at specific times and in a specific order. Action plans embody particular strategies for achieving strategic objectives. They are, therefore, the product of all prior steps.

PARTICIPANTS IN THE COOPERATIVE MANAGEMENT TEAM

Who's Involved?

As citizens and voters, we elect politicians to act on our behalf in the policymaking process. Cooperative members act similarly by voting for board members to act on their behalf over the affairs of the cooperative.

Ultimately, the members of the cooperative are responsible for the entire organization and its performance; however, it is impossible for all members to be involved all the time in all management decisions that affect their cooperative. Hence, members delegate some important managerial duties and responsibilities to the board, which they hope will be carried out in a manner that reflects their aspirations and needs. The board will then, in turn, hire professional managers to help them achieve cooperative objectives.

To guarantee that management will serve the members' best interests, cooperatives use the articles of incorporation, which may be viewed as a contract between the board and the cooperative membership. The articles are the basic legal document that defines the purpose of the organization and mandates the statutory obligations the board must fulfill in operating the cooperative. The articles may also specify the conditions for distribution of assets upon dissolution of the firm or state other policies that the membership wishes to make clear in advance.

In addition to the articles of incorporation, bylaws are usually developed to help govern the cooperative. Bylaws further specify limits on the board authority and practices that the cooperative must follow. Bylaws are procedural policies of continuing importance to the membership that are not expected to change frequently. Unlike the articles, bylaws generally do not include statutory requirements.[1] Rather, they are a set of rules that the membership, the board, and the management have agreed upon for operating the cooperative. Bylaws may cover such matters as selection of board members, length of terms, handling of net income, qualifications for membership, voting rights, and a host of other issues related to the operation of the cooperative.

The board of directors, then, has broad management power over the business affairs of the cooperative. It is, for practical purposes, the legally accepted decision-making voice of the membership, acting as a body within the constraints of the articles and bylaws.

Responsibility, Authority, and Chain of Command

The board, the general manager, and department managers play key roles in the management of the cooperative organization. If the cooperative is to function effectively, these groups must work together on behalf of the entire cooperative membership. Each group must clearly understand its responsibility and authorities, develop the necessary management skills, and work toward common goals if the cooperative is to run smoothly and effectively.

Although common to all corporate businesses, the board of directors plays a particularly important role in a cooperative corporation because of its importance to the cooperative control principle. By law, the management of a cooperative is the board's responsibility. Therefore, members of the board must act diligently with necessary skills and knowledge about the cooperative. If they do not, they may be liable for harm caused the cooperative by their neglect or misbehavior in office. Directors of farmer cooperatives may even be personally liable for conflicts of interest or for permitting management to violate a cooperative's laws or charter.

---

[1]There is one exception: IRS requires that the bylaws contain a statement that the member agrees to allow the cooperative to retain patronage refunds to capital.

Because directors are almost always farmers who patronize the cooperative, many states carefully restrict directors' dealings with the cooperative. Most states prohibit a director from entering into a contract with the cooperative that differs in any way from the business contracts accorded regular members or holders of common stock of the association.

Individual directors have no more authority in the cooperative than any other member. Individual directors have special authority only when they participate in making joint board decisions with other directors. Board committees have only the power conveyed on them by the whole board. Decisions made by a majority of the board in an official board meeting are binding on all directors. Individual board member objections to a board decision may be recorded in board meeting minutes, although the board member must abide by the decision of the

TABLE 17.3 The management team and responsibility, authority and accountability

Management group	Responsibility	Authority	Accountability to
Members	All responsibility for management of the coopera-tive	All authority not delegated in the articles and by-laws to directors	Themselves
Board	All responsibility for the manage-ment of the coop-erative not reserved for members in Art-icles and bylaws	All authority not delegated to the general manager as day to day operational authority	The membership and the state and federal government
General manager (CEO)	All responsibility for operational management not reserved for Board in bylaws and Board policies	All authority not delegated to the department heads	The board of directors
Depart-ment or section manager	All responsibility for operational management of department or section not reserved by the general manager in management policies	All authority not delegated to the departmental employees	The general manager

majority. An individual board member may also resign to express disagreement with a board decision.

Conflicts (or adversary relationships), lack of necessary management skills, or pursuit of different goals among the members of the management team may create serious problems. One key factor in preventing such problems is a clear understanding of responsibility, authority, and accountability in the cooperative. Managers cannot shed any of these simply by delegating them to subordinates (Table 17.3). For example, the board of directors may delegate to the general manager responsibility and authority to maintain equipment. However, should the general manager fail, the board remains responsible for the poor condition of equipment and may be held accountable by the membership. Similarly, the general manager may delegate responsibilities to department heads along with the authority to carry them out. But should the department heads perform poorly, the manager remains responsible for their performance.

Because they remain responsible for the actions of subordinates, delegating managers or boards sometimes are reluctant to give their subordinates much authority. This kind of problem is not unique to cooperatives. It occurs in other organizations as well. However, both responsibility and the authority to act must be delegated to subordinates if the organization is to function properly. Otherwise, subordinates cannot realistically fulfill their responsibilities. Further, once authority is delegated, interference by the delegating manager should be limited.

Who Decides What?

Since boards can act only as a body, and board members are generally engaged in managing their own farm businesses, it is all but impossible for them to perform the entire range of management tasks required to keep a cooperative running effectively. Therefore, they usually hire a general manager, to whom they delegate the responsibility and authority to make and carry out the operational decisions of the cooperative (Table 17.4). The general manager, in turn, is then usually free to further delegate responsibility and authority to department or section managers and to other employees within the cooperative.

The involvement of the cooperative management team in the planning process varies, depending on whether or not the planning is strategic or operational. The board of directors has a fiduciary responsibility to members as trustees of the cooperative's assets and must take the major responsibility for long-range planning. The board has a major involvement with planning in the goal-setting process. However, this does not mean that the board should do the planning independently. In fact, the board rarely, if ever, attempts this task alone. In this stage, cooperative boards usually give more input than IOF boards. In the IOF, a board representing stockholders is less concerned about business activities

**TABLE 17.4 Responsibility and involvement of the board and profes-
sional CEO or general manager in management functions
in a cooperative**

Management function	Professional CEO or general manager	Board of directors
Planning		
Operational	Major responsi- bility	Little or no involve- ment
Strategic	Heavy involvement	Major responsibility
Organizing	Total responsibility	No involvement
Directing	Major responsibility	Only as applied to general manager
Staffing	Total responsibility	Only as relates to personal expense
Controlling	Responsibility for hired employees and operation	Responsibility for overall goals and strategic objectives

than about long-run return on investment and preservation of capital. Cooperative boards, on the other hand, are concerned about business activities, because
these activities partly determine the profitability of members' farming operations.

Unlike IOFs where profit for investors is the overriding long-run objective, cooperatives must be concerned with the dual objective of satisfying members as investors and as patrons. The cooperative must earn an
adequate return and be efficient. At the same time, it must contribute to
member profitability. The latter may at times conflict with the maximization of long-term coooperative net income. Hence a delicate balance between board and management interests must exist, with neither
dominating. Therefore, responsibility in the goal-setting and planning
process is usually shared as follows.

Boards and top management typically establish strategic objectives in
the planning process, with top management presenting proposals and the
board having the final decision. Then professional managers and
employees, usually without board involvement, will establish operational
objectives to support the longer-range objectives of the cooperative.

As a rule, top professional management is heavily involved in the strategic
planning process that occurs next. Department heads, employees, and sometimes members or professionals from outside the cooperative organization will
participate as well. However, responsibility to ensure that planning occurs and

to make the final decisions rests with the board of directors and cannot be abdicated.

Operational planning typically does *not* involve the board. The CEO, department heads, and employees at lower levels are responsible for short-range or operational plans. These plans should support, be consistent with, and contribute toward the realization of long-range goals and plans.

In some cases, the board may want to review and approve operational plans. However, since board members may act only as a body, and since as full-time farmers they may not be familiar with operational details, it is not recommended that they participate directly in operational planning.

Professional cooperative managers generally assume the responsibilities for goal setting among the employees within the constraints of long- and intermediate-range board goals. The skills required to motivate employees, to establish effective reward systems, and to manage conflict, change, and stress within the cooperative are very similar to those required in the IOF.

TABLE 17.5 Management decisions in a cooperative

Membership	Boards	Professional Management
1. Change articles of incorporation	1. Selection and compensation of general manager	1. Maintenance of fixed assets
2. Change bylaws	2. Purchase of major fixed assets	2. Selection and compensation of employees
3. Consolidation with another cooperative	3. Credit policy	3. Interpretation of government regulations.
4. Merger with another co-op	4. Selection of auditor and attorney	4. Day-to-day purchasing of products
5. Liquidation	5. Levels of long- and short-term debt	5. Marketing of products for members
6. Selection of board	6. Lines of business activity	6. Allocation of expenditures within budget accepted by board
7. Recall directors	7. Policy on affiliation with regionals	7. Inventory management
8. Sale of majority of fixed assets	8. Authority given management	8. Personnel assignments and promotions
	9. Selection of sources of supply	9. Credit decisions (within board policy)
	10. Long-range plans	10. Setting pricing and margins within budget and policy
	11. Bottom line versus service to members	11 . Accounting and management information
	12. Budget level for various types of expenses	12. Patron complaints.
	13. Equity retirement	
	14. Cash patronage refunds	
	15. Sources of short- and long-term credit	
	16. Long-term leases/contracts	

In addition, organizing, directing, and staffing in a cooperative are almost exclusively the responsibilities of the chief executive officer (CEO) or general manager. It is recommended that the board not become involved in these functions, because past experience has shown that performance suffers when boards of directors become involved directly in organization and supervision of personnel. Rather than using direct intervention, boards should address their concerns about organizational structure, performance, or coordination of personnel to the general manager.

Other examples of specific decisions typically made by each participant in the cooperative management team are listed in Table 17.5. Of course, these decision-making duties will vary from one cooperative to another. In large cooperatives, professional management will typically make more of the decisions than the professional management of small cooperatives. It is also possible for the decisions listed to be made jointly by both the board and the manager. Board decisions often require input from professional management, and in turn, managers in many situations seek board input before making a decision.

SUMMARY

Making decisions in a cooperative about what goals to pursue and how best to attain them is the essence of cooperative management. Planning, organizing, directing, staffing, and controlling are the major functions or activities that are involved. The planning activity can be divided further into operational planning and strategic planning. The membership, the board of directors, and the hired managers are all involved in the cooperative management process. Each of these groups needs to understand its responsibilities and authorities and to work together as a team in order for a cooperative to be successful.

DISCUSSION QUESTIONS

17-1. What are the five major functions or activities of cooperative management? Give an example of a specific decision for each general function. Is any one of these functions the most important in your opinion? Why or why not?

17-2. What are the major differences between operational planning and strategic planning in a cooperative? Give examples of operational decisions that cooperative representatives might make. Give examples of strategic decisions that cooperative representatives might make.

17-3. Cooperative mission statements are not very useful in the planning process because they are nearly always very general with little detail. Explain why you agree or disagree with this statement.

17-4. What three main groups are involved in cooperative management? What is the chain of command among these three groups?

17-5. Distinguish between responsibility and authority in cooperative management.

17-6. Identify some of the typical cooperative management decisions made by members. Which of these is the most important in your opinion, and why?

17-7. Identify some of the typical cooperative management decisions made by the board of directors. Which of these is the most important in your opinion, and why?

17-8. Identify some of the typical cooperative management decisions made by the general manager. Which of these is the most important in your opinion, and why?

17-9. Managing a cooperative is easier than managing an IOF. Explain why you agree or disagree with this statement.

17-10. The manager of a cooperative should not get involved with decisions that typically are made by the board of directors. Explain why you agree or disagree with this statement.

17-11. Compare and contrast boards of directors in cooperative versus IOF organizations.

REFERENCES

ANDERSON, CARL R., *Management: Skills, Functions, and Organization Performance.* Dubuque, IA: Wm. C. Brown, 1984.

BEIERLEIN, JAMES G., KENNETH C. SCHNEEBERGER, AND DONALD D. OSBURN, *Principles of Agribusiness Management.* Englewood Cliffs, NJ: Prentice-Hall, 1986.

DOWNEY, W. DAVID, AND STEVEN P. ERICKSON, *Agribusiness Management.* New York: McGraw Hill, 1987.

DRUCKER, PETER F., *Management: Tasks, Responsibilities, Practices.* New York: Harper & Row, 1974.

DUFT, KENNETH D., *Principles of Management in Agribusiness.* Reston, VA: Reston, 1979.

GREENE, CHARLES N., EVERET E. ADAM, JR., AND RONALD J. EBERT, *Management for Effective Performance.* Englewood Cliffs, NJ: Prentice-Hall, 1985

ROY, EWELL PAUL, *Cooperatives: Development, Principles, and Management,* 4th ed. Danville, IL: The Interstate Printers & Publishers, Inc., 1981.

18

Directors
and Managers

Ron E. Deiter,
Iowa State University

Roger G. Ginder,
Iowa State University

INTRODUCTION

In describing what cooperative management is, we have identified (1) *who* are the participants on the cooperative management team, (2) *what* general functions or activities are their responsibility, and (3) examples of *which* specific managerial decisions they ultimately make. We now turn our attention to organizational and operational concerns in cooperative management. The emphasis here will be on some issues related to *how* the cooperative management process is or should be carried out in order for a cooperative to obtain and use resources effectively, efficiently, and in accordance with the organization's objectives.

BOARD OF DIRECTORS

The directors of a cooperative board are elected by the members to be their representatives in the management process. Directors should be intimately attuned to the needs and wants of patrons, because they too are patrons of the business—this is usually not the case for investor-oriented firm (IOF) board members. The managerial effectiveness of the cooperative board ultimately depends on the composition, experience, and talents of each board member, which in turn, depends on how the members elect, train, and compensate their directors.

Board Member Qualifications

Cooperative statutes usually require certain director qualifications and election methods. Members must follow these statutes carefully when they elect board directors. For example, most cooperative statues require that board directors be elected by members or stockholders from among their own number. Because of the important responsibilities that directors have, before casting their votes, members should also weigh carefully and objectively the strengths and weaknesses of each nominee. Board member elections should be more than popularity contests. Directors do not merely serve in an honorary position, although some board members unfortunately view this as their role.

Fulfilling the duties and responsibilities of a director requires much conscientious thought and hard work. The job carries with it tremendous legal obligations, as many of the corporate powers of the cooperative are vested in the hands of the board. In addition, the actions of the board ultimately determine the overall management character of the cooperative. If the board is well qualified and actively involved in managing the organization, there is more apt to be a proper balance of management by the board, the members, and the general manager. On the other hand, if the board is either indifferent about or incapable of functioning properly, the manager must control the cooperative, resulting in a potentially dangerous one-person organization with a so-called rubber-stamp board.

In assessing a candidate's qualifications to serve as a director, members can ask a number of questions. These include questions about the following characteristics of the candidate:

1. *Business judgment.* What evidence is there to indicate that the candidate possesses sound business judgment? Has the candidate been able to manage other business affairs prudently and successfully? How much related experience or educational training in business does the candidate have?

2. *Leadership capabilities.* To what extent has the candidate demonstrated leadership skills in the cooperative or in the community? Does the candidate elicit feelings of confidence, trust, and respect from fellow members? Is the candidate articulate? Does the candidate have well- founded opinions about what the cooperative should or should not be doing?

3. *Work habits.* Is the candidate likely to be an active, hard-working director? Does the candidate have the time, energy, and ambition necessary to serve as a director? Has the candidate demonstrated an ability to get along with and work with others as a team?

4. *Personal character.* Does the candidate have a reputation of integrity, honesty, and respect for the law? Has the candidate demonstrated loyalty to the association and to cooperative principles in general? Does the candidate patronize the cooperative fully and not have any conflicts of interest? To what extent is the candidate assertive yet not brash, persistent yet not uncompromising, and optimistic yet not unrealistic?

5. *Knowledge of cooperative principles.* Does the candidate understand and appreciate the uniqueness of the cooperative way of doing business? Does the candidate have a grasp of the opportunities as well as the limitations facing the cooperative? Is it clear that the candidate understands or is willing to learn the role of a director on a cooperative board?

Not all of the director candidates can be outstanding in each of these areas. Nevertheless, every candidate should be at least satisfactorily or sufficiently qualified in each area before being seriously considered worthy of election to the board.

When members recognize differences in candidate qualifications, they may wish to consider voting for those candidates who have special skills or talents that perhaps current board members do not have. For example, electing to the board someone who has unique financing or marketing skills may result in a more balanced, complementary, and effective set of directors. Of course, members may also simply and understandably vote for directors who they think will best represent their constituency in managing the cooperative.

Members can remove directors, as well as choose them. Most state laws outline the removal procedure, requiring members to conduct a meeting to consider removal, to give the director an opportunity to be heard, and to cast a deciding vote regarding the director's status. When a director's position becomes vacant for reasons other than an expired term, most statutes allow the remaining board members to fill the vacancy by electing another director according to a majority vote. Special rules may apply if directors are chosen by district.

Size of the Board

The board's effectiveness may also depend on its size rather than on each member's innate skills. If the board is too small, it may not be able to perform all its duties or to represent all major member viewpoints. On the other hand, if the board is too large, decision making may be cumbersome and slow, not to mention the greater difficulty and expense associated with scheduling board meetings. In setting board size, members should carefully consider these effects.

Between 65 and 75% of the agricultural cooperatives in the United States have an estimated 6 to 15 directors on their boards (Table 18.1). About 10% of the cooperatives, most of whom are regionals or large locals,

TABLE 18.1 Cooperative board of director size by size of cooperative (percent)

| Board size | Size of co-op | | | Total |
	Small	Medium	Large	
5 or less	32	10	3	19
6–9	58	53	32	51
10–15	10	28	33	21
More than 15	0	10	32	10

Source: Biggs.

have more than 15 board members, usually 16 to 35. The average size distribution of cooperative boards seems to be similar to that for all IOFs. However, there is some indication that cooperatives usually have more directors in proportion to their size than do IOFs (French et al.).

One possible explanation for this is that as cooperatives have grown in size through mergers, members of the merging firms have been unwilling to give up their representation on the board. As a result, the boards have often grown larger as they have been consolidated. In addition, large cooperatives have large boards because of their attempt to maintain geographical or commodity representation on their boards. For example, nearly two-thirds of the large cooperatives elect board members on a district basis, whereas more than two-thirds of the other cooperatives usually elect directors on an *at-large* basis, meaning that all members cast votes for candidates for all vacancies (Biggs).

Most cooperative statutes require a minimum number of directors. The most common number is five and three is second most common. Some statutes also limit the maximum number of directors from nine to 13.

State laws often indicate how a board of directors may organize itself to carry out its duties. In most states, the board of directors is responsible for choosing its own cooperative officers, including a president, one or more vice-presidents, a secretary, and a treasurer. Some offices may be combined. Usually, the president and at least one vice-president must be board members. Boards often use committees, and many statutes mention them, indicating that bylaws may provide for an executive committee and may allot to such a committee all the functions and powers of the board of directors, subject to the general direction and control of the board.

Director Tenure

In some cases, members establish cooperative policies that determine the conditions under which one can serve or be eligible to serve as a director. These policies establish qualifications for board membership and limit the length of a director's term, as well as the number of terms that a director can serve. A few statutes limit director terms to a maximum of three, although most do not specify any time limits. A number of statutes specifically suggest a staggered term plan, but few require it. Most cooperative associations (74%) have directors elected to three-year terms, while the bulk of the remaining associations elect their directors to either one- or two-year terms (Table 18.2). Members should continuously consider whether or not these policies are in their best interests.

Placing limits on the number of terms that a director can serve results in an automatic rotation of board membership, because the cooperative is forced to replace a director whose length of service reaches the maximum set by the members. Although fewer than 15% of all farmer cooperatives in this country currently limit directors' length of service, there are potentially some managerial advantages associated with this policy of automatic rotation. First, automatic rotation facilitates the automatic removal of an ineffective, unproductive, or inefficient director. Without automatic rotation, an incumbent director is sometimes difficult to unseat. Second, with new directors, cooperative boards can gain new ideas and new perspectives without a corresponding increase in board size. Third, a larger number of cooperative members can share the privileges and responsibilities of serving as a director, thus becoming better informed about their cooperative and more loyal to it as a result.

Of course, the possible advantages associated with automatic director rotation (due to length of terms served, age, or whatever) should be weighed against potential disadvantages. The biggest disadvantage is that such a policy may force a cooperative to replace the most experienced, capable, competent directors. As a result, unfortunately, the emphasis may be on length of service rather than on quality of service. To deal with this problem, cooperatives should stagger the terms of office so that only a portion of experienced directors leave the board each year. Some cooperatives use nonvoting associate or junior board memberships to give prospective

TABLE 18.2 Length of cooperative director terms by size of cooperative (percent)

Term length (years)	Size of Co-op			
	Small	Medium	Large	Total
1-2	24	23	19	23
3	72	75	79	74
More than 3	4	2	2	3

Source: Biggs.

new directors an opportunity to sit in on current board meetings and observe the board in action. In addition, new directors can enroll in director training schools and workshops to gain valuable experience and knowledge.

Finally, some cooperatives attempt to gain new perspectives by electing to the board certain nonvoting public representatives (e.g., tax consultant, banking officer, college professor) whose task is to deal with current issues confronting the cooperative. About half the states make a specific exception to the typical requirement of an all-member board, permitting the appointment of nonmember directors whose duty is to represent primarily the interest of the general public. These directors have the same powers and rights as other directors, although statutes typically limit their numbers to one-fifth of the total directors on the board.

Director Training and Compensation

For the board of directors to play an active role in the cooperative management process, there is usually a need to educate and train directors, especially new ones, regarding their duties and responsibilities. Approximately 30 to 40% of all cooperative directors participate in educational meetings, workshops, seminars, and short courses (Biggs). These training sessions often cover such topics as the scope and operations of the cooperative, the distinctions between board and manager responsibilities in the cooperative, methods of cooperative financing, and contracting.

Most orientation programs and even many of the training sessions are conducted by the cooperative itself, with contributions from management staff, experienced (past or present) directors, and outside people (e.g., from the Banks for Cooperatives, universities, law firms, etc.). Many of these programs are offered in the cooperative's office, although there may be fewer distractions and interruptions if the programs are conducted elsewhere. Directors should be exposed to as many viewpoints and educators as possible in the training process so that they can develop a broad and unbiased perspective of managerial issues with which they will have to deal on the board.

Most directors receive a meeting fee plus travel allowance (Biggs). There can be substantial variation among cooperatives in the amount of annual director compensation, depending on the number and cost of meetings as well as on the cooperative's philosophy. Usually, however, cooperative directors receive less compensation than do IOF directors (French et al.). Directors should not determine their own salaries or compensation. Rather, members should determine an amount that sufficiently compensates for time and expenses and does not discourage qualified members from serving on the board or being a part of the cooperative management team.

MANAGERIAL STAFF

Management Selection

Management selection is usually the most important decision that a board will ever make. Management selection typically presents a more difficult problem for boards in farmer cooperatives than in IOFs, many of whose board members are hired executives themselves. Most cooperative board members are farm operators who generally have had little experience in top management or executive selection. Moreover, some local cooperatives may not have a written plan or even an up-to-date job description for the position of general manager. If the previous general manager performed well over a long period of years, the board may have given little thought to management development or succession.

The board in these situations may not have a clear concept of what the manager really does. Because board members lack knowledge about the skills necessary for good manager performance and about proper techniques for evaluating candidates, they have nothing to work with except intuition, hunch, and hope. In such circumstances, board members have made serious errors in management selection and have been responsible for cooperative failure in several cases.

Although the possibility of management selection error may never be totally eliminated, the board can take several actions to reduce such errors. The most important of these is to create clear job descriptions stating the responsibilities of each management position. Cooperatives that have not gone through a managerial change for a number of years need to reevaluate and make an effort to keep the job description up to date.

A second valuable tool for management selection is an up-to-date plan. In particular, clearly stated strategic objectives and strategic action plans are helpful in assessing potential candidates. The plan is also helpful to potential candidates as they attempt to determine whether or not they will be comfortable with the job. Such self-evaluation by candidates helps to avoid selection errors. Cooperatives that do not have well-defined plans are at a severe disadvantage in articulating and communicating the expectations of the board and membership to potential candidates.

A starting point for preparation of management standards is the job description itself which outlines the manager's responsibilities. Job descriptions are, however, very short documents. To every job description boards should add performance standards that specify what the board considers to be adequate performance of the responsibilities (Table 18.3).

Finally, a clear set of performance standards can be very helpful in the selection process. Like the job description, performance standards communicate the expectations of the board to potential candidates. Ideally, each major responsibility listed in the job description should be accom-

TABLE 18.3 Comparisons between responsibilities in manager's position description and manager's performance standards

Position description	Performance standards
1 Tend to stress overall responsibilities	1. State results that will exist when responsibility is fulfilled
2 Tend to be general	2. Tend to be specific
3 Answers the general question "What is to be done?" Why, when	3 Answers specific questions as to how much, how well, in what way, when
4 Usually, one statement for each major responsibility or function	4. Usually, several statements for each function

panied by one to five performance standards. Board members should remember that performance standards specify the minimum, not the ideal, performance required; they should be written to reflect minimum job requirements, not individual capabilities of those who hold the position, and they should *not* be changed simply because they have been exceeded.

Management Compensation

Compensation levels for management positions are extremely difficult to establish. Members of a cooperative tend to be more conscious of management compensation and, in many cases, critical of high levels of compensation than shareholders in competing IOFs. Farmers may be willing to do business with an IOF without questioning the level of compensation received by its management, but they expect their own cooperative management to accept minimal compensation to avoid higher costs or lower patronage refunds.

This creates a problem for the board of directors. As we have seen, good general managers or CEOs are essential for the cooperative's very survival. Consequently, it is important for the board to recruit competent management by offering compensation comparable to that given in competing IOFs. At the same time, however, the board often must answer to a membership that is aware of and critical of high compensation in the cooperative.

Two major tools have been employed by boards to deal with this problem. Boards have increasingly used salary surveys to determine base levels for management compensation. Salary surveys, combined with the second tool, carefully written performance standards, allow the board to arrive at a base level of compensation for standard performance and to establish a range above and below the base for superior or substandard performance.

Salary surveys determine the compensation levels for managers in a number of similar cooperatives and in IOFs. Generally, firms and positions selected for the survey require skills and management responsibilities similar to those used by the cooperative in question. For example, a survey would ideally include firms with similar product lines, sales, types of assets, trade territories, and numbers of employees or patrons. The cooperatives then use average compensation (including noncash prerequisites) for general managers in these firms to establish a range of compensation levels for their own managers.

It is also possible to vary salaries according to years of experience, number of employees supervised, sales volume, or other similar factors by using statistical regression techniques. However, some have questioned the validity of this approach, because they believe the results may tend to perpetuate failure of boards in the sample cooperatives to address compensation properly. For example, in one cooperative salary survey, the manager's salary was found to be negatively related to the years of service in the cooperative. Consequently, if a cooperative were to base its salary schedule on this survey, its experienced managers would be systematically underpaid, resulting in an unnecessary turnover of good talent.

Salary survey information can be useful to the board in establishing realistic salary ranges, but the board should not rely heavily on them in determining the compensation of individual managers within those ranges. A better alternative for this task is to evaluate managers according to a previously established job description and performance standards.

AN APPRAISAL

Managing a Regional Versus a Local Cooperative

Managing a regional cooperative is similar in many ways to managing a local cooperative: similar participants with similar skills must perform similar managerial functions. However, some rather dramatic differences exist because the regional cooperative's operation is usually larger in scope, magnitude, and complexity than is the local cooperative's.

One major difference occurs in the delegation of authority and responsibility. As firms increase in size, it becomes physically impossible for one person to keep track of all operational details for all phases of the business. As a result, regional associations usually divide their activities into smaller, more manageable units called departments or divisions (e.g., finance, member relations, insurance, administration, marketing, agricultural services, personnel, etc.).

In some cases, regional cooperatives have formed subsidiaries to provide specialized products or services that they own or control. These various operating units are sometimes larger in size than most local cooperatives and are usually headed by a representative of the cooperative's management staff, such as a departmental head or vice-president. For these second-line managers to be able to carry out their managerial functions effectively, they need to have the corresponding authority and responsibility delegated to them. They, in turn, may also have to do a substantial amount of delegating.

Of course, local cooperatives sometimes have departments, too, but locals usually do not have as many as regional cooperatives. As a result, regional cooperatives also have more layers of management and, subsequently, more need to delegate. Department leaders are active participants in the management of the regional cooperative. They report to and are accountable to the president of the cooperative (analogous to the general manager of a local cooperative). Consequently, the management process in a regional is not likely to be as autocratic as it can be in a local cooperative. Also, the manager of a regional cooperative is likely to devote more time to major policies, problems, and planning and to delegate operational details to the department managers.

The value of information used in decision making is a second major difference between regional and local cooperative managements. Because of the larger number of dollars, products, people, and assets involved, there usually is more at stake when regional cooperatives make major decisions than there is when local cooperatives do. The cost of making a mistake, as well as the return associated with doing something right, can be magnified several times over for a regional cooperative. Thus it is usually advantageous for a regional cooperative to devote significantly greater amounts of time, staff, and money to obtaining information that can be used by management in decision making. As a result, larger cooperatives rely more on staff specialists to provide advice and counsel to management than do small local cooperatives, whose managers often must rely largely on their own experience, judgment, and knowledge to make a decision. These staff specialists advise regional managers about matters such as personnel, legalities, planning, product development, financing, engineering, and economics. Trade associations and state councils also frequently assist in providing regionals with this kind of information. One can view the management of the regional as one that involves a larger team effort. Because regional cooperatives so strongly value the extra information they use in decision making, they frequently put more emphasis on managerial training schools and seminars as well.

The regional cooperative's communications system also differs from that of the local cooperative. Whenever management has more people to work with, there will be greater difficulties in keeping everyone adequately informed about cooperative plans, policies, and strategies. Regional cooperatives must coordinate the activities of managers, employees, or patrons who are involved in planning, organizing, recruiting, hiring, training, directing, evaluating, compensating, and serving. Obviously, communication problems will often arise in such a firm. To solve those problems, regional cooperatives hold additional management staff meetings, and they employ communication specialists to help keep communication lines open.

Communications between management and members will also be more difficult. Not only is there a larger number of members to communicate with, but they are also spread over a wider geographic area. It is not as easy as it is in a local cooperative for members to talk directly with the general manager or department heads. Members also may live farther from the board member serving them. The communication process in a regional, then, may be accomplished through local or branch management, through field representatives who have direct contact with the members, or through public relations efforts such as newsletters, magazines, and area meetings.

In regional cooperatives greater communication problems for members are usually accompanied by similar communication problems for the board. Like the membership, the board often is larger in size, and thus as we have discussed, harder to assemble and work with efficiently. Therefore, regional cooperatives will use an executive committee of the board, rather than the whole board, to handle many policy matters.

The fourth major difference between regional and local cooperative managements lies in establishing equitable operating policies. It is more difficult for regional cooperative management to adopt and implement uniform operating policies, because a larger membership is likely to be more diverse or heterogeneous. As a result, cooperative desires to treat members equitably or fairly may in fact lead to different policies for different members, especially when many products or services involving substantial value-added activities are involved. A similar concern for regional cooperative management is how to establish fair operating policies among departments or among branch and local outlets when those units face different forms and levels of competition and have different operating costs. It can become very difficult to establish policies that all deal fairly with pricing, patronage refunds, managerial compensation, and subsidizing losses across units. This is not to suggest that local cooperative managements are free of these problems, but rather that regional cooperative management must deal with these matters more often and on a larger scale.

Finally, large cooperative management may differ in the scope of activities involved. Because of their greater size, area of service, human resource base and access to capital, regional cooperative management often has to make decisions dealing with special opportunities that are perhaps more available to them than to locals or with special requirements that pertain to them more than to locals. For example, usually only large companies such as regional cooperatives can expand or change the nature of the cooperative's business on a large scale by merging with or acquiring other companies, by expanding product lines or services, by developing export markets, or by integrating further backward or forward into the marketing channel.

The regional cooperative management may also be expected to represent cooperative interests legislatively and to provide leadership for smaller cooperatives. Local cooperative management often looks to regional management for guidance in such areas as recruiting, training, providing a varied and guaranteed supply of quality products and services, conducting research, maintaining programs, developing new products and new markets, and providing economic information.

In addition, regional cooperatives are often subject to greater scrutiny and regulation by governmental agencies, and they must also comply with a myriad of laws dealing with such issues as price and nonprice competition, taxation, security registration, environmental protection, and occupational health and safety.

Management Failures in Cooperatives

The effectiveness of management is one of the most important factors in determining the success or failure of any agribusiness firm, whether it be a cooperative or an IOF. One study cited by Downey and Erickson attributes at least 88% of all business failures to ineffective management. A growing number of grain elevator insolvencies (mostly IOFs were surveyed) has been reported in an eight-state region of the upper midwest over the period 1974-1982 (Casey et al.). This study also attributed most of the firm failures to mismanagement. Management failures in cooperatives are probably not any more prevalent than management failures in other types of businesses. In fact, Schrader et al. conclude that the recent performance of cooperatives is equal to or superior to that of IOF agribusiness firms on the basis of several performance measures, including efficiency, prices paid and received by farmers, services offered, innovation, and growth.

Specific failures in cooperative management can usually be attributed to any one of the groups in the tripartite cooperative management team—the members, the board, or the manager. In some instances, management problems can result from a lack of teamwork, cooperation, or communication among these groups. The patrons and the board play a greater role in managing a cooperative firm than they do in an IOF and thus are often more important in determining the overall managerial success or failure of a cooperative than an IOF. We will now give some specific examples of failures or mistakes that have been made by the members, the board, and the manager in managing a cooperative.

1. *Members* fail to (a) select qualified board members, (b) patronize their cooperative fully, or (c) adequately support their cooperative financially.

2. *Boards of directors* fail to (a) establish appropriate plans for their cooperative, especially for long-term investments and capital acquisition, (b) hire effective managers and compensate them accordingly, (c) implement effective member-communication programs, (d) hold managers accountable to them for their decisions, or (e) adopt sound operational policies or guidelines for managers to abide by (e.g., they overextend credit, make poor pricing or patronage refund decisions, or create a faulty product mix or marketing strategy).

3. *Managers* fail to (a) keep the board and members adequately informed; (b) properly manage the cooperatives' resources, including inventory, equipment, buildings, and employees; or (c) rely on sound record-keeping and accounting practices.

If cooperative members, directors, and managers purposely try to avoid some of these managerial pitfalls that have beleaguered some other cooperatives in the past, their chances for success will be much greater. Of course, the ultimate responsibility for effective management rests with the members, for it is they who elect the board members, who in turn select the managers.

SUMMARY

A cooperative must deal effectively with a number of important organizational and operational issues if it is to be successful. Establishing appropriate policies and procedures regarding board size, as well as director selection, training, compensation, and tenure, can ultimately determine the success or failure of a cooperative. Cooperative managers also need to be appropriately selected, evaluated, and compensated. Managing a regional cooperative is often more difficult than managing a local cooperative for various reasons, including greater complexities in delegating authority, handling information, communicating with members, and establishing equitable operating policies. When there have been failures in cooperatives, there have usually been corresponding failures on the part of at least one of the groups involved in the management of the cooperative--the members, the board, or the manager.

DISCUSSION QUESTIONS

18-1. What qualifications should a person have to be considered for a local cooperative director position? Of these, which is the single most important qualification in your opinion, and why?

18-2.   Should cooperatives elect directors on a *district* or *at-large* basis? Why?

18-3.   Suppose that a majority of the members of your cooperative (you are also a member) are proposing a policy for your cooperative that would prohibit anyone from running for a board-of-director position after reaching the age of 65. Would you support such a policy? Why or why not?

18-4.   What is meant by a policy of automatic rotation of directors? What are some of the potential advantages and disadvantages of such a policy?

18-5.   What are some things that a board of directors can do to minimize the chances of making a mistake in hiring a manager?

18-6.   How much money should the board of directors of a local cooperative pay their general manager? What factors should be considered in determining a general manager's salary?

18-7.   How is managing a regional cooperative different from managing a local cooperative?

18-8.   What is the biggest management mistake, in your opinion, that a member can make? Why?

18-9.   What is the biggest management mistake, in your opinion, that a director can make? Why?

18-10.   What is the biggest management mistake, in your opinion, that a manager can make? Why?

REFERENCES

BIGGS, GILBERT W., *Farmer Cooperative Directors: Characteristics, Attitudes.* Washington, DC: USDA ESCS, FCS RR 44, Feb. 1978.

CASEY, RICHARD P., DENNIS M. CONLEY, AND JOHN W. AHLEN, *Grain Elevator Insolvencies and Bankruptcies in Eight North Central States, 1974-1982.* Illinois Legislative Council Memorandum File 9-391, Mar. 1984.

DOWNEY, W. DAVID AND STEPHEN P. ERICKSON, *Agribusiness Management.* New York: Mc-Graw Hill, 1987.

FRENCH, CHARLES E, JOHN C. MOORE, CHARLES A. KRAENZLE, AND KENNETH F. HARLING, *Survival Strategies for Agricultural Cooperatives.* Ames, IA: Iowa State Univ. Press, 1980.

MATHER, J. WARREN, GENE INGALSBE, AND DAVID VOLKIN, *Cooperative Management.* Washington, DC: USDA ESCS CIR 1, Sec. 8, Apr. 1980.

RUST, IRWIN W., *Should Co-ops Rotate Directors?* Washington, DC: USDA FCS Reprint 383, Nov. 1971.

SCHRADER, LEE F., E. M. BABB, R. D. BOYNTON, AND M. G. LANG, *Cooperative and Proprietary Agribusinesses: Comparison of Performance.* Purdue Univ. Agr. Exp. Sta. Res. Bull. 982, Apr. 1985.

19

Communications

Richard H. Vilstrup,
University of Wisconsin-Madison

Frank W. Groves,
University of Wisconsin-Madison

INTRODUCTION

Education has historically been a fundamental part of a cooperative organization and success. The Rochdale pioneers in 1844 especially emphasized the importance of member education and information, stating that cooperatives should allocate a definite percentage of net income to education and give frequent financial reports to members.

In 1966, the International Cooperative Alliance reiterated the need for cooperative education by stating: "All cooperatives should provide for the education of their members, officers, employees, and the general public in the principles and techniques of cooperation."

These cooperative information programs have usually involved several distinct areas: member relations, employee relations, and public relations. Sometimes cooperatives also offered training programs for directors, management staff, and employees.

Every aspect of cooperative business and activity requires involvement and communication with people. It has been estimated that typical managers spend over 75% of their time in communicating—talking, writing, reading, and listening—because every major leadership or manage-

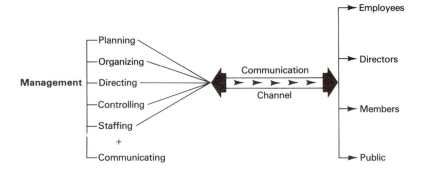

Figure 19.1 Communication —a key to effective communication.

ment decision or action depends on successfully getting a message through to members, directors, employees and customers (Figure 19.1).

The scope and importance of communication efforts can be illustrated by efforts of corporate management of Land O'Lakes and CENEX to relay important information quickly to management and to members of hundreds of local cooperatives located from Wisconsin to Washington and Oregon. The boards of directors of the two organizations decided to combine their feed, fertilizer, and fuel operation in a joint venture. To complete the action, they had to conduct a multistate informational meeting and successfully gain the approval of the memberships of the two organizations. To achieve projected savings, they also had to gain support of changes in identification, advertising, distribution, and personnel. Thus they had to use effectively nearly every communication channel and method available to complete the joint effort successfully.

This type of extensive communication can be difficult to achieve. It is thus essential for cooperative management to understand both the fundamentals of the communication process and the particular barriers to, as well as strategies for, effective communication within a cooperative structure.

COMMUNICATION MODELS

A thorough understanding of communications theory requires in-depth study. Therefore, we will illustrate only the basic and modified models. However, we encourage further study of references that include advanced communications models. Early models were advanced by Shannon, Weaver, Osgood, Westley, and MacLean (Shannon and Weaver). More sophisticated convergence models were later added by Kincaid and Schramm (Rogers and Kincaid, 1981 and 1984).

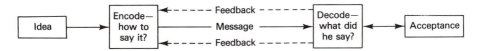

Figure 19.2 Communication feedback model.

System and Process

Communication is frequently described as a process of passing information from one person to another. It is the essential link or bridge of understanding between two people—the *sender* and the *receiver*. A person can initiate the message, but only the receiver can complete the communication cycle by giving sufficient feedback or return signals to prove that the receiver understands the message. The sender may be successful in making people listen, but intensive interchange and feedback may be necessary for the receiver to fully understand the sender's message (Figure 19.2).

The basic communication system consists of an idea that is transformed into a message and sent through a channel of communications to a receiver (Figure 19.3). This process appears simple, but in reality, true communication is difficult to achieve.

Barriers and Interference

Communicators need to recognize and deal with the barriers to communication (Figure 19.4). Three types of barriers in particular distort messages and thus prevent successful communication. These include (1) physical barriers that interfere with accurate communication, including noise/distraction and physical distances; (2) personal barriers caused by social or psychological factors, including value judgments, emotions, desires, and attitudes; and (3) semantic barriers arising from different meanings and uses of words.

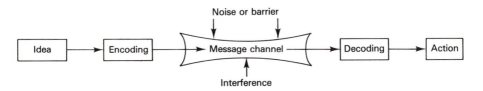

Figure 19.3 Basic communication process.

Figure 19.4 Communication barriers and interference.

ADAPTING THE COMMUNICATIONS MODEL TO COOPERATIVES

Unique Cooperative Needs

All cooperatives have special requirements in their bylaws or state cooperative laws that mandate communications with members. These often include requirements for holding an annual meeting, for notifying members of special meetings and board meetings, and for informing members about significant changes in financial operation or structure. Members are often required to vote on key issues, including mergers and consolidation or major changes in the articles or bylaws.

Because of their unique organization, cooperatives involve a wide range of people in the decision-making processes. Consequently, cooperatives have many unique challenges in communicating effectively. For example, members must be well informed because they own and control the cooperative. They must be prepared to weather economic and competitive pressures in order for their cooperatives to provide them with the benefits they expect from their membership.

Cooperatives must also provide more training to managers than do investor-oriented firms (IOFs). Directors are generally selected from the membership and patrons of the cooperative. Thus many new and inexperienced directors need to be informed and educated about the cooperative business. On the other hand, because cooperative managers are usually hired and are not personally involved in investment or ownership, they also need to be informed about the unique structure and purpose of the cooperative.

In addition, cooperatives frequently must communicate and decide major policy actions in a public environment rather than behind closed doors. Consequently, cooperatives must educate the public in order to avoid opposition that results from misunderstanding about cooperative objectives.

Consider the many cooperatives that have been forced in the past few years to close a plant or elevator. When this occurs, cooperative members, cooperative employees, and the public all need to understand the basic reasons for the operational and policy change by board and management. For example, when Farmland Industries closed beef processing facilities, and when Land O'Lakes sold its beef processing plants to Cargill, it was essential that members and ranchers patronizing the cooperative facilities understood the reasons behind those actions.

To achieve this type of extensive communication, the cooperative communications team formulates a communication plan or strategy to inform members of necessary policy changes or of future plans. Each step and issue can be outlined within the context of a communication system. Every aspect of the informational program should be designed to win acceptance and support of the people involved in the cooperative. Therefore, management should consider the economic and social environment of the cooperative and allow adequate feedback from members and employees (Figure 19.5). Furthermore, cooperatives need to gain the support of members in order to maintain membership participation and thus ensure the volume necessary to capture potential savings envisioned from economies of size.

Increased Emphasis Needed

Presently, dramatic changes in the business environment in which cooperatives operate are creating an even greater need for efficient and effective member education and communication than has existed in the past. It is becoming increasingly difficult for members to comprehend their interaction with aspects of the cooperative.

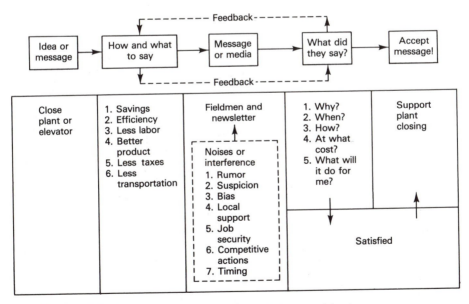

Figure 19.5 Cooperative communications strategy model.

Most cooperatives were relatively small and single-product organizations when they first were organized. It was simple for members to visualize the impact of their decisions on the cooperative and to understand how the cooperative contributed to their economic well-being. However, as cooperatives become larger and more complex, members must know more about their cooperative than they needed to know in the past, in order to understand and fulfill their role. Further, the actions of one member have a smaller impact on the entire cooperative; as a result, members feel more distant.

The increased complexity of finance and redemption plans also make these issues more difficult to understand. Margins are generally smaller than they were in the past, so the benefits are not so obvious. The heterogeneous needs and characteristics of the members in a large cooperative often require a variety of services and pricing policies. This in turn magnifies information requirements and raises concern over fairness. In addition, advanced technology in agricultural production also increases the need for information and makes it difficult for members to visualize benefit.

Finally, greater risks resulting from unstable government policies and domestic and international supply and demand create more stress for members and cooperatives alike. Also, the financial losses that result for producers and cooperatives lead to misunderstanding and confusion. Clearly, cooperatives must emphasize education in the future, even more than they have in the past, to overcome these problems.

Feedback

In addition to educating members, cooperative management also needs to learn from members. Each management function (planning, organizing, directing, staffing, and controlling) requires effective communications and an open channel to all parts of the cooperative. Dynamic cooperative leadership programs are based on clear directives, strategic information flow, and adequate and reliable data based on feedback from members and employees (Figure 19.6). This feedback is an important source of reliable information that can guide management in making tough decisions. For example, feedback information from members is valuable to a livestock marketing cooperative such as the midwestern Central Livestock Association in planning new market services, to Rice Growers Association (RGA) in California planning product manufacturing and processing facilities, or to a rural electric cooperative in trying to assess future power needs. Because member education and member feedback are so interdependent, a new term, *cooperative communications,* is now used to describe all cooperative information and education programs.

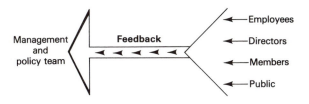

Figure 19.6 Feedback for decision making.

PREREQUISITES FOR SUCCESSFUL COMMUNICATIONS

To create a successful communications program, cooperatives need a board and management committed to a communications program, to good business management, and to excellent products and services. The program itself must be well planned and should include opportunities and continuous evaluation based on the feedback of members, employees, public citizens, many agencies, and other cooperatives (Figure 19.7).

Cooperatives—such as Sunkist, Gold Kist, Harvest States, and Agway—which have effective communications programs have developed skilled teams of communicators, along with strong budget support, to achieve their communication goals. Wisconsin Dairies Cooperative, which has resulted from a combination of many smaller dairy cooperatives in Wisconsin, Minnesota, and Iowa has effectively utilized telephone com-

Figure 19.7 Components of a cooperative communications program.

munications systems which link management through multiple meeting locations with members and provided successful feedback with surveys in its monthly publications. Their skilled communications staff have accelerated member involvement and management support and helped build the cooperative into a Fortune 500 company.

Commitment

Management and the board of directors must be committed to a cooperative communications program if it is to be successful. They must provide an adequate budget as well as moral backing; a well-trained staff; and guidance in planning, control, and evaluation. These programs should operate continually, not just when the cooperative has a problem or a crisis. Further, periodic evaluation should be required for all ongoing programs. Evaluation is one of the most important areas in communication. It should not be neglected, as it has been in the past.

Every cooperative that takes cooperative communications seriously is confronted with decisions on the amount and type of information to give members. In making this decision, management should consider whether benefits from dissemination or the information exceed costs and thus help attain the objectives and goals of the cooperative while keeping within policy guidelines. If the answer is positive, management may develop a program to transmit the information to the members.

Quality Products and Services

Communications programs are not a substitute for poor performance any more than they are for poor management. Cooperatives can develop effective communication programs only when they also develop high-quality products and services that meet member needs and have a value equal to or better than that offered by competitors.

ESTABLISHING A COMMUNICATIONS STRATEGY

Building a Data Base

Before starting a communications program, cooperative management needs to gather information about the cooperative's members, market, competition, and general economic situation. For many years Growmark Cooperative has used a special member survey panel that would give management quick reactions regarding member concerns to critical issues or priority questions. Communications writers have noted that careful internal analysis based on systematic research can provide the basic strategy and foundation for a successful public relations program (Cutlip and Center; Vilstrup and Groves, 1976).

Age distribution of membership

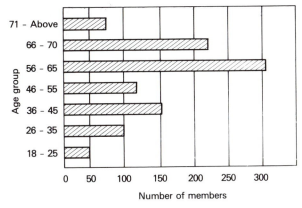

Panel 1

Age distribution of employees

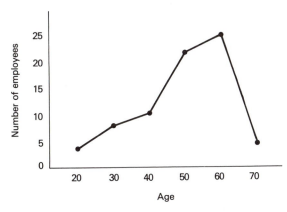

Panel 2

Figure 19.8 Examples of demographic information.

Member Profiles

An important part of the cooperative's data base is the member and employee profile. This profile is a factual analysis of data, including a member's age, income, education, and geographic location. All of this information can be used as a basis for planning member relations and education programs (Figure 19.8). More information on survey research is available in the reference by Hogeland.

Member Knowledge

Cooperative researchers and communicators also need to establish a base level of cooperative knowledge. A 1986 study by the National Rural Electric Cooperatives (Cooper and Secrets Associates) revealed that many

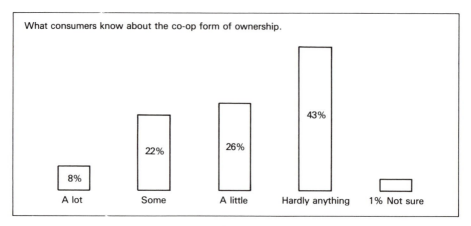

Panel 1

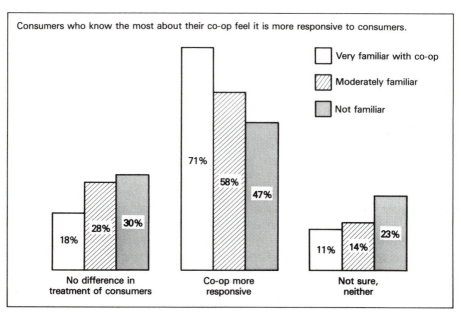

Panel 2

Figure 19.9 Consumer understanding of cooperatives. *Source*: Cooper and secrests Assoc.

consumer members knew little about the ownership of their electric cooperative and demonstrated the need for education and communication with the cooperative (Panel 1, Figure 19.9).

Questions to Consider

In determining what members know about their cooperatives, management can ask the following questions:

1. Do members know and understand the distinguishing features, objectives, goals, policies, and philosophies of their cooperative? Do they know what legislation enables cooperatives to organize and to operate? What restraints are imposed upon cooperatives?

2. Are members familiar with the organizational structure and the operation of their cooperative? Do they know where to get information and where to take their problems? Do they understand how they are represented on the board of directors?

3. Do members understand the cooperative's financial statement? the policy on equity formation and redemption? the member and cooperative tax obligations?

4. Are they familiar with the background and history of their cooperative? with the different member programs offered by the cooperative? Do they know the potential benefits and limits of cooperatives?

5. Are members getting correct information about varieties, expected yields, fertilizer levels, and recommended pesticides? To what extent do members use the cooperative as a source of technical information that is readily available from other sources? Can arrangements be made with other agencies to distribute their technical information in order for the cooperative to reduce distribution costs?

6. What about the general business climate? Do members understand the effects of business trends on the cooperative? Are they aware of general government policies and regulations that could affect their business? Is this information readily available elsewhere, or should the cooperative provide it?

7. As for consumer information, do the members understand how to be good shoppers? Are they aware of stores in which cooperative labeled products are available?

Member Attitudes

In addition, studies of attitudes toward cooperatives can provide valuable information. Informed members feel that the cooperative is more responsive than IOFs to consumer-members (Panel 2, Figure 19.9).

Educational Publications

Finally, developing interesting and imaginative programs requires an intensive search for new material and an innovative interpretation of educational publications. New information and a novel approach serve not only to capture members' attention initially, but to involve them in a review of the basic facts, programs, and policies of the organization.

WHAT TO COMMUNICATE

Once sufficient data have been gathered, management can analyze the data to determine what types of information members need most. In one study leadership groups were given several topics and were asked to indicate which ones they needed for decision making. We have listed the total responses for each group in Table 19.1. The types of information most frequently requested are given in Table 19.2.

TABLE 19.1 **Ranking of most needed cooperative information**

Topic	Young leaders	All members
Cooperative principles	1	1
Taxation	2	3
Bylaws	3	2
Capper-Volstead Act	4	5
Articles of incorporation	5	4
Parliamentary procedure	6	6

Source: Vilstrup and Groves (1981).

Communicators often list all the types and kinds of cooperative, economic, and technical information that would be helpful to members. They then use this list as a basis for measuring the members' information level and determining future communication programs. Educational efforts should be increased in deficient areas; however, priorities must be set because the amount of information management wishes to communicate

TABLE 19.2 **Ranking of financial information needed**

Rank	Financial information
1	Patronage refund
2	Net profit
3	Current liabilities
4	Accounts receivable
5	Member equity
6	Net margins
7	Current Ratio
8	Working capital

Source: Vilstrup and Groves (1981).

will usually be greater than the amount that existing communications staffs and budgets are able to handle and that members want to receive.

We normally think of communicating about positive accomplishments or how the cooperative operates. But basic operating difficulties should not be concealed from members. Members need the facts to understand change and provide support.

METHODS OF COMMUNICATION

The next step in the communication process is to determine who will communicate important information to members and how. There are many ways to communicate with people. One can make personal contact, write messages, and send messages through electronic devises such as radio, TV, audio- and videocassettes, and computers.

The most effective form of communication is one-to-one personal contact. Small-group discussions are slightly less effective. In both cases there is opportunity for instant feedback and reaction. Even in moderate-sized lecture discussions, there is some opportunity for instant feedback. Unfortunately, in most cooperatives personal contact is limited to that between employees and members because the general manager and his staff are often located many miles away.

Media Channels

As cooperatives have grown larger and the distance between members and management has increased, cooperatives have expanded their communication efforts. Media channels available to cooperative communications personnel may be classified as face-to-face, audio/audiovisual, and print channels (Figure 19.10). One of the most effective is the newsletter, which may be a single sheet or a magazine of several pages. Whatever the form of the newsletter, all information should be clearly and concisely presented, should interest the members, and should be sent out regularly. In a number of research studies, members have listed the newsletter as their most important source of information about their cooperative. For example, in the study mentioned earlier, a specific inquiry was directed at young leaders to determine what they felt was their best source of cooperative information (Table 19.3). The emerging leaders indicated that they received cooperative information from the organization through a variety of sources, ranking newsletters (distributed by both regional and local cooperatives), local cooperatives, and directors as the three top choices.

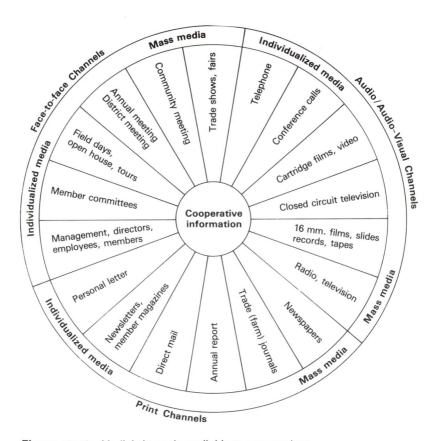

Figure 19.10 Medial channels available to cooperatives.

**TABLE 19.3 Ranking of best sources of
cooperative information**

Rank	Source of information
1	Newsletter
2	Local cooperatives
3	Directors
4	District meetings
5	Newspapers
6	Employees
7	Radio
8	Television

Source: Vilstrup and Groves (1981).

Meetings

One of the best opportunities for the cooperatives to communicate with membership is the annual meeting. A well-planned annual meeting can provide a showcase for cooperative activities for the year. Conducting a successful annual meeting requires planning, notifying members and public citizens, developing an effective agenda, organizing an interesting meeting, selecting a satisfactory location, and providing necessary services.

An integral part of every annual meeting includes the preparation and presentation of the annual report of the cooperative. Annual reports can be simple financial statements listing the current balance sheet and statement of operations; or they can be elaborate accounting reports by the board chairman and management; pictures of new facilities, employees, and directors; and graphic presentations of the past year's business activities. An adequately designed annual report can provide an effective tool for informing the members and the public about the cooperative.

Meetings directed toward local audiences are often more effective than large annual meetings. Consequently, most cooperatives that cover a fairly large area divide their territory geographically, and each division usually has its own annual district meeting. These meetings present a good opportunity for members to meet with management, to give feedback directly to management, and to become informed about their association. For more information on developing effective communications through meetings, see the references on meetings listed at the end of the chapter.

Other Methods of Communication

Sixty-one hundred leaders in 41 states who were also actively farming were also asked what they felt to be the best methods of communicating with the cooperatives (Table 19.4). The most common response was the telephone. It was apparent that members liked the convenience and

TABLE 19.4 Ranking of the best method or place for members to communicate with their cooperative

| | Ranking | |
Method	Husband	Wife
Telephone	1	1
Director	2	2
Employee	3	3
District meeting	4	5
Annual meeting	5	4
Letter	6	6

Source: Vilstrup and Groves (1981).

availability of making phone contacts. Consequently, many cooperatives are improving this channel of communication with in-watts lines, offering 24-hour service with telephone recorders during hours when the organization is closed.

A few cooperatives also are effectively using TV, VCRs, and movies in their communications programs. Several large regional organizations have video-recorded part of their annual meetings for later presentations to members who could not attend. Recently, in fact, Southern States Cooperative transmitted its annual meeting on closed-circuit TV by satellite to different geographic areas and involved members directly through an interaction conference broadcast (Duffey).

Similarly, some cooperatives have made good use of mass media—radio, TV, and newspapers—in their communications programs. Speed and relatively low cost per exposure are prime advantages of mass media. The media can (1) keep the public informed about the cooperative, (2) keep members aware of what the cooperative is doing, and (3) build up the interest of nonmembers.

Employees can be one of the most important groups in any effective cooperative communications program. In most cooperatives, employees often make the greatest impression on members. This is true because the member usually has frequent person-to-person contact with one or more employees. In fact, it might be the checkout person, the person who pumps the gas, the clerk in the credit union, the milk hauler, the field representative, the tank truck driver, or the general manager—who is the cooperative in the eyes of the individual member.

Consequently, educating employees with cooperative as well as technical information cannot be overemphasized. Every organization needs good employees, but cooperatives need especially good employees who understand cooperative principles and practices, possess adequate knowledge about products and services, and have the ability to transmit this information to members.

Cooperative Image

One of the most important reasons for having a communication program is to maintain a favorable cooperative image. Unfortunately, cooperatives have not always had a good image in the eyes of the public. Some of this has been due to deliberately misleading information given to the public by those who are against cooperatives. Some has resulted from cooperatives failing to keep the public informed.

WHEN TO COMMUNICATE

The board, management, and communications staff are caught between opposite positions in disseminating information to members. On the one hand, members need current information about the cooperative and impending policy decisions. On the other hand, information that could help competitors should remain with the board and management until final decisions are made and are ready for membership discussion and approval.

Although each cooperative has to decide when to release information, two guidelines may help:

1. If the information will help competitors, relay it only to the board, the management, and the key staff (unless critically needed for membership decision making).
2. The more controversial an issue is, the greater the need for the cooperative to provide factual and reliable information to members. Otherwise, the cooperative can develop a negative public image due to misunderstanding about the cooperative's objectives.

Adding a new product or changing store hours may require only a routine announcement in the cooperative's newsletter. However, a proposed merger, a closing of existing facilities, or a change in the way members are represented may require that cooperatives communicate much additional information to members before the issue is resolved. To minimize controversy and to ensure a smooth transition, those affected by the change need facts to dispel rumors, some of them started by competing firms, and fears. Cooperatives also must provide members enough information to make intelligent decisions.

Recently, previously competing cooperatives such as AMPI and Morning Glory or Land O'Lakes and MidAmerica Dairies have had to convince their members of the advantages of combining many of their manufacturing operations for efficiency and improved marketing programs. On the other hand, Farmland Industries has had to help their members in Farmarco understand the need to sell facilities to Union Equity, a successful wheat marketing cooperative with worldwide outlets.

BALANCE AND PAYOFF

Money spent on well-planned and carefully evaluated member communications programs is a good investment. Good will and understanding can be stored in membership as money is deposited in a bank, and can be drawn on in time of crises when understanding and support are needed.

However, a continuing problem for cooperatives is the balance between adequate education programs and the demand for funds elsewhere in the organization. When margins are adequate and business is growing, the problem is minimal, but when times are tough and a cooperative is retrenching and readjusting, the cooperative often sacrifices education funds. Unfortunately, this is the very time it should expand education programs. New and young members must continually be told why the cooperative is there, what conditions were like before the cooperative was established, and what conditions would be like if the cooperative were eliminated.

Because of the unique need for education programs in cooperatives, these businesses are at a disadvantage when competing with IOFs. The owners of the IOF know why they have invested—to maximize the return on their investment. On the other hand, to the owners (members) of a cooperative, the benefits are often more obscure and must therefore be reiterated. As we noted above, cooperatives can bank the benefits of a good education program and use them to help the business over inevitable rough spots.

Sixty national cooperative leaders identified the following characteristics of a well-informed member. They said that a member who understands the organization's structure, policies, and actions generally will remain more loyal, have fewer complaints, and take a greater interest in the cooperative. They will continue to patronize the cooperative when the going is rough, and they will offer more constructive criticism and suggestions. They will inform their neighbors about the organization in terms they understand and serve as effective salespersons for the organization, informing the community of the cooperative's contribution to the local economy. They can help promote new products and services, and they are easier to do business with. They will meet their obligations and pay their bills to the cooperative. Educated members will help stop rumors; defend the cooperative; and develop a favorable climate of understanding between members, employees, and directors. They will promote a progressive attitude and build member confidence and pride in the cooperative and its management (Vilstrup and Groves, 1976).

OTHER EDUCATION PROGRAMS

To this point we have discussed mainly programs conducted by cooperative businesses on their own. However, cooperatives also contribute to and are supported by other organizations that provide education and training. Some examples include the following:

1. *State cooperative councils.* While the primary mission of most state cooperative councils is to represent their members before legislative and rule making bodies, many also have active education programs. Often these programs are carried out in cooperation with the state extension service and/or the farm credit district.

2. *Land-grant universities and 1890 universities.* Several state university systems have active programs for cooperatives. These include teaching cooperative

courses as part of the resident instruction program, conducting research about cooperative businesses, and providing education and training, usually through the Cooperative Extension Service. Some universities have formally organized their cooperative education programs. Examples include the Graduate Institute for Cooperative Leadership at the University of Missouri, the Arthur Capper Cooperative Center at Kansas State University, the University Center for Cooperatives at the University of Wisconsin-Madison, and the California Center for Cooperatives at University of California-Davis.

Federated regional cooperatives often conduct education programs for their member cooperatives. *Farm Credit districts* frequently conduct education programs for their borrowers. These programs are often cosponsored with state councils and/or with state extension services. The role of the *national cooperative organizations* is described in Chapter 6.

SUMMARY

The unique structure and ownership of the cooperative requires effective communication with member owners and users of the cooperative. Communicating is a major responsibility of both the cooperative board and management.

The changing business environment and growth of cooperative has accelerated the need for member information. Successful cooperatives hire skilled staff and provide adequate budgets to ensure the adequate flow of business and organizational information.

Cooperatives need to use modern communications technology and methods to compete successfully. The availability of video, satellite, VCRs, teleconferencing, and computers provide new opportunities for cooperative communicators as they combine them with successful conventional communication techniques.

Skilled cooperative leaders will need to research the needs and concerns of cooperative members in planning organizational strategy. Winning commitment from members in mergers, consolidations, and acquisitions requires high levels of member understanding. Communicating with new members, potential members, government agencies, financial and business institutions, and the public will continue to be a major challenge for board and management.

DISCUSSION QUESTIONS

19-1. How would you determine the member concerns and needs in the cooperative?

19-2.   Why are the communications programs in a cooperative more difficult than in an IOF organization?

19-3.   Who has the major responsibility for member communications in a cooperative?

19-4.   Why do members rate cooperative publications higher than meetings as a method of getting information from their organization?

19-5.   Does a cooperative need to budget more money for member communication during good times or difficult times?

19-6.   How important is a good public, member, or business image to a cooperative business?

19-7.   What type of member information is needed if a cooperative plans to merge, consolidate, or start a joint venture?

19-8.   Is the responsibility of the communications director different in a cooperative than in an IOF?

19-9.   When may the flow of information to members need to be delayed until the board has completed their deliberations?

19-10.   How would you evaluate the success of a member relations program?

REFERENCES

COBIA, DAVID W., AND LUIS A. NAVARRO, *How Members Feel about Cooperatives.* Dept. of Agr. Econ. Res. Rep. 86, North Dakota State Univ., 1972.

COOPER AND SECRESTS ASSOCIATES, *A Profile America's Rural Electric System's Manager and Directors and Consumer-Members.* Washington, DC: National Rural Electric Cooperative Association, Oct. 1985.

CUTLIP, SCOTT M., AND ALLEN H. CENTER, *Effective Public Relations,* 5th ed. Englewood Cliffs, NJ: Prentice Hall, 1978.

DUFFEY, PATRICK, "Satellite Links Southern States Delegates at 10 Outlying Sites," *Farm. Coop.* (Jan. 1986):4.

GROVES, FRANK, AND DICK VILSTRUP, *Cooperative Ideas That Work.* Univ. Center for Cooperatives, Handbook UW 3, Univ. of Wisconsin, 1976.

HOGELAND, JULIE A., *A Guide to Survey Research for Local Cooperative Management.* Washington, DC: USDA ESCS ACS CIR 24, 1980.

HOLLAND, GARY. *Running a Business Meeting.* New York: Dell Publishing, 1984.

INTERNATIONAL COOPERATIVE ALLIANCE, Minutes of committee report on Rochdale Principles 1966 Version, Geneva, Switzerland, 1966.

KIRKMAN, C. H., *Mr. Chairman,* Washington, DC: USDA FCS Info. 6, Rev. Oct. 1976.

ROGERS, EVERETT M., AND D. LAWRENCE KINCAID, *Communications Networks.* New York: Free Press, 1981.

ROGERS, EVERETT M., AND D. LAWRENCE KINCAID, *The Convergence Model of Communications and Network Analysis, Communications Networks.* New York: Free Press, 1984.

SCHAARS, MARVIN A., *Cooperatives, Principles and Practices*. Univ. Center for Cooperatives A1457, Univ. of Wisconsin–Madison, 1971.

SHANNON, CLAUDE, AND WARREN WEAVER, *The Mathematical Theory of Communications*. Urbana, IL: Univ. of Illinois Press, 1949.

VILSTRUP, DICK, AND FRANK GROVES, *Cooperative Communications Techniques*, Handbook UW 2, Univ. of Wisconsin–Madison, Reprint 1976.

VILSTRUP, DICK, AND FRANK GROVES, *Cooperative Leadership Dimensions–Research-Guidelines*, Handbook UW 2, Univ. of Wisconsin–Madison UW 5, 1981.

Part VII
Structural Dynamics

20

*Structural
Dynamics*

Richard H. Vilstrup,
University of Wisconsin-Madison

David W. Cobia,
North Dakota State University

Gene Ingalsbe,
Agricultural Cooperative Service, USDA

CHANGING ENVIRONMENT

Dynamic changes in the economic environment often generate the need for significant changes in cooperative structure. New technology can alter economies of size, substantially improve transportation, or otherwise drastically change operations, thus completely reshaping industry cost structure.

The dairy industry characterizes the kind of change possible. After World War II, perishability was substantially reduced, and the market area increased several times by the transition from can to bulk handling and from small delivery truck to semitrailer distribution. This in turn reduced by nine-tenths the delivery costs for packaged fluid milk. These innovations pushed cooperatives into capital-intensive supply balancing and made manufacturing activities possible only through large-scale organizations. Consequently, a period of rapid consolidation began and continues at a slower pace today. The decline in the number of dairy cooperatives nationally from 1,930 in 1950 to 385 in 1986 is illustrative of the dramatic restructing that has taken place.

Farm productivity, government policy, and a variety of market forces have also precipitated structural change. Examples include rail deregulation; new farm programs; shifts in the degree of antitrust enforcement; wide swings in demand and supply, characterized by exploding export demand in the mid 1970s and sudden retrenchment in the first half of the 1980s; changing tastes and preferences in domestic markets; technological changes in farm production, creating the need for new services and the end to old ones; uneven productivity changes on the farm, such as increases in the feed conversion rates for poultry exceeding that of beef; sudden withdrawal by major agribusiness firms [e.g., Gulf Oil (geographic) and U&I Sugar (a Northwestern sugar beet processor's total withdrawal)]; bankruptcy or liquidation creating competitive vacuums; and entry by domestic and foreign companies, creating a more vigorous competitive environment.

All of these changes cause instability and uncertainty, wrenching old institutions and relationships out of their traditional framework. All change is disruptive; in fact, often the older structure must be destroyed before the new structure can be developed. The impact of change, however, will vary. Some sectors will be continually plagued by forces that disrupt operations, while in others conditions will remain relatively stable.

Generally, though, the rate of change is increasing. Thus cooperatives must develop methods to profit from changes by adapting more quickly than they have in the past or by directing the kinds of changes most beneficial to the cooperative. Growth brings problems, but decline may be more difficult to confront. Therefore, cooperatives can best serve their members by choosing the most beneficial methods of and adjustments to growth.

ADJUSTMENT ALTERNATIVES

Agricultural producers may respond to dynamic forces by creating a new cooperative, by expanding or liquidating an existing one, or by creating alliances with other cooperatives or investor–oriented firms (IOFs). Cooperatives can expand or change the mix of their services by internal or external methods of growth. For example, a cooperative may construct new facilities (internal growth) or acquire an ongoing operation (external growth). (Do not confuse these terms with internal and external financing.) Many cooperatives also have successfully combined internal and external growth. Generally, either type of growth or expansion will change the financial and control structure of the cooperative. The appropriate choice of alternatives depends on competition, initial resources, location, and potential savings.

Taking advantage of some opportunities may require the creation of new cooperatives. But more often than not, the economic need served for generations by many cooperatives has largely evaporated. These cooperatives must then liquidate or merge to conserve their members' equity, especially if they have insufficient capacity, volume, or finances to compete.

In other cases, as market areas expand because of lower transportation rates and greater economies of size, cooperatives find that they are no longer isolated. Their market areas overlap with other cooperatives. They are competing for the same business with other cooperatives and there is unnecessary duplication of equipment and effort.

TABLE 20.1   Number of farmer cooperatives removed from and
added to the ACS list, 1970-1985[a]

| Year | Reasons for removal | | | | | Additions |
	Out of business	Merger or consoli- dation	Acqui- sition	Other	Total	
1970	224	144	108	48	524	567
1971	137	60	39	17	253	458
1972	139	48	41	22	250	52
1973	70	26	12	28	137	194
1974	56	27	24	53	159	60
1975	79	16	13	22	130	20
1976	31	22	18	48	119	9
1977	64	28	24	714	830	31
1978	196	51	39	131	417	281
1979	132	16	19	90	257	102
1980	71	17	17	98	203	51
1981	68	43	8	86	205	123
1982	51	65	36	77	229	143
1983	86	57	17	54	214	78
1984	59	49	53	56	217	10
1985	86	41	38	53	218	61
1986	139	57	61	23	280	24
Total	1,688	767	567	1,620	4,642	2,264

Source: ACS USDA
[a]The actual change may have occurred a year or more earlier. The year-to-year variation is partially due to the reporting mechanism.

These conditions have accelerated the number of consolidations, mergers, acquisitions, liquidations, and joint ventures (Table 20.1). During the recent surge of cooperative reorganization, many innovative and pragmatic restructuring plans have been used to retain cooperative service and membership in rural communities (see Table 20.2 for specific examples). Even though membership and cooperative numbers have been declining, cooperatives have been selling more supplies to farmers and marketing more of their products (see Chapter 3).

TABLE 20.2   Selected structural changes made by cooperatives

Cooperative	Action	With IOF	Year
Midland Cooperatives	Merged with Land O'Lakes		1982
Farmland Industries	Sold 3 insurance companies to Nationwide Insurance		1982
California Canners & Growers	Went bankrupt; assets bought by Tri/Valley Growers		1983
North Pacific Grain Growers	Merged with GTA to become Harvest States Cooperatives		1983
Ag Processing, Inc.	Formed from merging soybean processing facilities of Land O'Lakes, Farmland, & Boone Valley Co-op. Proc.		1983
Landmark	Merged with Ohio Farmers to become Countrymark		1985
Midwest Breeders Cooperative	Merged with Minnesota Valley Breeders to form 21st Century Genetics Cooperative		1985
MFA Oil	Bought 47 retail gasoline outlets from Midstate Oil	X	1985
Farmland Industries	Sold Far-Mar-Co wheat and grain sorghum facilities to Union Equity Co-op Exchange		1985
Northeast Dairy Co-op Federation (NEDCO)	Went bankrupt		1985
GROWMARK	Formed ADM/GROWMARK joint venture with ADM	X	1985
Agri Industries	Formed Agri Grain Marketing joint venture with Cargill	X	1986
Capitol Milk Producers	Sold assets to Southland Corporation	X	1986
FCX, Inc.	Went bankrupt; assets purchased by Southern States and Gold Kist		1986
Producers Rice Mill	Bought Pioneer Foods Division of Pillsbury	X	1986
Gold Kist	Formed IOF subsidiary Golden Poultry, Inc.	X	1986
CENEX & Land O'Lakes	Established 3 joint ventures in farm supplies		1987
GROWMARK	Purchased 32 retail fertilizer outlets from IMC	X	1987
Plains Cotton Cooperative	Purchased denim plant from American Cotton Growers		1987

GROWTH AND ADJUSTING TO CHANGE

Benefits and Reasons for Growth

Growth is usually considered a sign of a healthy, successful business—and often for good reasons. Most advantages to growth are associated with economies of size. For example, new technology is frequently linked with higher fixed costs and requiring larger volume. Pecuniary economies, such as quantity discounts, especially in transportation and advertising, are often substantial. Other benefits of economies of size come from the ability to spread fixed costs from research; from qualified management; and from special training programs available for management, employees, and members over a larger number of units. In addition, size often permits a cooperative to attract and keep qualified and motivated management and employees. (Of course, large businesses also *need* better management simply because they have more human and material resources to coordinate.)

Other incentives to grow include investing idle cash as well as to achieve marketing and bargaining power, political power, legislative influence, and financial strength. A growing organization is frequently healthy and presents an image of achievement that is almost universally admired.

Disadvantages of Size

Cooperatives need to expand to provide new market outlets and accumulate more volume for efficiency. But as cooperatives grow, the cooperative's service, location, representation, leadership, image, and logo frequently change. Members often feel that the cooperative is more impersonal, distant, and unresponsive. Members find that they are in a giant cooperative spread across several states and serving multiple markets. Although many members realize large cooperatives need to move in new directions, members also miss the personal relationships they had in the past with staff, management, and leadership. Some are suspicious of bigness and change, because the communication system has grown too complex for them to understand (see chap.19).

Many cooperatives more than 50 years old have outstanding records of achievement and service. These cooperatives were frequently started by a handful of members in local town halls, schoolhouses, and churches—with limited funds and maximum member involvement. The cooperative was next door, and its leadership resided in the community. Members generally voted directly on issues rather than through an intricate delegate structure. Members saw the management, and employees frequently were their personal friends. The lines of communication were short and the opportunity for involvement was readily available.

Now, however, many small, local cooperatives have had to merge with other cooperatives to compete successfully and satisfy member needs. The resulting large regional cooperative frequently requires an even stronger member commitment but offers less opportunity for direct member involvement. Because a large cooperative often serves a large geographic area, individual contact between managers and members is limited.

Larger organizations have an intricate system of districts, delegates, and directors. Board members often represent entire districts, states, and regions and need a local delegate structure, as well as an adequately informed field staff, to assist in member communications of board policies.

In addition, multiple levels of management and staff may inhibit the free upward flow of member feedback in large organizations. The repeated transmission of messages and directives downward, through an involved organizational structure, can result in faulty interpretation, misunderstanding, and dangerous omission of essential facts and information.

Large growth cooperatives may serve members directly through member associations or a combination of membership plans. However, direct contact with producers may still be limited to massmedia, written communication, or invitations to local annual meetings. Finally, members of regional cooperatives often have less in common with each other than do members of local cooperatives. Regional cooperatives often serve members who have widely varying interests in commodities and who, therefore, may compete against each other in the marketplace. Most members have vested interests in the location of facilities and processing plants, but this is the extent of their common interests. The following list summarizes the disadvantages regarding membership in a growing cooperative: (1) less direct member contact, (2) fear of reduced sensitivity to member needs, (3) less local identification, (4) frequently less member involvement, (5) more complicated voting structure, and (6) extra layers of management. Despite these disadvantages, most cooperatives must expand in size to capture economies associated with size.

Direction of Growth

The *direction of growth* is the relationship of the cooperative's original line of business to the expanded activity. A cooperative may direct its business line in any one or more of three directions: horizontally, vertically, or conglomerately.

Horizontal Horizontal growth is the expansion of an existing line of business, that is, the combining of like activities or functions. For example, an elevator firm may purchase a neighboring elevator, or a farm supply cooperative may expand its fertilizer mixing capacity. The major advantage for a cooperative that confines itself to horizontal growth is that member needs will be more uniform, and management can specialize in one endeavor. The cooperative can more easily achieve economies of size.

Vertical Vertical growth occurs when businesses enter into successive vertical stages of business, extending into new stages of producing or marketing the same product (Figure 8.1). Vertical growth is further classified as forward (or downstream) and backward (or upstream). Forward vertical growth results from extending production or services closer to the ultimate consumer (e.g., an apple-packing cooperative building a canning facility). Backward vertical growth results from gaining control over sources of supply (e.g., a federated supply cooperative manufacturing fertilizer). Farmers using the marketing cooperative is forward integration, and using the supply cooperative is backward integration.

Through vertical integration in their cooperatives, farmers gain or are guaranteed access to supplies of inputs (e.g., fuel and fertilizer) and to markets for outputs. In addition, cooperatives can coordinate quantity and quality more readily, thus reducing costs and risks. Capturing monopolistic profits of IOFs operating in this stage may be an objective. (See Chapter 8 for further explanation of these issues.)

Farmers must remember, however, that integrating vertically to eliminate the middle man in the marketing channel (and to claim the advantages listed above) does not eliminate the need for functions such as transportation and processing nor the associated capital requirements and risks. Further, a cooperative that integrates vertically may need to develop sophisticated management and marketing skills, as well as to increase investment. Another drawback is the possible elimination of pricing points.

Conglomerate Conglomerate growth occurs when businesses add activities unrelated to existing lines of business. For example, in addition to milling durum wheat, Domain Industries in the Midwest produced turkeys and manufactured feed, electronic equipment, plastic toys, and packaging equipment. Clearly, some product lines are more closely related than others. However, all are unrelated enough that it would be misleading to refer to this kind of expansion as horizontal growth.

Cooperatives may become conglomerate but to a limited degree. Diversification is congeneric in that products handled and services rendered are related to needs of agricultural producers. For example, Farmland Industries manufactures feed, sells and services computers, manufactures car batteries, formulates pesticides, constructs steel buildings, and processes a variety of pork products. Agway, Gold Kist, Southern States, and others have become conglomerate in a similar way.

Sometimes conglomerate growth is hard to avoid. A cooperative may almost by accident pick up an unrelated business in an acquisition. For example, Agri-Industries acquired Pickett Brewery as part of an acquisition of grain river-terminal property.

Cooperatives can use diversification to stabilize seasonal patterns, to spread risks, to provide one-stop shopping, to achieve economies in record keeping, to share common fixed costs such as management and warehousing, and to avoid the cost of establishing several cooperatives in the same area.

On the other hand, diversification can still be costly for cooperatives. The more diverse and wider variety of services and products handled, the more diverse—that is, less homogeneous—will be the members. Consequently, the cooperative will have more difficulty achieving business at cost and harmonizing the sometimes conflicting needs of members.

Member Education

Restructuring often has painful implications for members, employees, and the community. Cooperative leaders contemplating a structural change need to design an education program carefully to make the change a positive transition. Member support will be based on adequate understanding and sound information. Ken Duft highlighted this issue by writing:

> Members are sometimes led to expect too much from the newly formed organization. ... the member should not be led to believe that cooperatives can solve all problems, right all evils, or adjudicate all grievances. They must view their cooperative as part of a free and competitive capitalistic economic system and not as a reform agency, a welfare institution, or a charitable society. They must assess the merger and the resultant new cooperative within a realistic criteria where efficiency of operations and service to patrons become the important factors (Duft, p. 4).

Members generally vote against reorganization proposals that they do not understand. Successful cooperatives have used letters, meetings, personal contact, and the media to assist in the informational and educational program.

SUMMARY

Our economy is in a seemingly unending spiral of change. Technology, government policy, world trade, peace and war, markets, and farms all contribute to powerful forces that propel all businesses, sometimes at breathtaking speed and at other times at a ponderous pace, toward changing structure (size, function, product line, and linkages within and outside the organization). A cooperative can provide better services at lower cost by growing to achieve economies of size. The cooperative may grow horizontally by expanding its major line of business, vertically backward toward sources of supply, forward to the ultimate consumer, or conglomerately by adding somewhat unrelated activities. Growth has its disadvantages, especially for cooperatives. Most of these disadvantages relate to how members perceive and relate to their cooperative. More specifically, growth results in more complicated and less direct voting and communication and in loss of local identification.

Farmers can respond to environmental changes by creating new cooperatives, by promoting growth of existing cooperatives, or even by liquidating. They can forge new relationships with other businesses or reorganize the cooperative internally. These options are discussed further in the next two chapters, emphasizing unique aspects as they relate to cooperatives, briefly mentioning a few procedural steps to start a cooperative and some elements that contribute to its success.

DISCUSSION QUESTIONS

20-1.　Identify changes taking place at the farm level regarding such factors as size distribution of farms and technological capabilities. Identify the probably effects of these changes on cooperatives.

20-2.　Contrast the impact of horizontal with vertical growth for cooperatives.

20-3.　What are the benefits and costs of growth for cooperatives?

20-4.　How can unrealistic expectations of members lead to failure? What can be done to avoid these unrealistic expectations?

REFERENCES

DUFT, KENNETH D., "Cooperatives Combinations and Firm Growth," *Agribusiness Management*. Pullman, WA: Washington State Univ., May 1980.
FRENCH, CHARLES E., JOHN C. MOORE, CHARLES A. KRAENZLE, AND KENNETH F. HARLING, *Survival Strategies for Agricultural Cooperatives*, Ames, IA: Iowa State University Press, 1980.

21

Starting
a Cooperative

Gene Ingalsbe,
Agricultural Cooperative Service, USDA

INTRODUCTION

Too many cooperatives may exist in some areas, commodities, or functions, calling for consolidation. In other areas, it may be necessary to create a cooperative to capitalize on a new or different opportunity. As cotton and tobacco production shifts out of the Southeast, many farmers are turning to fruit and vegetable production and, consequently, forming new cooperatives to market their products. In other parts of the country, farmers have considered forming cooperatives to handle products and services as diverse as mushrooms, dairy goats, orchids, wine grapes, wild rice, fish, greenhouse plants, and recreation. In addition, opportunities of another kind arise when an existing investor-oriented firm (IOF) withdraws from the market and leaves a vacuum.

The following situations characterize the circumstances that have led to the creation of cooperatives:

1. In 1876, an Iowa company bought all the patents on barbed wire and immediately raised the price 40%. Farmers responded by forming the Iowa Farmers'

Protective Association and began making their own barbed wire. Their action saved the state's farmers between $5 million and $6 million in one year.

2. In the early 1970s, owners abruptly closed their California canning plant because they were dissatisfied with returns on their investment. Growers supplying the plant lost their market for fruits and vegetables, which jeopardized their entire farming investment. To regain their link with consumers, the growers bought the plant through the formation of Pacific Coast Producers in 1971 and continue to operate the processing cooperative successfully.

3. Minnesota corn farmers attempting to increase their income committed the price of a pickup truck for a membership share in a cooperative (Minnesota Corn Processors), that would enable them to market corn as starch and syrup. Their corn wet-milling plant began operating in 1983.

4. A change in government commodity-support programs prompted peanut growers to accept more risk to gain greater income benefits. They formed the Virginia-Carolina Peanut Farmers Cooperative Association in 1985 to shell and market their peanuts.

INITIAL CONSIDERATIONS

Elements of Success

Starting a cooperative is fraught with difficulties, such as finding sources of financing and qualified management. Consequently, the failure rate of new businesses is high. The following factors must be present in order for a cooperative to succeed in contributing to the well being of its members: economic need, financial feasibility, adequate volume and member support, qualified management, adequate financing, and membership prepared for the rigors of competition. Absence of any one of these can result in failure. Therefore, in the organizing process, those attempting to start the cooperative should determine the extent to which these elements are present.

Steps in Forming a Cooperative

Organizing a cooperative is a complex and painstaking undertaking. Leaders must demonstrate a combination of leadership, expertise, enthusiasm, practicality, dedication, and determination to see the project completed. Unfortunately, cooperative organizing efforts often get started with inadequate information and planning, with little or no economic analysis, and with insufficient producer support. Consequences are false starts, delays, and all too often, eventual failure.

Creating a cooperative is a systematic, time-consuming process. As a rule of thumb, the time necessary to form a cooperative from idea to open

STEPS TO FORM A COOPERATIVE

1. Discuss with small groups of producers the perceived economic need that formation of a cooperative might fulfill.

2. Hold an exploratory producer meeting. Vote whether to continue. If affirmative, select a steering committee.

3. Conduct an investigation to determine feasibility of cooperative, if so work out details of proposed cooperative. This step may require several intermediate steps such as:

   a. Conduct a producer survey as a basis for determining cooperative feasibility.
   b. Hold second general meeting to discuss results of producer survey. Vote whether to proceed.
   c. Conduct a market, or supply, and cost analysis.
   d. Hold third general meeting to discuss the results of the market, or supply, and cost analysis. Vote whether to proceed, this time by secret ballot.
   e. Conduct a financial analysis and develop a business plan.

4. Hold general meeting to hear results of the feasibility study. Vote again whether to proceed. If affirmative, vote a second time on whether the steering committee should remain intact or changes should be made.

5. Draw up necessary legal papers and incorporate.

6. Call a meeting of charter members to adopt the bylaws. It is a good idea to invite all potential membership to ratify the bylaws. Elect a board of directors.

7. Call the first meeting of the board of directors and elect officers. Assign responsibilitied to implement the business plan.

8. Conduct a membership drive.

9. Acquire capital, including developing a loan application package.

10. Hire the manager.

11. Acquire facilities.

12. Start up operations.

Figure 21.1 Sequence of events to form a cooperative.

house is one to two years. Figure 21.1, based on field experience, summarizes a suggested sequence of events to follow to create a successful cooperative.

Individuals organizing a cooperative should also consult bulletins such as those of Ingalsbe and Goff to follow details of the steps in Figure 21.1 and thus avoid costly delays, violation of statutes, and unnecessary adversary relationships. However, even taking these measures will not guarantee success. Organizers will usually need to consult with special advisers to deal with unique problems.

Leadership and Advisers

A compelling need and a few leaders can spark the idea of forming a cooperative. Usually, organizers are producers who recognize a common economic need they believe a cooperative can fulfill. Responsibility for creating a cooperative rests with this leadership group. They should begin by discussing their idea at one or more small group meetings with other key producers. If they conclude these key producers and others support their idea, they should next seek the advice of someone familiar with cooperatives.

Specialized help is necessary throughout the various stages of forming a cooperative. Leaders need someone familiar with the process to work with them step by step concerning legal, economic, and financial aspects. Specialists with an understanding of cooperatives are usually available; for example, land grant universities usually have an extension economist. Other sources of help include a state department of agriculture, a state cooperative council, a Bank for Cooperatives, or an established cooperative. Agricultural Cooperative Service, U.S. Department of Agriculture, has a small staff whose sole purpose is to help develop agricultural and rural cooperatives by conducting feasibility studies, by providing educational services, and by helping with implementation. Other agencies, such as the Small Business Administration and state and regional development agencies that provide assistance to business in general may also be helpful.

Cooperative organizers need attorneys familiar with state cooperative statutes to draw up or check the organization papers, to acquire property before starting, to make capitalization plans, to borrow money, and to write agreements and contracts. Legal counsel will be needed on a continuing basis after the cooperative is operating, to ensure that it is conforming to laws applied to businesses. References for such an attorney can be obtained through the contacts mentioned in the preceding paragraph.

In addition, organizers should seek early financial counsel from a financial institution to anticipate capital needs and methods of financing. These institutions can help design the feasibility study to meet requirements of a lending agent, and they have staff personnel who are specialists in finance and accounting matters. Cooperatives should also employ an independent accounting firm (with cooperative accounting experience, if possible) prior to selling any stock or collecting or handling members' money. The cooperative needs an accountant to establish the bookkeeping system, the tax records, and an equity redemption plan, as well as an outside accounting firm hired by the board for the annual audit.

Finally, technical advice may be necessary, depending on the type of cooperative organized. For example, advice from someone with experience in marketing livestock would be important for farmers organizing a livestock marketing cooperative.

Steering Committee Function

The steering committee, with the help of one or more advisers, has a two-part responsibility. First, it judges whether the key elements of success are present. Second, if the proposal passes this test, the committee prepares a specific, detailed business plan for the new cooperative.

Economic need or justification (see Chapter 8) is fundamental to the formation and successful operation of a cooperative. To determine need, the committee should identify what service functions the cooperative could provide, whether they would be available from other sources, and whether the cooperative could offer lower costs and better quality than those other sources. It is usually best for a new cooperative initially to provide one or a few services not requiring elaborate or costly facilities. Then the cooperative can expand after a successful start. The committee should also consider intangible functions such as preserving a market, stabilizing prices, or encouraging more orderly marketing.

The committee should carefully consider alternatives to starting a cooperative and determine whether similar services could be provided through membership in a cooperative nearest the area, either directly or by the establishment of a local branch. Even if forming a new cooperative is the best alternative, the committee should assess the advantages of either establishing membership links with regional cooperatives that could extend marketing, purchasing, and service benefits, or encouraging IOFs to provide necessary services, thus avoiding added finance and leadership burdens.

CRITICAL ISSUES

Economic Feasibility

The steering committee and adviser must next identify the requirements for the proposed cooperative's markets or supply sources. Alternative methods of gathering this information include any one or a combination of the following:

1. Use industry research publications and trade association reports.
2. Develop and conduct a market or supply survey. Although the adviser should be primarily responsible for the survey, this phase should be a joint effort. Committee members should contact potential buyers or suppliers to determine their needs and requirements.

3. Request state and/or federal agricultural offices (such as the Cooperative Extension Service or Agricultural Cooperative Service), universities, commodity organizations, or private consulting firms to do the research for you.

4. Analyze the results of the market survey in relation to the findings of the producer survey. This process is the adviser's responsibility and may determine the scope of the cooperative's activities. Then the adviser or steering committee members should contact engineers, equipment dealers, real estate agents, and others who can provide ballpark estimates of both the costs to establish the cooperative's physical facilities and operating costs based on the probable range of activities.

If producer support is questionable at this stage—or any stage thereafter—the committee should require producers to make a token interest investment and to sign a premembership agreement. Producer investment should be in proportion to intended use of the cooperative, but at this time it should be a minimal amount, such as 10% of potential equity needs. The committee should meet this goal before continuing organizational efforts.

The emerging picture of the size and scope of the cooperative permits the development of basic operating assumptions regarding facilities needed, operating costs, capitalization, and financial requirements. The steering committee may have to contract with engineering firms or equipment dealers in order to get estimates of facility, equipment, and labor costs and then, in turn, to estimate operating costs realistically.

If facilities needed include land, buildings, and equipment, the committee bases estimates of these needs on the expected business volume by the probable members, along with some allowance for future expansion. The committee should evaluate the cost of buying or leasing an existing plant, the cost of building a new plant, and the cost of new versus used equipment. In addition, the committee should seek the advice of engineers or similarly skilled technicians both in determining the need for new facilities and in assessing the value of existing facilities being considered.

Operating costs include salaries of the manager and other employees; costs of utilities, taxes, depreciation, and interest; and costs of office and other supplies needed. Cooperatives marketing bulk commodities such as grain need only determine whether they can lower costs to meet competition. On the other hand, cooperatives marketing and processing nonbulk commodities must also estimate their market potential. Unfortunately, many of these cooperatives (such as International Cooperative, Inc., a defunct potato-processing cooperative at Grand Forks, North Dakota) have tragically overestimated their potential. Therefore, steering committees must carefully assess the probability of penetrating markets given the anticipated resources. If the operating revenues for the estimated volume of business are not much higher than estimated costs, the committee may want to estimate the volume needed to produce acceptable margins. In most businesses, per-unit operating costs tend to decline as the volume increases. For a cooperative to have the lowest possible operating costs, its members must furnish the maximum amount of business the cooperative can handle.

Producer Survey

Potential membership and volume are best estimated by formal survey techniques (see Hogeland for information about these techniques). An adviser usually drafts the producer survey questionnaire for the steering committee to review and make comments. Information collected should include estimates of potential volume, based on the volume of supplies and/or products marketed during the most recent or typical year. Surveyors should collect information about characteristics of producers, such as size, enterprise combinations, location, familiarity with and use of cooperatives, and willingness to join, finance, and use a cooperative.

Estimates of both membership and volume should be conservative. Not all persons interested will join. Not all who join will do so at the outset. And not all members will make fullest use of the cooperative's services.

Results of the producer survey should reveal how extensively economic need is perceived and the degree of interest in a cooperative to fulfill that need. The survey should indicate whether the expressed support in terms of business volume and financial commitment is sufficient to organize the cooperative and to operate it successfully.

Competitive Reaction

Estimates of anticipated volume should take into account competitive reaction. Depending on the situation, a new cooperative may be welcomed with enthusiasm or met with vigorous competitive opposition. If the latter is the case, leaders must be prepared to react to various strategies of competitors, including price changes to divert potential cooperative members' business; better contract terms or canceled contracts; attempts to influence lenders against providing credit; and even negative publicity, misstatements, and rumors attacking the cooperative business concept.

Management

Experienced and qualified management is particularly critical when a new organization and facility begin to operate. Overcoming unforeseen difficulties, establishing procedures, hiring—all require qualified management not swayed by personal preferences of steering committee members. At this point, the cooperative should not proceed unless such management is available and cooperative members are willing to pay for it.

Capital Availability

Next, the committee must decide whether adequate equity and debt financing is available. Initial capital requirements can be determined by projecting the cost of facilities, number of members, and volume of business.

Members' share of initial equity capital should be large enough for them to want to protect their financial stake in the business by providing an equity base for the cooperative. Investing initial capital is a basic member responsibility. It is desirable that members invest in proportion to their expected use.

Once the committee has decided that adequate financing (including debt capital) is available, it should prepare a financial analysis report that outlines all assumptions and income and expense projections based on standard financing practices. The report should be reviewed and revised as many times as necessary to arrive at a realistic business plan that can be approved by potential members and implemented without significant change.

Decision to Organize

Results of this feasibility analysis should be reported at a general meeting of potential members. Discussion should cover the cooperative's purpose, goals, and economic functions, including assumptions and financial projections for startup and for at least the first three years of operations.

Specific topics include the following:

1. Volume projections
2. Membership fee and equity investment requirements—initial and continuing
3. Sources of and obligations associated with borrowed capital
4. Financing projections, including tables for monthly cash flows, annual pro forma operating statements, balance sheets, and source and use of funds
5. Grower payment schedules
6. Projected patronage refunds—cash and retained
7. Equity redemption plans
8. General implementation schedules

Membership approval to go ahead with organizational plans allows the steering committee (often thereafter referred to as the organizing committee) to arrange for incorporation and to carry out the business plan.

Generally, a minimum member volume and equity commitment is established, based on the business analysis. This threshold sign-up should be met before the committee hires an attorney to draw up the several legal documents involved—unless, perhaps, the committee needs help to draw up a premembership agreement.

Incorporation Statutes

Articles of incorporation Because cooperatives are usually corporate entities authorized by the state, prerequisites are found in their state legal statutes. After the cooperative is formed, the committee should prepare and file legal documents called articles of incorporation with the appropriate state official. These articles are a set of legally binding rules under which the cooperative must operate. When they are accepted by the state office, the cooperative is legally recognized as a legal person with all rights and obligations described in the state's laws.

Incorporation statutes require that articles of incorporation contain certain kinds of information about the cooperative, including its name, its purposes, its business location, and its term of existence, as well as the names of incorporators and initial directors, and the number and terms of directors. The articles should specify member property rights and information about capital stock, such as the number of shares authorized, par value, and preferred stock, if any. Provisions in articles must not conflict with statutory requirements.

The amount of detail required varies from state to state. Those forming the cooperative must make important decisions about the new organization as they draft these articles. In fact, this is often the first time that farmers must make specific decisions about the cooperative as a legal corporate business entity. In this drafting process, they should carefully consider the principles discussed in Chapters 1 and 2 that distinguish cooperatives from other corporate businesses.

When the state accepts these articles, they become, along with the corporation statutes, the cooperative's *charter*. The cooperative is granted legal status as long as it continues to meet the statute's requirements and the rules it established in its own articles. Articles are legally binding; cooperatives can amend them only by following statutory procedures and filing the amendments with approval of the members.

A cooperative may choose to incorporate under general incorporation statutes, thus gaining potential advantages but assuming associated risks (Roy). Advantages include no limitations to the amount of dividends paid on capital stock, greater freedom to do business with nonmembers, and more flexibility in transacting business with members. Associated risks are different treatment under antitrust and tax laws and additional compliance with Federal Security Acts.

Bylaws Bylaws are a set of written rules the cooperative uses to govern itself. Cooperatives have a fairly long list of rules already in the charter, including state cooperative statutes and articles of incorporation filed with the state. However, bylaws perform two additional functions. First, they implement charter rules by describing in more detail how cooperative affairs will be run to ensure that laws and articles will not be violated. Second, bylaws address topics not found in laws or articles but necessary to make the cooperative a successful business organization.

Bylaws are usually filed only with the cooperative, not with a state office. Nevertheless, they have legal significance. They cannot be treated as general operating guidelines to be used only when convenient. Bylaws, as the name suggests, are like laws often considered as contracts that bind members, directors, and management. Directors, in particular, are responsible for ensuring that the cooperative adheres to bylaw provisions. Many states let the bylaws themselves establish penalties for violation. Examples will be cited later on bylaw provisions that serve as member-cooperative agreements.

The committee should prepare these bylaws with the help of an attorney, making sure that the bylaws do not conflict with state statutes, articles of incorporation, or cooperative principles. Then the bylaws must be adopted and amended by members within 30 days after the cooperative comes into existence. Most states require adoption by either a majority of all members or a majority of members voting at a meeting. In a few states, the board of directors can adopt the initial bylaws, but these provisions normally allow members the opportunity to change them.

Bylaws usually describe in more detail the specific qualifications for membership in the cooperative; the limitation on transferability of membership stock; and the reasons for suspension or termination of membership, such as breach of bylaws, loss of eligibility, or nonpatronage. Bylaws usually specify the time and place of annual meetings and the procedures for special member meetings along with notice and quorum requirements. Bylaws contain any special voting rules, such as proxy and mail voting, ballot counting, and majority requirements.

In addition, bylaws mandate the appropriate number of directors, their qualifications, and their conduct. Bylaws describe terms of office (including staggered terms) and specify the number of times a director may be reelected. They describe director election procedures such as nominating, districting, and voting. They describe board actions to fill vacancies, to hold regular and special board meetings, and to give meeting notices. Bylaws may authorize the membership to set the amount of compensation for board members. Members may place special rules in bylaws to prevent conflicts of interest. Finally, bylaws explain the duties of directors and all other officers.

Other legal documents Other legal documents include membership applications, membership or stock certificates, marketing agreements, meeting notices, and waivers of notice.

If a marketing cooperative does not have a marketing contract, its bylaws should specify the obligation of the members to market their products through the cooperative, the terms and conditions under which the products will be marketed, and accounting procedures that should be used together with the amount of business done with members and nonmembers. These provisions are important factors for determining antitrust and tax treatment.

Membership application and/or marketing contracts are needed. A marketing contract allows the cooperative to function properly by ensuring sufficient control over the quantity, quality, and delivery conditions of products to be delivered. This is especially helpful in the first few years of operation when the cooperative is establishing its reputation as a going, responsible, and successful business. (See chap.11 for a more detailed discussion of contracts.)

Final Procedures

After preliminary actions have been completed, charter members adopt the charter and elect the board of directors. The board in turn selects its own officers and assigns to directors individual or committee responsibilities to implement the business plan. Members not on the board may be assigned to committees, but at least one board member should be on each committee for communication reasons. The board then begins to develop an implementation schedule, establishing target dates for important events, and developing plans to implement details of managing and running a cooperative.

SUMMARY

New businesses are started continually, and cooperatives may prove to be the appropriate structural form. Starting any business is difficult, and the risk of failure is high. Several essential ingredients are necessary for success. They include an economic need, financial feasibility, qualified management and adequate volume and financing.

Vigorous and persistent leadership is necessary to nurture the idea of a cooperative to the day it opens for business. Specialized help is needed in several areas of the forming process, such as legal, financial and market analysis, engineering, and cooperative organization.

Two principal legal documents that must be written are the articles of incorporation and bylaws. They must conform with state incorporation statutes relating to various types of businesses. Most states have statutes relating to various types of businesses. Most states have statutes written particularly for cooperatives. The articles of incorporation contain general information about the name of the business, the location, the purpose, and the incorporators. The bylaws are more specific, presenting a set of rules on how the cooperative will implement the charter rules and how it will govern itself.

Constant communication with potential users of the cooperative is necessary to develop understanding and support and to gain the information necessary to determine economic feasibility. Successful operation depends on educating members on their governance responsibilities and demonstrating the value of their patronage.

DISCUSSION QUESTIONS

21-1.   What are some conditions that might call for starting a new cooperative?

21-2.   What types of specialized help would a group of people need in the organization process?

21-3.   Name sources of special help.

21-4.   Discuss the function of the steering committee.

21-5.   Discuss some key factors that would help determine whether successfully organizing a new cooperative would be economically feasible. Explain how you would determine these factors.

21-6.   What is the strongest indication that a person would be an active participant in a new cooperative?

21-7.   Name and describe two important legal papers that must be drafted when a cooperative is incorporated.

21-8.   What is the final step in the organizing process?

21-9.   What are some critical stages in the organizing process that involve the entire potential membership?

21-10.   Name at least three elements that must be present for a new cooperative to have a chance of success.

REFERENCES

HOGELAND, JULIE A., *A Guide to Survey Research for Local Cooperative Management.* Washington, DC: ESCS CIR 24, 1980.
INGALSBE, GENE, AND JAMES L. GOFF, *How to Start a Cooperative.* Washington, DC: USDA ACS CIR 7, 1985.
ROY, EWELL P., *Cooperatives: Today and Tomorrow,* 2nd ed. Danville, IL: Interstate Printers & Publishers, Inc. 1969.

22

Adjustments by Existing Cooperatives

Richard H. Vilstrup,
University of Wisconsin-Madison

David W. Cobia,
North Dakota State University

Robert Cropp,
University of Wisconsin-Platteville

INTRODUCTION

Altering the scope of operations (growth or retraction) to accommodate changing conditions can be handled by internal or external changes, by forming or canceling alliances with other businesses, or by modifying internal organizational structure. Several alternatives exist. Internal changes include constructing new productive capacity or closing down operations. External changes refer to amalgamating existing companies or selling off ongoing operations. External growth may take place through mergers, consolidation, or acquisition. Joint ventures, marketing agents, holding companies, or contract agents are organizational linkages that require coordination with other independent businesses, either cooperatives or investor-oriented firms (IOFs). Organization of a cooperative may be modified by forming subsidiaries or changing structure between centralized and federated.

Many of these changes require a majority decision of membership. In addition, legal advice should be obtained during the planning phase.

INTERNAL GROWTH

Cooperatives can grow internally by constructing new facilities which add new productive capacity that did not previously exist. Examples of successful cooperatives that have grown largely from internal expansion are Ocean Spray Cranberries, Inc., Riceland Foods, Inc., and Sunkist Growers, Inc.

A cooperative may have several reasons for wishing to expand internally. This option has many of the fresh-start advantages of starting a new cooperative; for example, cooperatives can avoid the image of gobbling up competition and thus evade antitrust problems. As a result, public relations are made easier as well. In addition, bad habits and relationships of the past do not encumber operations. On the other hand, it is easy to establish new policies and procedures without destroying old ones. Finally, employee and membership morale and support tend to be high, as opposed to the anxiety often created by external growth such as mergers.

Building new facilities, however, does not make sense if excess capacity exists in the industry, unless the excess capacity is hopelessly obsolete or improperly located. Although this may appear to be intuitively obvious, many cooperatives have been guilty of redundant expansion. This has been particularly true in the rush to build train-loading capacity in grain- and oilseed-producing areas. Objective feasibility studies (discussed in Chapter 21) are a major tool needed to avoid this problem. Necessary restructuring of the industry might be sidestepped if growth is internal. Internal growth may also lead to a failure to capture opportunities in a timely manner, and it may lead to apathy among members, management, and employees.

Cooperatives generally achieve internal expansion by obtaining additional loans, by soliciting member investment, by allocating current income, or by using capital retains and leasing. Internal growth often involves less change in the cooperative structure but frequently results in slower growth. Adding large numbers of new members may require a review of the cooperative's governance plan to ensure adequate representation and control.

Cooperatives can use several methods of internal expansion. They can enlarge an existing facility, add new services, solicit new members by serving a wider geographical area, build new facilities or purchase new equipment and add new products or markets.

EXTERNAL GROWTH

External growth is growth by a company that has been achieved by uniting with other companies or parts of other companies. This method of growth is achieved by merger, consolidation, or acquisition. Merging means combining the assets of one or more organizations with those of the surviving cooperative. Consolidation is combining the assets of two or more organizations into a new organization. Acquisition is the purchase of all or part of the assets of one firm by a second firm. The acquired firm may be integrated into the buying firm or operated separately under the new ownership structure. The term *merger* is often used more broadly to include all forms of external growth, including consolidation and acquisition. Examples of cooperatives that have used external growth to grow are Agway (a consolidation of Grange League Federation, Eastern States Cooperative, and Pennsylvania Farm Bureau) and Harvest States (a consolidation of Grain Terminal Association [GTA] and Pacific Grain Growers—GTA also had previously acquired several IOFs). Merger, consolidation, and acquisition techniques have also been used when cooperatives have run into financial difficulty and have been forced to join forces with neighboring organizations.

Reasons for External Growth

Most writers on this subject confuse incentives for merging with incentives for growth. For example, a firm may wish to invest idle cash, to stabilize seasonal patterns, to achieve potential economies of size, or to gain control over markets. These objectives may be achieved by internal as well as external growth.

External growth methods are frequently selected by cooperatives wishing to expand quickly by adding new geographical territory or by acquiring new facilities or services. External expansion may be cheaper, faster, and less risky than building new facilities. External expansion does not result in intensified competition, and it may be the only way to acquire some controlled markets, patents, processes, and raw materials.

Several financial advantages may be present as well. Credit is often relatively easy to obtain for an established activity, and exchange of stock may require no cash. In addition, the acquisition of working capital or even losses for tax purposes may be attractive. Cooperatives have other incentives, in addition to those listed previously, which are not as valid for IOFs, that make external growth attractive. As market areas grow and economies of size increase, cooperatives also may consolidate operations to eliminate costs of overlapping market areas, competitive advertising, and duplicate facilities and record keeping.

Finally, an acquisition may be forced upon a federated cooperative because a member has become overly indebted to the federated cooperative. In this case, acquisition may be the most feasible way to rescue the operation and protect loaned funds.

Too many view mergers as a last alternative to liquidation. Merging should be considered as building an organization that will pay off for farmers. During recent years, some cooperatives have found themselves in a vulnerable position with insufficient financing or business and have been forced to seek a merger partner to protect the investment of the members. Even so, it may be better for the cooperative to consider merger prior to liquidation.

Disadvantages of External Growth

Sometimes there are barriers and disadvantages to external growth, especially for cooperatives. Intercooperative relationships become more complicated, for example, when two or more cooperatives merge and combine memberships, boards, and managements. This new cooperative may also have difficulty eliminating duplicate facilities, changing poor practices and policies, or releasing excess employees because of the hostility that inevitably results from such actions. In addition, external growth may weaken community identity. Contractual obligations and redemption plans, especially the current status, may be incompatible. These problems can rarely be solved without sacrificing some member interests. Finally, evaluation of equity is complicated by the lack of an open market for stock and the frequent deterioration of one of the cooperatives as a feasible business.

Several years ago, John P. Comstock, a Bank for Cooperative (BC) vice–president with considerable experience at merger attempts, illustrated the futility of negotiating otherwise sound mergers when members objected. Negotiators may overlook factors more important to members than money. Objections may stem from loyalty to community and community rivalry, from reluctance to relinquish local control, and from pressure exerted by special-interest groups. According to Comstock, members ask themselves if they can support a proposed merger without being disloyal to the community. Intense community rivalries reinforce community loyalty. Therefore, farmers have an affinity for what is theirs, whether it is worth much or not. As Comstock explains, "For many members, giving up what they are a part of becomes insurmountable." He then tells of a small, worn-out, obsolete elevator which had poor management and weak finances and which was about to go out of business. A strong neighboring elevator finally agreed to a merger, although the neighboring elevator had nothing to gain from the merger. After all the roadblocks had been cleared, a special meeting was held, during which the merger proposal was submitted. After some discussion a member rose to his feet and said:

> "My father helped organize this elevator at about the turn of the century. Ever since boyhood I have been able to look out our kitchen window and watch the slanting rays of the early morning sun paint the old red building with

warmth and welcome. The setting sun silhouettes it against the sky at nightfall. It is like an old friend and neighbor. It almost seems a part of my farm. I count every member of the elevator among my friends.

"I know that our old elevator is worn out and obsolete. I know that our cooperative doesn't have the money to provide the variety of services that we would like to have. Nevertheless, we own it and it is ours.

"As long as I can remember, every bushel of grain taken off our farm was delivered to it, and as far as I am concerned, I intend to keep right on delivering it there for the rest of my life."

Then he sat down. A vote was taken. It was overwhelmingly against the merger (Comstock, p. 17).

In some cases, it is the board or management that blocks an otherwise sound merger, because board members know they may have to surrender their positions. Normally, merged boards are smaller than the combined previously independent boards. This problem can be minimized by allowing all board members in on the new board and then reducing its size over time. Management, however, is more difficult. Managers tend to be, and in fact often need to be, natural leaders, egotistical and strong-willed and therefore unwilling to step down. Consequently, some mergers have been delayed until one of the managers has retired or left the cooperative.

Misunderstanding can also result from mergers because merging cooperatives often have widely differing equity profiles. They have acquired equity in different ways and employed different equity redemption plans, and they are at different stages of redemption. The bulk of one cooperative's equity may be in the hands of inactive members, while another cooperative is up to date in its redemption to members, providing equity according to use. Members of the latter cooperative are understandably reluctant to dilute their equity by assuming the redemption burden of the less financially secure cooperative. However, this dilemma can be resolved by creating new criteria based on the circumstances. Hatfield et al. give examples of how new criteria can be created to achieve a balance between filling the needs of the cooperative and current patrons and protecting the rights of inactive members.

There are other disadvantages of external growth that are common with IOFs, such as inadvertently acquiring outmoded equipment that needs repair. This is another reason for the failure of the International Cooperative, Inc. (Pederson and Scott).

Merger

Procedures and problems[1] The major steps inconsummating a merger are screening initial ideas, determining criteria, making initial contact with other

---

[1]We will raise only selected issues at this point. See Swanson for a detailed description and step-by-step procedures and Brigham and Gapenski for standard financial considerations.

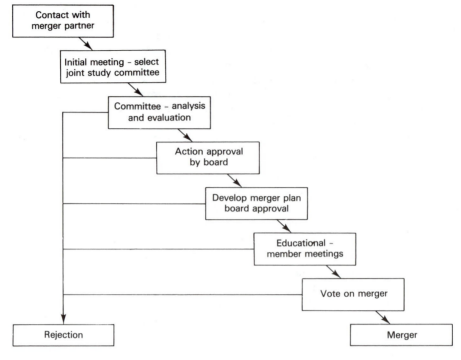

Figure 22.1 Typical merger procedure.

firm(s), conducting a feasibility study, and negotiating. We have listed the final steps in Figure 22.1.

Timing is critical in achieving successful mergers. Often, boards lack the foresight or courage to proceed when a merger is needed. Usually, the merger fails to receive serious consideration until it may be the only way out. By then, the cooperative has dissipated its financial resources, its goodwill, and its patronage base. Thus its bargaining power for achieving a merger on relatively favorable terms has evaporated.

Successful mergers are also prevented by competitors that raid members during negotiations. As a result, the merging cooperative must try to maintaining confidentiality, respond to rumors, and deal with resultant demoralized employees.

Legal issues Mergers and consolidation result in fundamental changes in the cooperatives involved; thus merging or consolidating cooperatives must meet certain legal requirements. In all cases the cooperatives must be authorized by appropriate statutes in order to proceed. The older state cooperative corporation statutes do not have merger provisions, so cooperatives operating in those states will refer to the business corporation statutes for guidance. However, many states have amended their cooperative statutes to include merger provisions. Most of these amendments describe the method that merging cooperatives must use to transfer owner equity.

Most merger and consolidation statutes require several steps in the process. First, the merger plans must be approved by the boards of directors of all cooperatives involved. The boards then must adopt a resolution and a written plan of merger and submit these to the membership for approval, often by two-thirds of the members. Finally, the merger plan is formally filed with the state office handling incorporation. The merger plan operates to modify the articles of incorporation and becomes part of the new organization's charter. In some states members who dissent from the merger have a right to withdraw rather than become part of the new organization; therefore, their equity withdrawal rights must be carefully analyzed by organizations considering merger or consolidation.

Consolidation

Consolidation, the creation of a new cooperative from previous independent cooperatives, has appeal when the image, name, geographic location, product line, and philosophical orientation of the parent cooperatives are incompatible with those of the new organization. Creating a totally new cooperative provides neutrality by eliminating former legal entities, including names, organizational structure, and so on. It also presents an opportunity to make a fresh start by adopting the strengths and shedding the weaknesses of parent cooperatives to an extent that would not be feasible with a merger. However, a consolidation also may be more expensive than merging because of extra legal fees and because of the loss of identity associated with predecessor cooperatives.

Cooperatives are usually consolidated when organizations are of equal size or strength and a new structure will provide more potential efficiencies, financial returns, and improved services. Some members prefer to use consolidations in forming a new organization because they appear to be equal partners and not part of an organization that has been swallowed up in a merger or acquisition.

Acquisition

Failure to dispose of excess facilities, to reduce the number of employees, and to establish pricing policies at satellite stations that reflect costs, in addition to problems associated with postmerger follow-through, often negate potential savings. Consequently, some cooperatives choose acquisitions instead of mergers or consolidations. This alternative minimizes the seriousness of problems associated with mergers and consolidations.

Restructuring cooperatives through acquisition instead of a merger is an alternative growing in popularity, because many of the drawbacks associated with external growth can be minimized through acquisition. The surviving cooperative has more freedom in changing management and employees and in eliminating duplicate facilities. In addition, the purchasing cooperative may complete the acquisition without having to seek approval of its members, as long as their member equity is not materially affected. Acquiring a financially weak neighboring or competing cooperative is often easier under cooperative law than merging or consolidating. Acquisition is also the route to acquiring an IOF.

Acquisitions can quickly allow cooperatives to selectively offer new services to members. They can purchase an ongoing organization with successful management and trained personnel. Through acquisition they can acquire a successful marketing operation with a substantial share of the market. Acquisition has been used by such regional cooperatives as Land O'Lakes, Farmland Industries, CENEX, and Harvest States to purchase manufacturing, processing, distribution, and retailing businesses.

Drawbacks of acquisitions include cash required to make the purchase and the difficulty in arriving at an equitable appraised value of the operation to be purchased. Finally, small cooperatives do not have the staff or experience to evaluate opportunities.

ALLIANCES

Cooperatives often have been criticized for their failure to cooperate among themselves to achieve and capitalize on economies of size, to eliminate duplication, and to gain market access. However, in recent years cooperatives have become involved in several traditional and innovative alliances. A cooperative can enter into several institutional arrangements to achieve objectives of gaining efficiencies in operation, entering into other activities, enhancing financial strength, gaining market entry, increasing market power, and reducing competition among cooperatives without the financial resources required for merger. These alternatives include federations, marketing agencies-in-common, joint ventures, holding companies, and subsidiaries. (See Chapter 3 for a description of these organizational linkages.)

Federations

Two or more independent cooperatives may create another cooperative or federation of which they become owner-patrons while retaining their independent status. Many of today's prominent federated cooperatives were originally organized this way.

Federations often are created by several independent cooperatives that desire a service whose economic operations they do not individually have the volume or resources to achieve. (See the section "Benefits and Reasons for Growth" in chap. 20.) The impediments to a merger may be too great, the advantages to continued independent operations, overwhelming; in addition, members or management may oppose a merger no matter what the circumstances. On the other hand, a federation might succeed when a merger clearly will not, because its prime activity may involve a relatively small portion of the member cooperative's line of service.

Marketing Agencies-in-Common

Fragmentation in the marketplace can be overcome by the common management of marketing agencies-in-common. These are particularly effective when a family of products can be marketed under central management control, because member cooperatives maintain operational independence while capitalizing on the economies and market power of a larger organization. An example is the combination of the marketing operations Sun-Maid Raisins, Diamond Walnut Growers, Sunsweet Prunes, and Valley Fig Growers into Sun-Diamond Growers of California. Cooperating Brands, Inc. is a marketing agency-in-common initiated by a fruit- and vegetable-processing cooperative for marketing juice. Members are classified by the brand name used for their major juice products: Citrus World for oranges, Welch Foods for grapes, Ocean Spray for cranberries, Tree Top for apples, and Tri/Valley Growers for tomatoes. Another example is Norbest, Inc., which markets turkeys for six member turkey-processing cooperatives. Marketing agencies-in-common have been used by dairy cooperatives in negotiation above federal milk marketing order premiums.

Cooperative Joint Ventures

Joint ventures among cooperatives enable cooperatives to expand services or products, to increase efficiency, and to control operating costs and margins (Frederick). They are also a convenient device for avoiding problems associated with external growth.

One successful joint venture occurred between Prairie Farms Dairy and Mid-America Dairymen. The two dairy cooperatives bought Hiland Dairy of Springfield, Missouri, on a 50-50 basis. Mid-America Dairymen now supplies the bulk milk, and Prairie Farms manages the fluid operation.

A joint venture of a different kind between Land O'Lakes and CENEX was formalized in December 1986. The main objective of their three-year agreement was to save an estimated $10 million by eliminating duplicate facilities in feed and petroleum. CENEX is to manage the petroleum; Land O'Lakes, the feed and seed facilities.

Cooperative-IOF Joint Ventures

Cooperatives enter into joint ventures with IOFs to gain access to resources and market positions controlled by the IOF. A recent cooperative-IOF venture was the 1985 sale of grain marketing facilities by Growmark in exchange for common stock and a position on the Archer-Daniels-Midland (ADM) grain subsidiary board. This joint venture gives Growmark producers access to worldwide markets through ADM markets. Another example is the 1961-1962 agreement between Florida Orange Marketers (FOM), a producer cooperative, and the Coca-Cola Company's Food Division, of which Minute Maid is a part. The agreement was made in an attempt to establish a fair transfer price for oranges. Coca-Cola Company now procures about one-third of its orange supply from FOM, while FOM producers enjoy price premiums resulting from quality control and from the popular Minute Maid brand image. In such arrangements, income from an IOF joint venture is essentially nonmember business and is treated as taxable income of the cooperative.

Joint ventures with IOFs can be formed as an IOF corporation, cooperative, partnership, merger, or lease. Each structure has its own unique tax, antitrust, and management consequences (Shereff et al.). If joint ventures between cooperatives and IOFs involve only contractual arrangements such as promises to sell or buy, the legal relationships and consequences may be manageable and fruitful. However, if joint ownership of a business entity is involved, the financial, legal, and taxation complications can be serious (Shereff et al.). If such joint ventures tend to reduce competition, antitrust action may result. For example, the joint venture of Allied Grape Growers with Heublein, Inc. in 1969 resulted in a 1972 Federal Trade Commission charge that section 7 of the Clayton Act was violated because this action eliminated actual or potential competition. Cooperatives need assurance that they are not being exploited by the firms with which they are dealing. Cooperatives, for example, may provide capital-intensive, low-margin operations while IOFs reap benefits of higher-margin, less-capital-intensive activities.

Holding Company

In recent reorganization talks among major cooperatives, the holding company approach has surfaced as another alternative for cooperatives wishing to benefit from consolidating selected functions. Regional purchasing and marketing cooperatives can own and control a holding company by means of a voting trust. In turn, the holding company has its own subsidiary corporations and operates businesses complementary to those of the member regional cooperatives.

The holding company can achieve economies of size in areas beyond operations such as accounting and financial administration, personnel management, or corporate planning. In addition, cooperatives may own holding companies so that they can consolidate selected operating, purchasing and marketing functions. Clearly, a benefit of owning a holding company is the ability to centralize certain functions when economies of size are present yet at the same time to maintain established business, although with a somewhat smaller degree of autonomy. Membership control is important. It needs to be maintained to make sure that the holding company continues to operate in the best interest of members of the cooperative.

Contract Agencies and Franchises

Several types of agency and franchise relationships can be created to meet a variety of cooperative needs. We have described these in Chapter 3. The basic principles used to examine the feasibility of the other structural forms should be used to evaluate these alternatives. Economic, legal, financial, member, board, and employee concerns should all receive careful scrutiny.

INTERNAL ORGANIZATIONAL LINKAGES

As a cooperative grows or contracts, a continual process of internal restructuring occurs. Cooperatives need to modify the relationships between various parts of the organization that relate to each other to accommodate changes in function, scope, and capabilities of management team. Legal options available include converting to a federated or centralized system and vice versa, creating subsidiaries or converting the cooperative to another business type.

Conversion Between Centralized and Federated Structure

The relative importance of the factors that originally prompted organizers to use the centralized or federated structure (see Chapter 3) may change, making a conversion to the other form desirable. Gold Kist went through such a change in 1979 when its local member cooperatives became branch stations and its patrons became direct members of Gold Kist.

Subsidiaries

A cooperative may organize and operate subsidiary corporations (1) to handle nonmember business, (2) to provide capital or financing of certain risky ventures, (3) to purchase or market certain goods, (4) to qualify for business in other states, (5) to sell a different brand name, or (6) to conduct export trade.

A subsidiary may be operated either as a cooperative or as an IOF corporation. The business and legal obligations of the subsidiary, however, are separate from those of the parent cooperative; that is, losses, liabilities, and so on, do not affect directly on the parent cooperative. Cooperative subsidiaries are frequently controlled by selected management and directors of the cooperative. This interlocking control provides direction to the subsidiary in serving members.

Two examples of subsidiaries organized by cooperatives are Welch Foods as a cooperative processor and merchandiser for National Grape Cooperative, and Gold Kist's formation of AgraTech Seeds, Inc., a subsidiary to market seed (Table 20.2).

Conversion of Corporate Form

Cooperatives can be converted to IOFs, and vice versa. Reasons for such transformations are not always clear. If benefits seem to be declining or the equity seems in jeopardy, some members may wish to sell their interests to other members who are interested in buying them and converting the organization to an IOF. On the other hand, owners of an IOF corporation may wish to convert it to a cooperative. A cooperative in North Dakota has made this transition three times.

LIQUIDATION

Recognition and honor is generally extended to those who had the foresight and skills necessary to create a cooperative. But it also takes foresight to recognize when a cooperative should liquidate—and it takes much courage to act on that realization. Typically, cooperatives are created when other businesses are not providing necessary services or when competition is not functioning. Sometimes competition provides new or superior services or prices and returns from cooperative activity evaporate. In such cases cooperatives should liquidate to stop erosion of equity. Even if equity is not eroding but returns are near zero, farmers can find a better use for their funds.

Recognizing the Need

It is critical that cooperative directors and managers frequently evaluate their competitive and financial position and determine the need to merge or consolidate while they still have sufficient financial strength. Feasibility studies should indicate when an activity should be dropped.

The difficulty is in determining the stage of a recent reversal. Is it a sign that the worst is yet to come, or is it the darkness before the dawn? In other words, management must decide whether the change is temporary or fundamental. In some cases the problem can be corrected and the cooperative will return to profitability. In other cases there may be no hope because the damage is so severe or the negative factors are beyond control of the cooperative. Or members can vote to liquidate, having determined that the cooperative has performed its task and is no longer needed.

Poor performance can be linked to poor management or the evaporation of the cooperative's economic justification. Management may be incompetent, having become involved in moral turpitude or perhaps having speculated unwisely in the futures market. In other cases the cooperative may have inherited a high-cost structure that it cannot easily change. A competing business with a lower-cost structure may be capturing substantial patronage. A fundamental decline in demand may also result in losses.

Alternatives

Feasibility studies should indicate when an activity should be dropped. Methods of cooperative reorganization through liquidation include the following (Turner; Vilstrup):

1. Sell assets of weaker cooperatives to a neighboring cooperative and thus retain essential services for members.

2. Sell just the nonproductive assets of the cooperative.

3. Sell all the assets of the local cooperative to the regional cooperative and have the regional provide member service, or convert the operation from a federated to centralized cooperative.

4. Liquidate the assets, dissolve the cooperative, and reorganize the cooperative under a new financial structure.

5. Sell the assets to the regional and lease back the facilities in order to retain local member service.

6. Liquidate assets and dissolve the cooperative to protect the remaining member equity.

Disposing of a vertical stage does not necessarily eliminate service to members. Divesting of horizontal facilities could leave some members without services if the facility is operating at the first level from members.

The type of major restructuring possible is illustrated by the action taken by major soybean-processing cooperatives facing declining exports of oil and meal. In 1983 plants were operating well under capacity and one was closed. Maixner describes how a restructuring improved the situation.

> Farmland and Land O'Lakes separated five crushing plants in Minnesota, Iowa, Missouri and Arkansas from their other . . . ventures. The crushing

plants were joined to the Boone Valley plant in a new regional cooperative, called Ag Processing, Inc., based in Omaha, Neb. The new cooperative drew in 326 local farmer co-ops. About 285 of the farmer groups are in Minnesota, Iowa and the Dakotas. The consolidated co-op has more than survived the extended slump in the soybean market: All plants are operating, and at a profit . . . . The cooperative's central marketing division sells all soybean oil for the plants, and coordinates all sales of flour and meal. [The volume generated] . . . gives the marketing office a lot of latitude in filling large-volume orders, and meeting customers' desires for shipping and time of delivery (Maixner, p. 30).

Legal Procedures

Dissolution or liquidation is the process by which a cooperative ceases to do business, liquidates its property, and distributes equity to shareholders and others. Liquidation may be voluntary or involuntary. Voluntary dissolution is initiated either by the board of directors or the members as a body, and member approval is required. Typically, members appoint trustees to carry out the dissolution process. They wind up the cooperative's business and liquidate the assets.

One of the more important tasks is to distribute assets. According to most laws, property remaining should first be applied toward debts and obligations of the cooperative. Any remaining surplus should then be apportioned among members. Although a few statutes give detailed directions for distribution, most place considerable emphasis on what the articles, bylaws, or agreements say about distribution on dissolution.

A sample legal document gives an example of a bylaw provision on dissolution distribution. It says:

> Upon dissolution, after (1) all debts and liabilities of the association shall have been paid, (2) the par value of stockholders' shares returned, and (3) all capital furnished through patronage shall have been retired without priority on a pro rata basis, the remaining property and assets of the association shall be distributed among the members and former members in the proportion which the aggregate patronage of each member bears to the total patronage of all such members, unless otherwise provided by law (Neely, p. 581).

Cooperatives may establish other priorities, depending on the intended plan of operation and limitations in statutes and other laws. Each of the plans takes careful study to protect member assets and to retain essential services. Most cooperative laws are carefully written to protect members in the liquidation process. These laws require that members be given adequate information for decision making and that a substantial majority of the members approve the liquidation plan.

SUMMARY

Existing cooperatives can be restricted in several ways. They can combine with other cooperatives, make alliances with other firms, liquidate, and reorganize internally. The appropriate scope and organizational structure is dictated by the environment, by the job to be done, and by the resources available. Within these constraints, what evolves depends on the management and leadership skills of those involved. They will succeed only if they have the ability to recognize and overcome problems.

DISCUSSION QUESTIONS

22-1. Why might a cooperative look for a merger partner rather than construct a new processing plant? Which factors are unique to cooperatives? Contrast these issues with reasons for growth in general given in Chapter 21.

22-2. Why might another cooperative decide to grow internally rather than externally?

22-3. Distinguish mergers from consolidations and acquisitions, and discuss why a cooperative may select one over the others.

22-4. Why would a cooperative want to form a joint venture with an IOF in light of the disadvantages of doing so?

22-5. Why might a joint venture between two cooperatives be a better institutional arrangement than would a complete merger or consolidation?

REFERENCES

BRIGHAM, EUGENE F., AND LOUIS C. GAPENSKI, *Financial Management Theory and Practice*, 4th ed. Hinsdale, IL: The Dryden Press, 1985.

COMSTOCK, JOHN P., "The Human Problem—Why Some Mergers Fail," *News Farm. Coop.* May 1963.

FREDERICK, DONALD A., *Successful Joint Ventures among Farmer Cooperatives.* Washington, DC: USDA ACS RR 62, Apr. 1987.

HATFIELD, HARLAN, et al., *Consolidation of Allocated Equity for Merging Cooperatives Previously Operating on a Revolving Fund.* Dept. of Agr. Econ. Misc. Report 96, North Dakota State Univ., Apr. 1986.

MAIXNER, ED, "A New Era: U.S. Farmer Cooperatives Adapt to Changing Times," *Agweek*, Apr. 14, 1986:30.

NEELY, J. MORRISON, *Legal Phases of Farmer Cooperatives.* Washington, DC: USDA FCS CIR 100, May 1976.

PEDERSON, GLENN, AND DONALD F. SCOTT, *International Cooperative, Inc.: The Evolution and Dissolution of a Potato-Processing Cooperative.* Dept. of Agr. Econ. Report 182, North Dakota Agr. Exp. Sta., Jan. 1984.

SHEREFF, HENRY D., et al., "Joint Ventures with Noncooperatives, Opportunities, and Problems," *American Cooperation 1982.* Washington, DC: American Institute of Cooperation, 1982.

SWANSON, BRUCE L., *Merging Cooperatives, Planning, Negotiating, Implementing.* Washington, DC: USDA ACS RR 43, Jan. 1985.

TURNER, MICHAEL S., "Cooperative Acquisitions and Other Forms of Restructuring," *Farmer Cooperatives for the Future,* ed. Lee F. Schrader and William D. Dobson, NCR-140, et al., workshop, St. Louis, MO; Dept. of Agr. Econ., Purdue Univ., Nov. 6, 1985.

VILSTRUP, DICK, "Cooperative Acquisitions and Other Forms of Restructuring: Discussion," *Farmer Cooperatives for the Future,* ed. Lee F. Schrader and William D. Dobson, NCR-140, et al., workshop, St. Louis, MO; Dept. of Agr. Econ., Purdue Univ., Nov. 6, 1985.

23

Antitrust
Laws

James R. Baarda,
Agricultural Cooperative Service, USDA

INTRODUCTION

In a world of multibillion dollar mergers and corporate takeovers, multi-national corporations, and incredibly complex intercorporate arrange-ments on every conceivable subject, it seems strange indeed that farmers need concern themselves with antitrust laws when creating and operating a cooperative for purely economic purposes. Antitrust concerns are, never-theless, important. Managers, directors, and advisors need a basic under-standing of rules applied to cooperative organization and operation, so antitrust laws will not be violated. Penalties may damage cooperatives and ultimately be paid out of farmers' pockets. Antitrust laws apply to all busi-nesses, including cooperatives, but special laws change the way that an-titrust policies apply to farmer cooperatives.

Why do farmer cooperatives need special attention? The answer lies in our basic national antitrust laws, which prohibit businesses from doing certain things that farmers, through their cooperatives, must do to make the cooperative a success. In this chapter we discuss antitrust laws to see how they affect farmers and then look carefully at special rules applying to farmers and their cooperatives. We will see what cooperatives can do and cannot do, and will note some current issues unresolved in cooperative antitrust laws.

FEDERAL ANTITRUST LAWS

In an economy based on free competition, rules are required to prevent behavior that is harmful to competition and to maintain a business structure in which competition can work. These rules are found in what are generally called antitrust laws. The U.S. Supreme Court, in *Northern Pacific Railway Co. v. U.S.* (1958) has described the philosophical underpinning of the Sherman Act, our premier antitrust law:

> The Sherman Act was designed to be a comprehensive charter of economic liberty aimed at preserving free and unfettered competition as the rule of trade. It rests on the premise that the unrestrained interaction of competitive forces will yield the best allocation of our economic resources, the lowest prices, the highest quality and the greatest material progress, while at the same time providing an environment conducive to the preservation of our democratic political and social institutions.

Farmers have a stake in antitrust laws because they are interested in smoothly operating markets. They must also consider how antitrust laws affect their own businesses. The Sherman Act, the Clayton Act, and the Federal Trade Commission Act are all of interest to farmers and their cooperatives.

The Sherman Antitrust Act

The first and still fundamental statement of national antitrust policy is the Sherman Act of 1890. Enacted with strong farmer support, the act established the overall approach to preserving a free competitive economy and indicated the way that rules would be applied to individual businesses to protect all business and its competitive environment.

The Sherman Act, written in broad language with sweeping implications, prohibits (1) "every contract, combination in the form of trust or otherwise, or conspiracy in restraint of trade or commerce" and (2) monopolization, attempts to monopolize, and combinations or conspiracies to monopolize trade or commerce. The substance of the act has not changed in the nearly 100 years since it first became the law of the land.

The Clayton and Federal Trade Commission Acts

The Clayton Act and the Federal Trade Commission Act, both of 1914, added to basic antitrust laws of the Sherman Act. They are more detailed than the Sherman Act, and each contains provisions important to farmer cooperatives.

The Clayton Act prevents mergers or the acquisition of assets of one company by another if the effect "may be substantially to lessen competition or to tend to create a monopoly." It also lets injured parties sue and receive treble damages if injured by antitrust law violation, and it holds directors and officers responsible for their corporation's criminal behavior if they approve or participate in illegal actions.

The Federal Trade Commission Act says "unfair methods of competition . . . and unfair or deceptive acts or practices . . . are hereby declared unlawful." Most of the act relates to the structure and activities of the Federal Trade Commission.

INTERPRETATION OF ANTITRUST LAWS

It is clear that the terms "monopolization," "attempt to monopolize," "restraint of trade," and "unfair methods of competition" must be applied to specific behavior. This is done on a case-by-case basis when a government agency (the Department of Justice or Federal Trade Commission) or a private complainant brings an action charging someone with antitrust law violation. A court must then decide how the law applies to a particular set of facts.

In the 100 years since the Sherman Act became law, the broad language of our antitrust laws has been interpreted by the courts in hundreds of specific cases. The operating rules developed by those cases generally divide antitrust rules into two kinds. Some actions by business are judged by a *rule of reason*, meaning that the actions are illegal only if they are done for improper reasons and have detrimental effects on other business. Other business practices are illegal no matter what their purpose or effect, because they have such potential for harming the economy. These are called —*per se* (pronounced "per say") violations and are not judged by a rule of reason.

One per se rule demonstrates why farmers need protection from antitrust laws when creating and using a cooperative, although it is not the only antitrust law with which they need be concerned. It is illegal for businesses selling a product to get together and agree on prices at which the product will be offered for sale. Price fixing is *per se* illegal. The size of business, the reasonableness of the price, and the power of the group to influence price are all irrelevant. No matter what the circumstances, such price fixing is illegal. This law creates a problem for farmers who join together in a cooperative marketing effort, because they typically do just what is prohibited under the *per se* rules. Thus they need special protection to make cooperative marketing legal. Before such protection existed, farmers cooperating together often faced legal problems. For example, in the early twentieth century, dairy farmers in some areas were prosecuted when they formed a cooperative and refused to deliver milk when prices offered were too low.

THE CLAYTON ACT, SECTION SIX

The earliest law to address special needs of farmer cooperatives was the Clayton Act of 1914. Section 6 states that nothing in antitrust law forbids the existence and operation of

> agricultural organizations, instituted for the purposes of mutual help, and not having capital stock or conducted for profit, or to forbid or restrain individual members of such organizations from lawfully carrying out the legitimate objects thereof; nor shall such organizations, or the members thereof, be held or construed to be illegal combinations or conspiracies in restraint of trade, under the antitrust laws.

Section 6 marked the beginning of legislative protection for cooperatives. However, it alone was not enough. Its lack of precise definitions and especially its restricted reference to nonstock cooperatives suggested that further legislation was needed.

THE CAPPER-VOLSTEAD ACT OF 1922

After debates in both the House and the Senate, the *Capper-Volstead Act* was passed and signed into law in 1922. Sometimes rather expansively called the *magna carta* of farmer cooperation, the Capper-Volstead Act explains in more detail than does Clayton Act section 6 the extent of cooperatives' special recognition in U.S. antitrust policy. The act is divided into 2 sections. Section 1 defines the extent, purposes, and organizational requirements for Capper-Volstead Act protection. Section 2 gives the Secretary of Agriculture authority to regulate a cooperative that "monopolizes or restrains trade . . . to such an extent that the price of any agricultural product is unduly enhanced by reason thereof." We will look first at requirements for Capper-Volstead Act qualification. Then, given the law and court decisions interpreting the law, we will summarize what farmers and cooperatives may and may not do. Later we will discuss the "undue price enhancement" problem.

Section 1 of the Capper-Volstead Act is so important that anyone who studies or works with cooperatives should be familiar with its language. It states:

> That persons engaged in the production of agricultural products as farmers, planters, ranchmen, dairymen, nut or fruit growers may act together in as-

sociations, corporate or otherwise, with or without capital stock, in collectively processing, preparing for market, handling, and marketing in interstate and foreign commerce, such products of persons so engaged. Such associations may have marketing agencies in common; and such associations and their members may make the necessary contracts and agreements to effect such purposes: provided, however, that such associations are operated for the mutual benefit of the members thereof, as such producers, and conform to one or both of the following requirements:

First. That no member of the association is allowed more than one vote because of the amount of stock or membership capital he may own therein, or,

Second. That the association does not pay dividends on stock or membership capital in excess of 8 per centum per annum. And in any case to the following:

Third. That the association shall not deal in the products of non-members to an amount greater in value than such as are handled by it for members.

Before we study the act in more detail to learn what it does, we should know what it does not do. The Capper-Volstead Act is not an incorporation law. Farmers who wish to form a cooperative association must do so under state laws. Neither is the Capper-Volstead Act related to cooperative taxation. How a cooperative is taxed depends entirely on federal and state tax laws. Finally, the act does not give anyone enforcement powers over internal cooperative affairs. With exceptions limited only to antitrust and Capper-Volstead Act qualification, no authority is given to any agency with respect to a cooperative's business affairs. No federal agency licenses or registers cooperatives or certifies them as being qualified Capper-Volstead Act cooperatives.

QUALIFICATION FOR CAPPER-VOLSTEAD ACT

Capper-Volstead Act protection is only available to organizations that meet criteria established in the act and further explained by court interpretation of statutory language. Qualification criteria can be divided into two parts. The first explains who can be a member, and the second explains what kind of organization is required. The Capper-Volstead Act mentions both kinds of qualifications, and courts have further interpreted the act's language.

Cooperative Membership

Who can be a farmer cooperative member? The act is specific in answering this question. Only persons "engaged in the production of agricultural products as farmers, planters, ranchmen, dairymen, nut or fruit growers" can qualify. The U.S. Supreme Court has interpreted this requirement strictly, and membership of any number of nonqualifying persons, no matter how small the number, will disqualify a cooperative from Capper-Volstead Act protection from antitrust laws.

Two U.S. Supreme Court decisions show how important the farmer-only membership requirement is. The first, decided in 1967, was *Case-Swayne* v. *Sunkist Growers*. Sunkist's membership was mixed. About 80% of its members were growers, 5% were corporate growers with their own packinghouse facilities, and 15% were investor-oriented firms (IOF) operating packing houses for profit. IOF packing houses had cost-plus-fixed-fee contracts with growers.

The Supreme Court ruled that the presence of IOF packing houses such as Sunkist members was not permitted under the Capper-Volstead Act. It stated:

> We deal here with "special exemptions to a general legislative plan," . . . and therefore we are not justified in expanding the Act's coverage, which otherwise seems quite plain. The Act states those whose collective activity is privileged under it; that enumeration is limited in quite specific terms to producers of agricultural products.

In 1978 the Supreme Court again addressed the producer-member issue in the case of *National Broiler Marketing Association* v. *United States.* The National Broiler Marketing Association was a poultry cooperative whose members were made up of integrated poultry-producing firms. Some of the integrated producers were engaged primarily in processing poultry, but not in growing poultry.

In this case, the Supreme Court noted that one purpose of the Capper-Volstead Act is to give farmers a chance to strengthen their market position with respect to buyers and processors. To permit others to join cooperatives would extend antitrust protection to those for whom Capper-Volstead Act protection was not designed and who did not need it. Therefore, the court ruled in favor of the United States, stating:

> We, therefore, conclude that any member of NBMA that owns neither a breeder flock nor a hatchery, and that maintains no grow-out facility at which the flocks to which it holds title are raised, is not among those Congress intended to protect by the Capper-Volstead Act. The economic role of such a member in the production of broiler chickens is indistinguishable from that of the processor that enters into a preplanting contract with its supplier, or from that of a packer that assists its supplier in the financing of his crops. Their participation involves only the kind of investment that Congress clearly did not intend to protect. We hold that such members are not "farmers," as that term is used in the Act, and that a cooperative organization that includes them—or even one of them—as members is not entitled to the limited protection of the Capper-Volstead Act.

The specific requirements of the words of the act and the strict interpretation by courts means that farmers must be very selective in who they join with in cooperative formation and who they accept on the membership roles.

Organizational Requirements

Once an association properly limits itself to farmer members, what kind of organization must it be? The word cooperative is not found in the Capper-Volstead Act. Requirements are found only by reading the act carefully. Some requirements for cooperative organization are specific and some are general. According to the act, associations must be "operated for the mutual benefit of the members . . . as producers." Although this requirement has not been carefully analyzed by any court decisions, it certainly will be satisfied by cooperatives adhering to basic cooperative principles and requirements of state laws previously outlined.

Cooperatives must also meet one of two options given in the act. One option is to restrict voting. The act permits two kinds of voting and prohibits another. A one-member, one-vote rule satisfies voting requirements of the Capper-Volstead Act, as does voting based on the amount of business done with the cooperative. (The common statement that the Capper-Volstead Act requires a one- member, one-vote rule is not correct, because not only is voting restriction an option, but patronage-based voting is not prohibited.) However, the act does not permit the typical corporate voting method of basing voting on the number of shares owned.

The alternative to restricting voting is to limit dividends on stock or membership capital to 8%. Of course, a cooperative may meet both voting and dividend limit requirements; in fact, most cooperatives do. The Capper-Volstead Act does not require this, however.

All cooperatives must also meet the nonmember patronage rule. They may not deal in products of nonmembers exceeding the value of business done with members—this is the *fifty percent rule.*

WHAT FARMER COOPERATIVES CAN DO

The Capper-Volstead Act, interestingly enough, does not specify what associations are authorized to do. Instead, it states that "persons engaged in the production of agricultural products" may perform certain activities collectively through associations. The distinction between persons acting together and associations is significant. As a business organization, a cooperative needs no authority to market, process, or otherwise engage in business. However, farmers, as noted previously, cannot combine to fix prices without specific permission. Therefore, when we describe what cooperatives may do under the act, we should keep in mind that the Capper-Volstead Act applies to farmers working together and therefore grants individual farmers limited protection from antitrust laws.

In the words of the act, farmer members may "act together" (the key to cooperation) in "collectively processing, preparing for market, handling, and marketing." Some of these terms have been explained in court decisions as a result of challenges to cooperatives' marketing activities.

Marketing and Bargaining

Marketing is a broad term. Many cooperatives purchase members' products and resell them to buyers, returning net income to members. This is clearly marketing. However, many farmer associations do not actually purchase members' products. They act instead as farmers' bargaining agents. The association bargains with buyers for prices and terms of sale; then farmers sell directly to the purchasers at the agreed prices and terms.

Two court decisions have established bargaining and other related activities as part of the marketing process protected by the Capper-Volstead Act. When two farmer bargaining associations sued potato processors in 1966 for antitrust violations, processors claimed that the associations had violated antitrust laws themselves. A U.S. Circuit Court of Appeals discussed the issues in *Treasure Valley Potato Bargaining Association* v. *Ore-Ida Foods* (1974).

Processors claimed that bargaining associations could not qualify for antitrust protection because they engaged in none of the functions enumerated in the Capper-Volstead Act. The Court disagreed, stating:

> We think the term *marketing* is far broader than the word *sell*. A common definition of "marketing" is this: "The aggregate of functions involved in transferring title and in moving goods from producer to consumer, including *among others*, buying, selling, storing, transporting, standardizing, financing, risk bearing, and *supplying market information*." Webster's New Collegiate Dictionary, 1953 Edition. [Emphasis added] The associations here were engaged in bargaining for the sales to be made by their individual members. This necessarily requires supplying market information and performing other acts that are part of the aggregate of functions involved in the transferring of title to the potatoes. The associations were thus clearly performing "marketing" functions within the plain meaning of the term. We see no reason to give that word a special meaning within the context of the Capper-Volstead Act.

In an action against Central California Lettuce Producers Cooperative, a Federal Trade Commission complaint said a bargaining association was not eligible for protection as a farmers' association because it did not bargain on members' behalf. The association did not grow, harvest, or ship lettuce in its own name. It did not negotiate directly with lettuce buyers, and it did not enter directly into sales agreements with buyers. It had no sales personnel in its own name, and it had no receipts from lettuce sales.

Instead, each member entered into separate arrangements with buyers for the sale of that member's lettuce. Individual members conducted all their own negotiations and sales. According to the administrative law judge, the cooperative served only "as a meeting ground for the lettuce producers to come together and agree on pricing policy." Members agreed to sell only at prices within ceiling and floor prices set by the association.

The Federal Trade Commission, on reviewing the case, ruled that the association action should be protected, rejecting claims that a cooperative must be a functioning corporate entity dealing directly with purchasers on behalf of farmers. The Commission stated:

> If . . . it is sufficient merely for the cooperative to unite producers in "collectively negotiating" over price, legal consequences should not attach if the cooperative presents the results of its decisions through each member rather than through a single agent representing each member. It would seem to make little sense, for example, to require that the employees of the growers become employees of Central only to thereafter go about their business of negotiating the sale on different terms of individual members' crops in their fields or cooling plants. As the Supreme Court said, in a different context, the Capper-Volstead Act does not lend itself to such an incongruous immunity-distinction . . . as that urged here.

Setting Prices

Court rulings have clearly established that cooperatives may set prices at which farmer products are to be sold through the association, and it is acceptable for pricing to be the only activity in which the cooperative engages on behalf of farmers. Of course, cooperatives also may do more than set prices, as stated in the Capper-Volstead Act. *Processing, handling, and preparing for market* are broad terms, and the variety of services farmers provide themselves cooperatively, discussed in other parts of this book, show the extent of those activities.

Cooperation by Cooperatives

Cooperatives can cooperate with each other. The Capper-Volstead Act authorizes farmers and cooperatives to use marketing agencies in common. It does not limit farmers to one cooperative or force local cooperatives to dissolve when their members wish to join marketing or processing forces with other groups of farmers.

Cooperatives can cooperate in several ways. Two or more cooperatives may simply coordinate their marketing or processing activities without forming another organization, as in the Treasure Valley case, although there are dangers associated with such an informal arrangement. On the other hand, cooperatives may form another cooperative to make a federated system. Local cooperatives become members of the federated structure, and for Capper-Volstead Act purposes, federated cooperatives are treated as associations made up of farmer producers as long as locals' members are farmer producers.

Cooperatives may coordinate their marketing by arrangements somewhere between informal coordination and fully federated structures. Marketing agencies-in-common are usually based on agreements to coordinate specific activities, leaving individual cooperatives to operate freely otherwise. Cooperatives may also make arrangements to share specific facilities or to supply items of produce. Formalities may include contractual agreements, joint ventures, corporate structures, or formal partnerships. Courts have interpreted the purpose of the Capper-Volstead Act as one of protecting a wide range of intercooperative agreements that permit farmers to market, process, handle, and prepare products collectively for market. The best way to analyze any arrangement is to go directly to the words of the Capper-Volstead Act and to compare what farmers are doing with the act's description of activities protected.

Making Contracts

The Capper-Volstead Act states that "associations and their members may make the necessary contracts and agreements to effect such purposes." These include contracts between members and their cooperatives and between cooperatives and those to whom cooperatives sell farmers' products.

Some cooperatives have agreements with members requiring members to market their products through the cooperative. Antitrust laws do not invalidate member agreements. For example, in the case of *Holly Sugar Corporation* v. *Goshen County Cooperative Beet Growers Association* (1984), sugar beet producers had signed such contracts with their cooperative. The cooperative and processor to whom beets were usually sold could not reach an agreement, yet the cooperative contract prevented members from selling directly to the processor. A Circuit Court of Appeals held antitrust laws would not prevent the cooperative from enforcing the contracts.

Cooperatives may also make contracts with buyers. These contracts have advantages and obligations for both parties and are often an integral part of a cooperative's marketing function. One kind of contract tested by antitrust laws is called a full supply contract. This commits the buyer to an agreement to buy all supplies from the cooperative and thus prevents purchases from anyone else. Such agreements must be carefully analyzed for antitrust implications. Generally, however, a cooperative can make a wide variety of agreements with buyers to meet its farmer-member needs.

Cooperative Size

Finally, the Capper-Volstead Act sets no size limits on cooperatives. Therefore, cooperatives need not limit membership, although some do because they lack capacity to serve all who would like to become members.

WHAT COOPERATIVES CANNOT DO

We sometimes hear the statement "cooperatives are exempt from antitrust law." This is an incorrect statement, and it is dangerous because it may mislead cooperative directors, management, and employees about antitrust limitations on cooperative actions. Another statement sometimes applied to cooperatives is: "The Capper-Volstead Act lets farmers form cooperatives; but, once formed, cooperatives are subject to antitrust rules just as other corporations." Although this is also not completely accurate (we have seen how cooperatives can coordinate pricing with other cooperatives, something IOFs cannot do), the statement comes closer to the overall position of farmer cooperatives under antitrust laws.

What are some of the kinds of things farmer cooperatives may not do under the antitrust laws, even though they qualify as Capper-Volstead Act cooperatives? Two kinds of activity are forbidden farmer cooperatives. One relates to involving nonproducers in otherwise protected activities; the other, to certain kinds of unacceptable behavior.

IOF, Nonproducer Dealings

We have seen that strict membership requirements limit cooperatives because the Capper-Volstead Act is designed to benefit only farmers, and nonproducer membership would allow joint marketing efforts by nonproducers. For similar reasons, once a cooperative is formed, it may not engage in activities with IOFs that would extend protection of the Capper-Volstead Act to nonfarmer, IOF businesses. Two important U.S. Supreme Court decisions discuss the principles and rules of law applied.

In *U.S. v. Borden* (1939), a dairy marketing cooperative was accused of involvement with a number of other parties in a widespread effort to illegally control prices and flows of milk. The Supreme Court limited the Capper-Volstead Act's protection in the following words:

> The right of these agricultural producers thus to unite in preparing for market and in marketing their products, and to make the contracts which are necessary for that collaboration, cannot be deemed to authorize any combination or conspiracy with other persons in restraint of trade that these producers may see fit to devise. In this instance, the conspiracy charged is not that of merely forming a collective association of producers to market their products but a conspiracy, or conspiracies, with major distributors and their allied

groups, with labor officials, municipal officials, and others, in order to maintain artificial and non-competitive prices to be paid to all producers for all fluid milk produced in Illinois and neighboring States and marketed in the Chicago area, and thus in effect, as the indictment is construed by the court below, "to compel independent distributors to exact a like price from their customers" and also to control "the supply of fluid milk permitted to be brought to Chicago." . . . Such a combined attempt of all the defendants, producers, distributors and their allies, to control the market finds no justification in Section 1 of the Capper-Volstead Act.

The standard applied by the Supreme Court has been applied in numerous cases since. The decision does not forbid cooperatives to have agreements or dealings with noncooperative firms. That, of course, would destroy the reasons for which they exist. However, cooperative-IOF arrangements generally must meet antitrust tests applied to arrangements among IOF firms under the usual rules of the Sherman, Clayton, and Federal Trade Commission Acts.

A second Supreme Court decision, made in the case of *Maryland and Virginia Milk Producers Association* v. *United States* (1960), applied reasoning similar to that used in the Borden case to another kind of activity involving IOFs—a merger with or acquisition of an IOF firm. The cooperative in Maryland and Virginia purchased the assets of a competing IOF dairy firm. The trial court had found several improper reasons for the cooperative's purchase of its competition. The Supreme Court called it a "classic combination or conspiracy to restrain trade" unless protected by the Capper-Volstead Act.

The Court denied Capper-Volstead Act protection for the acquisition, explaining:

> The contract of purchase here, viewed in the context of all the evidence and findings, was not one made merely to advance the Association's own permissible processing and marketing business; it was entered into by both parties, . . . because of its usefulness as a weapon to restrain and suppress competitors and competition in the Washington metropolitan area. We hold that the privilege Capper-Volstead grants producers to conduct their affairs collectively does not include a privilege to combine with competitors so as to use a monopoly position as a lever further to suppress competition by and among independent producers and processors.

Predatory Practices

Prohibited actions are not limited to dealings with nonproducer, IOF businesses. Another class of actions, sometimes called *predatory practices,* also falls outside protection extended by the Capper-Volstead Act. Predatory practices are nowhere defined. They are determined on a case-by-case basis in the context of all facts and circumstances of each situation. Generally, predatory practices are actions having no legitimate business purpose and intended to monopolize or restrain trade. They may, of course, involve agreements with IOFs, but they may also be unilateral acts by a cooperative. We can develop an understanding of predatory practices by looking at examples described by courts that have labeled conduct unacceptable and have ruled against protection by the act.

The court deciding the Maryland and Virginia case condemned interfering with shipments of nonmembers' milk, inducing a competitor to transfer marketing outlets outside the cooperative's market area, compelling a business to deal with the cooperative by member boycotts, using economic power to force others to deal with the cooperative, and as mentioned earlier, buying out competition at an exorbitant price to eliminate competition.

A district court in *Bergjans Farm Dairy Company* v. *Sanitary Milk Producers* (1965) also denied protection to the cooperative, explaining that the cooperative tried to compel buyers to deal exclusively with the cooperative, that it gave illegal rebates, that it tried to control and manipulate the market, and that it conspired with retail outlets to fix resale prices. The court also found the cooperative had a specific intent to control the market through various unlawful means.

That same year a circuit court of appeals in *North Texas Producers Association* v. *Metzger Dairies* found a cooperative compelled its customers to deal only with it and used a boycott to control the market. Practices described as predatory are not protected by the Capper-Volstead Act. Terms such as "predatory trade practices," "competition stifling practices," "monopolization," and "preying on independent producers, processors, or dealers intent on carrying on their own businesses in their own legitimate way" summarize unacceptable conduct, with or without protection of the Capper-Volstead Act, legitimate business practices or not. Of course, actions are interpreted only in the circumstances in which they take place, often in light of the cooperative's intent.

UNDUE PRICE ENHANCEMENT

Section 2 of the Capper-Volstead Act states that "if the Secretary of Agriculture shall have reason to believe that any such association monopolizes or restrains trade in interstate or foreign commerce to such an extent that the price of any agricultural product is unduly enhanced by reason thereof," the secretary may issue an order that the cooperative cease to monopolize or restrain trade. This section of the act has generated much comment and discussion, although no investigation has ever led to a finding of undue price enhancement or to an order against a cooperative.

The most interesting feature of the section is the question it raises about what constitutes undue price enhancement. A significant and entirely legitimate purpose of a farmers' association is to enhance the prices that farmers receive for their product. But at what point does the enhancement become "undue" enhancement? Although the question has been discussed by economists, no study in connection with an actual enforcement proceeding has answered that question either in general terms or for a specific cooperative.

The act also does not direct the secretary to issue an order regarding the unduly enhanced price. The conduct prohibited by section 2 is monopolization or restraint of trade (traditional antitrust standards). It is that conduct only against which the secretary may issue an order. The section also requires a cause-and-effect relationship between monopolization or restraint of trade and the unduly enhanced price. Section 2 gives no authority to act if prices are "unduly enhanced" through means other than monopolization or restraint of trade, or if monopolization or restraint of trade occur but do not result in unduly enhanced prices.

ROBINSON-PATMAN ACT

The Robinson-Patman Act of 1936 prohibits certain kinds of price discrimination. A seller may not discriminate in prices charged different purchasers if the result of that discrimination "may be substantially to lessen competition or tend to create a monopoly in any line of commerce, or to injure, destroy, or prevent competition with any person who either grants or knowingly receives the benefit of such discrimination, or with customers of either of them." One way of discriminating among customers may be to charge initially the same price to each but later give rebates to some and not to others.

Robinson-Patman Act rebate prohibition does not apply to patronage refunds paid by cooperatives. The act says that nothing prevents a cooperative "from returning to its members, producers, or consumers the whole, or any part of, the net earnings, or surplus resulting from its trading operations, in proportion to their purchases or sales from, to, or through the association."

AGRICULTURAL FAIR PRACTICES ACT OF 1967

Forty-five years after the Capper-Volstead Act defined farmer cooperatives' position with respect to basic antitrust laws, Congress addressed problems cooperatives face when economic pressure is used to undermine membership and effective marketing. In the Agricultural Fair Practices Act of 1967, Congress said that the marketing and bargaining positions of individual farmers are adversely affected unless farmers are free to join together voluntarily in cooperative organizations authorized by law. It said "interference with this right is contrary to the public interest and adversely affects the free and orderly flow of goods."

The act lists practices in which purchasers, contractors, or agents purchasing agricultural products may not engage as they deal with farmers. They may not coerce farmers to join or refrain from joining a cooperative or refuse to deal with a farmer because of the farmer's right to join or not join a cooperative. Nor may they discriminate against a farmer with respect to price, quantity, quality, or other terms of purchase because of membership in or a contract with a cooperative. They cannot coerce or intimidate producers in connection with a membership agreement or marketing contract with a cooperative, and may not give a farmer an inducement or reward for not belonging to a cooperative. The act does not prevent handlers and producers from selecting their customers and suppliers, and it does not force a handler to deal with a cooperative.

The act may be enforced by private individuals or the government. Producers who believe that they are victims of discrimination or prohibited treatment may ask a court to stop practices not allowed by the act. Also, if the Secretary of Agriculture has "reasonable cause to believe" that someone has engaged in a prohibited act, the secretary may request that the Attorney General bring a civil action to restrain the prohibited acts and grant other appropriate relief. For a number of reasons the Agricultural Fair Practices Act has not turned out to be a very successful way to address many of the difficult problems that occur when farmer producers, their cooperatives, and handlers bargain for farmers' products.

STATE ANTITRUST LAWS

States have their own antitrust laws, varying from state to state. Some are similar to federal antitrust laws, and some differ considerably. If the kinds of activities that farmers engage in through their cooperative are prohibited by state antitrust laws, farmers need state-level protection as well. Section 6 of the Clayton Act and the Capper-Volstead Act offer protection only from the federal antitrust laws, so only states can protect farmers from state antitrust law.

Most states have adopted cooperative antitrust provisions, usually as part of the cooperative incorporation statute of the state. Two related provisions are found in most statutes. A statute may state that formation of a cooperative marketing association is not in itself illegal under the state's antitrust laws. The most typical provision says:

> Any association organized hereunder shall be deemed not to be a conspiracy nor a combination in restraint of trade nor an illegal monopoly; nor an attempt to lessen competition or to fix prices arbitrarily or to create a combination or pool in violation of any law of this State; and the marketing contracts

and agreements between the association and its members and any agreements authorized in this act shall be considered not to be illegal nor in restraint of trade nor contrary to the provisions of any statute enacted against pooling or combinations.

In addition, state statutes usually protect intercooperative cooperation from antitrust laws. A typical statutory provision explains:

Any association may, upon resolution adopted by its board of directors, enter into all necessary and proper contracts and agreements and make all necessary and proper stipulations, agreements and contracts and arrangements with any other cooperative corporation, association or associations, formed in this or in any other State, for the cooperative and more economical carrying on of its business or any part or parts thereof. Any two or more associations may, by agreement between them, unite in employing and using or may separately employ and use the same personnel, methods, means and agencies for carrying on and conducting their respective businesses.

As in federal law, state antitrust protection is not guaranteed in the form of a blanket exemption for every activity of the cooperative. Even cooperatives properly organized and incorporated under appropriate state laws are prohibited from engaging in certain kinds of activity, depending on the antitrust law requirements of the states in which they operate.

CHALLENGE AND RESPONSE

The main provisions of the antitrust statutes have changed very little in the many years since they were first enacted, and section 6 of the Clayton Act and the Capper-Volstead Act have not changed at all. Other conditions, however, have changed considerably. Court interpretations constantly seek to apply the broad language and underlying principles to a vast array of situations. More important, the economic circumstances, cooperatives, and farming itself have changed. Consequently, new issues must be addressed. In this section we summarize some areas of possible future change in the application of antitrust law to cooperatives.

Farmer Members

We have discussed the critical importance of farmer membership in preserving benefits of the Capper-Volstead Act to farmers. The National Broiler Marketing Association case demonstrates the challenge to cooperatives as the very nature of the farming operation changes. Farming operations are sometimes complex organizations with various functions of the overall farming system conducted by separate entities. Consider this situation: an operation may use one entity to purchase and apply inputs, another to cultivate and harvest, another to store supplies, another to make market decisions, another to finance operations, another to manage operations, and another to own and lease machinery. In addition, ownership of the members' product may be in the hands of the buyer from the time the product is planted—all on land owned by yet another entity. This is a rather extreme example, but it is possible. In this case, which of these entities should be granted Capper-Volstead Act protection to join other producers in collective marketing?

Clearly, there are no easy answers. The Capper-Volstead Act contains no restrictions on the size of the farm or its organization, whether it consists of one individual, a family, a partnership, or a corporation. Size cannot be a criterion, nor can organizational form. An economic realities test is difficult to apply. For example, in the National Broiler Marketing Association case, the lower courts identified risk bearing as an important key, explaining that cooperative members bore the true risk of growing broilers. But the Supreme Court did not accept the conclusion of the lower courts on that issue. Rather, the Supreme Court discussed the act's purposes in relation to relative disadvantages in the marketplace. No criteria exist, however, to include or exclude membership based on market position.

The challenge to cooperatives and their advisors is to make careful judgments about cooperative formation and membership. Penalties for improper membership are severe.

Business Arrangements with IOFs

A number of cooperatives have found it beneficial to join in various kinds of joint ventures and other business arrangements with IOFs. However, as we have seen, the Capper-Volstead Act protection ends with the cooperative, and a cooperative will lose its own protection if it engages in activities with IOFs otherwise prohibited by antitrust laws. The complexity of some cooperative IOF joint arrangements makes it difficult to judge clearly the limits beyond which the coordination of activities may not go without losing protection.

Size, Power, and Monopolization

Some cooperatives are very large, representing a significant portion of products in the market. As we have mentioned, the Capper-Volstead Act places no limits on the size of a cooperative. On the contrary, courts have ruled that cooperatives may grow by voluntary membership until all producers of a product belong to the cooperative, giving it a 100% share of the market. To many, however, the relative size of cooperatives is an indication of their market power, and the ability to gain market power is related to special protection by the Capper-Volstead Act. Thus the challenge to cooperatives is to demonstrate that for a voluntary membership organization marketing the kinds of products produced by farmers, market share does not have the same meaning as market power.

Market size and power relate to a somewhat unclear area of antitrust law for cooperatives protected by the Capper-Volstead Act. We have seen that cooperatives may not engage in predatory practices or combine with IOFs to violate antitrust laws. But a question remains: Can a cooperative that does neither still violate antitrust laws?

The problem lies in defining monopolization for cooperatives. Can a cooperative monopolize a market without engaging in predatory practices, given that it can grow to represent all producers of a product? Recent court decisions have cast the issue in some doubt, although most cooperative scholars feel the predatory practices rule is still the best. The challenge to cooperatives, particularly to large cooperatives, is to monitor cooperative activities under the economic circumstances in which they operate. The ultimate answer, however, may be determined by judicial decision.

Antitrust Enforcement

The U.S. Department of Justice and the Federal Trade Commission enforce antitrust laws, and the Secretary of Agriculture enforces section 2 of the Capper-Volstead Act. The Department of Justice and the Federal Trade Commission jurisdictions overlap in some areas, so the two agencies coordinate their enforcement. Both agencies have brought antitrust suits against farmer cooperatives.

Issues have been raised about the effect of the Secretary of Agriculture's role under the Capper-Volstead Act on Department of Justice and Federal Trade Commission actions against cooperatives. It is generally held that the secretary's role does not exclude concurrent jurisdiction by the other agencies. The U.S. Supreme Court in *Borden* discussed the relation of section 2 and the Sherman Act in these words:

> The Capper-Volstead Act contains no provision giving immunity from the Sherman Act in the absence of a proceeding by the Secretary. We think that the procedure under section 2 of the Capper-Volstead Act is auxiliary and was intended merely as a qualification of the authorization given to cooperative agricultural producers by section 1, so that if the collective action of such producers, as there permitted, results in the opinion of the Secretary in monopolization or unduly enhanced prices, he may intervene and seek to control the action thus taken under section 1.

In recent years Congress has placed certain restrictions on Federal Trade Commission investigations and on prosecution of cooperative activities qualifying for Capper-Volstead Act protection. This does not mean that antitrust laws enforced by the Federal Trade Commission do not apply to cooperatives, only that prosecution must meet certain standards.

Challenge to Exemption

Criticism of antitrust protection for farmers is occasionally raised. The arguments usually focus on three areas. First, the Capper-Volstead Act was enacted in 1922 to respond to a particular set of economic circumstances. Opponents of the act argue that those circumstances have changed, particularly with the growth in size and sophistication of farming. A second reason for suggesting changes is that cooperatives, especially in some commodities, have grown in size and power to a point never contemplated by Congress when it passed the Capper-Volstead Act. The Capper-Volstead Act was designed only to protect the small and helpless, and should not be available to protect farmers no longer in that position. Finally, some see the treatment of farmers as unfair to competition because competitors do not have the same protection.

In 1978, Secretary of Agriculture Bob Bergland addressed a number of such issues before a commission investigating several aspects of antitrust law. He summarized responses to Capper-Volstead Act challenges as follows:

> The exemption from antitrust laws given to farmer cooperatives is a limited, carefully drawn exemption, narrowly construed by the courts to achieve only the purposes behind the exemption, and useful only to farmers in the difficult task of moving food to the consumer.
>
> Cooperatives have been and will continue to be examples of efficiency in the food system. They help bridge the gap between the farm gate and the consumer's table. They bring efficiency to the system because they themselves are efficient. And they force others in the system to be efficient to compete. They serve both the producer and consumer well as the yardstick by which effective marketing can be measured.
>
> The public is protected under present law, as are the farmers. The protection is balanced. The presence of cooperatives in the economy has had untold benefits for us all. In fact, without strong cooperatives, I cannot imagine an economic system as healthy, as productive, as responsive, or as efficient as the one providing us all food at a reasonable price.
>
> My own view, I believe well supported by history, experience, and research, is that the Capper-Volstead Act and our marketing order system are in no need of statutory modification.
>
> Actions to modify these agricultural provisions may be intended to increase competition, but they may in fact weaken competition. The buyers' side of the agricultural product markets has gained tremendous strength because of buyers' size. But the producers' side is still made up of individual farmers. Their only realistic hope of some equity in the market is effective cooperation.
>
> A false step based on a mistaken view of competitive forces in agriculture could well lead to increased Government regulation. This country cannot allow, and will not permit, the efficient producers of our most basic needs to be pushed to the edge of failure. If the protection given these individual farmers is weakened, we may be forced more deeply into market intervention by Government. The whole purpose of our national antitrust policy of preserving individual opportunity would be lost for the farmer, and if it fails for the farmer, we will all pay a terrible price (Bergland).

This statement incorporates much of what we have learned about basic U.S. antitrust policy, farmers' special needs, laws designed to address these needs, and the limits of protection extended to farmers for cooperative marketing.

SUMMARY

In the United States, antitrust laws make some kinds of business practices illegal, including agreements among competitors to fix prices, or behavior designed to destroy competition. The laws can apply to farmers who join together to price and market their products jointly. In 1922 the Capper-Volstead Act gave farmers permission to form cooperatives without violating antitrust laws. The act lets farmers market, bargain, price, and otherwise jointly organize their efforts. It does not fully exempt cooperatives from antitrust laws, and farmer cooperatives may not be used by nonfarmers to fix prices. Cooperatives cannot engage in predatory practices harmful to others. Several issues and problems face cooperatives under current law. Challenges to farmers' right to form cooperatives require students of cooperatives to be informed and articulate in explaining what cooperatives are and what the Capper-Volstead Act means, including the limits it imposes.

DISCUSSION QUESTIONS

23-1.   The Supreme Court has said that competitive forces will "yield the best allocation of our economic resources, the lowest prices, the highest quality and the greatest material progress." How do farmer cooperatives contribute to these same goals even though they have exemptions from the antitrust laws? How could these desirable attributes of our economic system be damaged if farmers could not form marketing cooperatives?

23-2.   A group of farmers and representatives of two companies that process and sell farmers' products explain a coordinated marketing plan to you. They can prove prices to consumers will be lowered and stabilized and a better product will result from coordinating marketing efforts. What advice can you give them?

23-3.   You receive requests from the following different groups for help to set up marketing cooperatives: Christmas tree growers, catfish growers, muskrat trappers, cattle-feeding operators who feed other people's cattle on a contract basis, two broiler integrators that own chicks but contract out growing to independent broiler house owners, a group of poverty-level sharecroppers who own no land and are at the pricing mercy of the only buyer in the area, and landowners who rent their land to tenants on a cash basis. What do you consider in each case, what questions do you ask, and what is your advice to each group?

23-4. You are approached by three irate farmers whose fresh produce was just refused at a buyer's plant because the buyer's nephew was caught stealing farmers' watermelons last summer and was turned over to the sheriff. The produce spoiled because it was too late to ship it elsewhere. The farmers have a long list of concerted activities against the buyer, suggested by their previous advisor, Rambo, some of which involve weapons purchases. What do you advise?

23-5. A local elevator owner complains to the newspaper that the cooperative competing with the elevator has unfair advantages because it is exempt from antitrust law and does not have to pay taxes. The owner claims that the cooperative consistently pays farmers less for products but uses contracts to force farmers not to leave the cooperative or to sell to whom they choose. The newspaper comes to you for your opinion. Use your knowledge of the cooperative form of business enterprise to explain cooperative principles and how they relate to antitrust laws.

REFERENCES

BERGLAND, BOB, testimony submitted by the Honorable Bob Bergland, Secretary of Agriculture, National Commission for the Review of Antitrust Law and Procedures, July 17, 1978.

JESSE, EDWARD V., ed., *Antitrust Treatment of Agricultural Marketing Cooperatives, NC Regional Res. Pub. 286, NC 117 Monograph 14-15, Univ. of Wisconsin-Madison, Sept. 1983.*

MANCHESTER, ALDEN C., *The Statuts of Marketing Cooperatives under Antitrust Law.* Washington, DC: USDA ERS-673, Feb. 1982.

NATIONAL COMMISSION FOR THE REVIEW OF ANTITRUST LAW AND PROCEDURE, *Report to the President and the Attorney General,* Washington, DC, Jan. 22, 1979.

24

Future Structure, Problems and Opportunities

Lee F. Schrader,
Purdue University

INTRODUCTION

Most agricultural cooperatives, as well as farmers and other businesses serving agriculture, experienced a major change in their economic environment from the 1970s into the 1980s. In 1974, cooperative leaders believed that their major challenge was to expand quickly enough to meet the needs of their patrons. Farmers saw no end to the demand for their products. By 1983, however, excess capacity was obvious nearly everywhere in agriculture. As a result, survival, rather than expansion, became a major concern. Further changes will be needed on farms and in cooperatives that serve farmers. The nature of cooperatives and their role in the future of agriculture are being shaped by factors outside the control of cooperatives or their patrons and by cooperatives' responses to changing conditions in agriculture. Major external factors beyond the cooperative's control include the emerging biotechnology and information technology, the internationalization of agricultural markets, and national agricultural policy. Major internal factors include the limitations of being a cooperative, capital acquisition, and choice of organizational structure.

TECHNOLOGY

Changing technology continues to affect farm structure. The emerging biological and informational technologies are likely to further trends toward larger and more specialized farms that were established by the preceding mechanical and chemical eras in farming. The number of U.S. farms as defined by the U.S. Bureau of the Census decreased from a peak of 6.8 million reported in 1935 to 2.3 million in 1984. Five percent of the largest accounted for 50% of sales in 1982. In the 1970s, the rate of decrease in number was largest for the smallest farms, while the number of large farms increased. In 1982, 45% of farmers indicated that their primary occupation was something other than farming. Rural residences (farms with sales of less than $10,000 in 1982) represented 49% of the country's farms but only about 3% of total output.

There appears to be a tendency toward a bimodal distribution of farm size: the small and large. Harrington and Manchester (p. 25) see the cause as follows: "Changes in technology and specialization of production have encouraged the formation of very large, highly capitalized farms and very small, part-time farms. The small family farm finds itself too large to allow full-time, off-farm employment and too small to yield adequate income from farming." There has also been a trend toward greater specialization on remaining farms. Economies-of-size for individual enterprises is a major factor in this change.

Technological advances are seldom size neutral; usually, they favor large-scale operation. Technologies presently being developed (such as plant and animal growth regulators, embryo transplants, and genetic engineering) all suggest large and specialized operations, unless new service functions are provided by off-farm firms for farms that are too small to provide their own economically. If the moderate-sized operation is to survive, more technical services must be available at costs comparable to those for the large units providing their own services. The small part-time or semiretirement operation may well survive by using less than state-of-the-art technology and used equipment and by accepting lower returns to capital and labor. Even so, the share of total production accounted for by the largest farms is likely to increase further.

The U.S. Congress Office of Technology Assessment (OTA)(p. 20) concluded that the effect of present farm policies, combined with biotechnologies and information technologies expected before the end of this century, "could be the development of a farm structure composed of three agricultural classes." They are (1) a large farm segment composed of "as few as 50,000 farms producing as much as three-fourths of the agricultural production," (2) a "struggling moderate-size farm segment—trying to find a niche in the market and survive," and (3) a "small, predominantly part-time farm segment" obtaining most of its net income from off-farm sources. OTA concludes that the moderate-sized farms are least likely to survive.

These same developing technologies suggest a further adaptation of crops and animals to produce products designed for specific end uses. Today's varietal differences may be only a sample of what is to come. For example, the production of waxy maize for starch production and white corn for certain foods is only a hint of the type of specialization within a plant or animal species which may be advantageous.

The prospect of modifying plants and animals to fit specific market needs suggests a much greater benefit from close coordination of production and processing. Products will also be adapted to fit specific processes. Both of these changes imply that markets for given specification will become narrowed. For example, with only a few exceptions, corn is corn in today's market; however, if varieties were developed specifically for feed, for wet milling, for dry milling, and perhaps for new uses as well, the number of alternative outlets for any one type of corn would be much smaller than the number that is now available for the commodity corn. Clearly, the coordination/allocation problems would be more complex. As the OTA (p. 12) report states, "biotechnologies will have the greatest impact because they will enable agricultural production to become more centralized and vertically integrated".

More complex coordination leads to more complex production control and generally, to a greater need to fine tune all stages of the food and fiber system. All of which calls for a higher level of management at all levels of the system. These trends suggest there will be less reliance on spot markets and increased importance of planning and developing methods to improve coordination of farm units with other operations in the food and fiber system. Some of the necessary decision making at the less concentrated levels of the system (farm, first handler, and supply retailer levels) will be provided by consultants or computerized expert systems. As the income differential between the well-managed and appropriately coordinated farms and traditional independent farms increases, fewer of the latter will survive. The question is not whether agricultural production and marketing will be more closely coordinated but whether the job will be done by farmers who choose to organize the system cooperatively or by other types of firms not controlled by farmers.

Key issues confronting cooperatives are (1) whether existing cooperatives will serve both large and small farmers, (2) how to deal with the bio and information technologies, and (3) to what degree the coordination function can be performed cooperatively. The size disparity among farmers challenges existing cooperatives to serve patrons with very different needs. For example, occasionally buying clubs have been organized among large farmers. These farmers pool their orders, pay in advance, and agree to accept products at the same time. The designated member(s) then invite(s) bids to supply the entire amount. Such volume, no-frills buying has resulted in materially lower prices to club members. This creation of new cooperatives is remarkably similar to actions of progressive farmers more than 60 years ago. The method may be feasible only when there is idle capacity in an industry. Nevertheless, present cooperatives will need to either devise the means to serve these patrons on their no-frills terms in a manner that is equitable to all, or else relinquish this business to the buying club type of cooperative. A cooperative's election to serve only the traditional medium-to-small farmer would indicate acceptance of a diminishing role.

Biological sciences have also produced technologies which, when adopted, will change the nature of agriculture. Existing cooperatives do not have the resources to conduct developmental research on their own. If all the development occurs in the private sector, cooperatives may have a role as a licensee to distribute products and to provide technical assistance to the farmer-user. If the development work is done by public entities, the opportunity for cooperatives to participate will be much greater. In either case, it will be important for cooperatives to win a role in the distribution of these new inputs and in the provision of management services; otherwise, they will probably not be able to retain their role in traditional input supply. If cooperatives and their patrons fail to accept the need for a new level of coordination of the production and processing functions, other firms may take the dominant position in an integrated agriculture. The success of cooperatives in the closely coordinated poultry and fruit and vegetable systems illustrates that cooperative coordination is possible, even perhaps with the major field crop and livestock systems, which will present the greatest challenges for them.

Cooperatives may also want to find some way of competing successfully in international markets. The export boom of the early 1970s placed U.S. agriculture squarely into an international market, and the midwest regional cooperatives made an effort to take a role beyond that of first handler in the grain trade. Their success was minimal, however, because their effort was late and fragmented. Farmers were not committed to the cooperative exporting effort, because they were not willing to forgo opportunistic behavior in order to build a long-term relationship. It is a challenge to explain to midwest producers how their cooperative's selling Argentine corn to a European buyer is in the U.S. producers' best interest even though that might be necessary from time to time to be a credible international supplier.

To a substantial degree the game is now back in the hands of the multinational trading firms. It may be more productive for cooperatives to develop products adapted to the needs of foreign buyers. Stable markets for unique products may be more appropriate targets than the commodity markets in which the lowest price is the major factor in a sale. The payoff from development of international markets for high-value products, while potentially large, is uncertain and may be realized only after years of investment.

BEING COOPERATIVE

Being a cooperative will also remain a challenge in the future. Olson has shown that the mutual benefit of some group achievement is not sufficient reason to expect all members of a group to act to achieve an objective. He states: "Indeed, unless the number of individuals in a group is quite small, or unless there is coercion or some other special device to make individuals act in their common interest, rational, self-interested individuals will not act to achieve their common or group interests" (p. 2). To illustrate this point, one can show easily that the income of all farmers could be increased through collective action to control the flow of products to market or to coordinate production and processing. But in nearly all cases the individual can achieve the greatest gain by acting counter to the group's best interest, provided that all or nearly all others in the group do act in the best interest of the group. When individual members of a marketing cooperative react to their per-unit return from the cooperative as their marginal revenue even though the marginal revenue to the group may in fact be much lower, they fail to exploit a monopoly position. To the extent that the benefits of cooperative action cannot be limited to participating members, the cooperative can expect free riders to diminish that action's effectiveness.

The determination of management rewards and incentives also presents a challenge for cooperatives. For the investor-oriented firm (IOF), profit is usually the clear objective. Thus rewards and incentives related to profits of the firm commonly are offered to managers and employees. Although not without problems, such incentives are usually appropriate and effective. The absence of such a clear and measurable objective for a cooperative complicates the process. Net income may not be relevant. Prices to members may be arbitrary. Members are concerned with the full return for their products (or net cost for inputs), including the value of patronage refunds. Changing initial prices for products marketed for farmers in turn changes savings and patronage refunds, but it does not change the final value to the member. Finally, the maximum value to members may not be achieved by operating the cooperative to maximize its net income.

Measurement of management performance and the reward system must be structured to keep the objective of maximizing patrons' welfare as a first priority. This task, however, is easier said than done. It is easy to say that a cooperative's objective is to work in the interest of its patron-members. The diversity of current patrons has been discussed, but there are others who have an interest in or might be served by a cooperative. Consider the set of all farmers in the cooperative's area. There are current patrons and others who could be. The nonpatrons include former patrons and those who were never patrons. The cooperative also has an obligation to its equity holders. This set includes current and former patrons some of whom are no longer farming. Almost every alternative action or policy is likely to affect these groups differently. Actions and policies that can be decided by an IOF based solely on the earnings of the firm have equity implications for a cooperative.

The situation is further complicated by the fact that cooperative members often may expect a higher level of service from their cooperative than they expect from others—at the same price. A wait for service accepted at another business may be considered too long at the cooperative where the patron is an owner. The farmer who usually stores grain on the farm may be critical if the cooperative has no space for the farmer's excess in a large crop year. These factors, unless resisted, tend to leave the cooperative with more capacity and personnel relative to turnover than other businesses have.

A cooperative faces similar problems when it needs to close a facility. Other firms can write off their mistakes and move on. The user of the closed facility or eliminated line of service has little choice but to go elsewhere. However, the cooperative member-patron is also part owner of the cooperative. Thus if the facility or service a member uses is discontinued, so is the benefit from the member's investment. The affected member also has a vote and will probably vote to keep the facility open, even if it is not profitable to do so. However, the more nearly a member's investment corresponds to a member's business with a cooperative (by line of business and facilities), the less of a problem this is. For example, if the cooperative handles only one product through one facility, the problem does not exist. Members must balance their desire for the service with the return to (or loss of) their equity in the organization. The better the correspondence between an individual member's equity and his or her volume of business, the more likely the appropriate decision will be made.

Obtaining equity capital will also remain a problem for cooperatives. Outsiders have little incentive to invest when returns are limited and losses are not. It is likely that the major source of equity will remain the patrons themselves. Their desire to be paid high cash patronage refunds lengthens the time that cooperatives must retain funds and makes noncash patronage refunds worth even less. A current base capital plan with equity well aligned with patronage and a short period allowed to redeem the capital of those no longer using the cooperative has the potential to capture the attention and loyalty of patrons. On the other hand, the cooperative that elects to retain all or nearly all net returns as tax-paid, unallocated capital neither operates at cost nor differentiates itself well from the operation of an IOF. Once a revolving fund has been allowed to lengthen, it is very difficult to recover. The patron has already begun to heavily discount allocated equity.

Finally, cooperatives will have some problems because their affairs tend to be more public than those of other firms, and the process of informing members means revealing information that may be useful to competing firms. The swift coup is rarely possible if the idea of member control is honored. However, if the membership is very small, this may not inhibit cooperative action.

If the cooperative exists for sound reasons, these problems may be annoying but hardly an excuse for failure. If the reasons for existence are marginal or absent, the problems of being a cooperative could be a deciding factor in its longevity. Cooperation as a business form can be used to perform any legitimate business function. One should recognize, however, that it carries some extra baggage that renders its use questionable outside situations for which it is specifically adapted.

STRUCTURAL ALTERNATIVES

Most existing agricultural cooperatives were formed to serve relatively homogeneous farms. The commodity cooperative organizations grew or shrank with the commodity or changed with their members. Cotton Producers became Gold Kist, and poultry replaced cotton as a major commodity. Arkansas Rice Growers Cooperative became Riceland Foods, with four times as many soybean grower-members as rice grower-members. The product mix of farm supply cooperatives also changed with the pattern of use by their members. But the increased specialization of patrons has rendered some patterns of equity accumulation and computation of patronage refunds out of date. Averaging over several lines of business for computation of net income brought little distortion when nearly all farms had a similar product mix. But now these procedures must also change.

The differences in size and needs between large and small farms have increased, and cooperatives have had difficulty adjusting to this change. Their further challenges are to provide the services, appropriately priced, that these different farmers need and to develop an appropriate allocation system for member control. If the one-member, one-vote principle is to be the rule, it may not be feasible for one cooperative to serve both large and small farms. The large volume user (and the large investor) may be unwilling to belong to a cooperative controlled by a large number of small users who are not willing to have prices fully reflect the lower costs associated with large-volume purchases. On the other hand, the organization of separate cooperatives for large and small users would be an inefficient use of management and facilities. A better alternative may be to organize separate cooperatives by product line, location, and so on, in order to cater to special needs while retaining a common management. Present regional cooperatives could provide management services to specific-purpose cooperatives on a cooperative basis. In this case the new type of local cooperative would be separately capitalized and would have its own board but would gain from management functions of and from coordination with other units under the same management. In this way, cooperatives could fulfill the need for separate facilities, procedures, and policies, yet still benefit from shared management. There is no reason that more than one organization cannot share one management or even some facilities. The structure of cooperative organizations must change to keep up with changing farm structure.

The organization and product/service array must also be adapted to the emerging technology. A key question is whether cooperatives will be able to deliver these technologies to the farms such that the moderate- sized family farm will be a viable participant in this new organization. Will the way be found to cooperatively coordinate the more complex input/production/processing systems which seem likely to emerge without forcing the moderate-sized farms out of business?

Federated systems will face the challenge of answering these questions by defining the role at each level. The larger farmers will require fewer local services but a higher level of management to handle the services they do need. Consolidation of (or outright demise of) some locals will continue. The role of the regional is already in question as more of the consolidated locals or superlocals become capable of providing for themselves products and services they once received from the regional. Recent consolidation of regionals has not affected the local-regional relationship appreciably.

However, local control of local units in a federated system does have disadvantages. First, innovations at the federated level must be sold twice—once to the local and again to the patron. Second, local control can impede changes needed to keep the system efficient. Sometimes changes need to be made by the whole system in order for them to be effective. This can be problematic for a cooperative, because often a few locals may choose not to change, particularly if they have invested recently in the technology or product line to be replaced. Finally, competition among locals is counterproductive. The coordination of production and marketing is more difficult if the processing/marketing function is at the federated level. In many cases it may be more efficient for the regional cooperative to deal directly with the patron. For example, the no-frills feed buyer may not need any of the services provided by a local cooperative. On the other hand, direct selling may not be politically possible in the federated system. Although complete reorganization may not be necessary to deal with this problem, regional cooperatives will need to find the means to deal directly with patrons on some products or services.

Some cooperatives have succeeded at this task through joint ventures. Joint ventures between cooperatives are a variation on the federated structure. Ventures with IOFs have been used to gain the advantage of skilled management and know-how in new areas and, at times, to acquire capital for an undertaking. Ventures with IOFs depend on well-thought-out contracts for mutual success. Often events that were unforeseen at the time the venture was initiated change the balance of material benefits a great deal.

RELATING TO MEMBERS

Member relations are and will continue to be a challenge for cooperatives. To many cooperative members and other persons, the idea of cooperation is firmly associated with equal treatment—equal prices, equal voting, equal time, and so on. Even though the principle of business at cost implies unequal prices when the costs of serving patrons vary, there is often a confusion of equity with equality. The need to distinguish clearly between equity and equality increases as the variety of patrons becomes more diverse. To be equitable in the sense of offering business at cost, cooperatives must price according to cost. A system of pricing based on volume of business and services actually used is necessary. Some unbundling of products and services will be needed. If a patron group can be served at lower cost because of differences in size or services required, and their cooperative does not reflect that lower cost in pricing, that group will form a new cooperative or patronize a firm that does recognize the difference.

Provision of equity capital is a closely related issue. Cooperatives, like other businesses, must offer returns to capital. Equitable treatment requires the provision of an equitable share of capital needed to operate the business. Further, if a patron's capital is commensurate with patronage, the distribution of patronage refunds rewards capital equitably. Revolving funds that do not revolve inspire neither loyalty nor respect. When the benefits to patrons are less than those necessary to justify the required investment, this may be a signal that the market can be served more adequately by another type of firm.

The debate regarding the wisdom of one-member, one-vote control also goes on. Some argue that democratic control means equal votes to each member. Others contend that equal voting is inequitable because the larger patron (investor) is under-represented (Knutson). They suggest that voting should be proportional to patronage. In the end, they claim, it is the volume of patronage that counts because if a group having a large volume is treated inequitably, they will vote with their feet by patronizing another organization. Therefore, it may be impossible to obtain equity capital in proportion to patronage without a similar allocation of control.

Cooperative members also continue to argue about the trade-off between risk sharing and business at cost. The buy-sell relationship typical of supply and grain marketing cooperatives involves a minimum of risk sharing, limited to averaging costs and sharing results of the organization's investments at other stages in the commodity system. When equity capital is not revolved or otherwise not kept proportional, the risks of loss are taken to a substantial degree by former patrons, while the benefits, if any, accrue to current patrons—hardly a defensible position.

If technology drives a commodity system toward closer coordination, it is likely that risk sharing will be increased. That is, as a coordinating agent (the cooperative) assumes responsibility for a greater portion of decisions at the production (farm) level, more of the quality and quantity that affect value will be outside the control of individual members. The member who gives up the option to pick the day to sell will be likely to prefer a pool average price to that of accepting the price offered on the day his or her produce happened to be sold. Cooperatives will have to continue to search for the right balance between offering farmers insurance and insisting that they accept the consequences of their own efforts, despite the continued risks and uncertainties. Perhaps different risk levels from which the member can choose will serve well in some cases.

Member education will continue to be a concern of cooperatives. The need to keep members informed was never more evident than during the period of adversity experienced by farmers and their cooperatives alike. A failure of farmers to recognize a cooperative as their own business has been particularly evident in the Farm Credit System, whose farmers in many cases have referred to their relationship with cooperatives as "we" (farmers) versus "them" (cooperatives). Incidences of farmers taking their own cooperatives to court are also indications of the failure of members to recognize and accept their role in the cooperative.

The concentration of agricultural production into fewer hands will result in larger individual investments in cooperatives. These farm managers may be expected to demand a more complete accounting by their cooperative managements. Involvement in cooperative decisions will also become more important than a free meal and entertainment at an annual meeting. Members who are more informed than is currently typical about the performance of their cooperative will be able to react rationally rather than emotionally to actions taken by the organization. An informed member who understands the cooperative's role is more likely to make patronage decisions based on his or her long-run best interest, rather than simply on the best price available at the moment.

RELATING TO THE PUBLIC

Cooperatives must maintain a favorable image in the eyes of the general public. It is important for them to retain income tax status and antitrust treatment that recognizes the unique character of cooperatives.

To take advantage of the exclusion of patronage refunds from the taxable income of the cooperative, cooperatives must make sure that their income computed for tax purposes corresponds to income used to compute patronage refunds. IOFs are allowed to make adjustments to income reported to stockholders in order to take advantage of any available more favorable alternatives possible under the tax code. The ability to lower the effective corporate tax rate through the use of these tax preferences has lowered the value of the single tax status afforded cooperatives. Exclusion of dividends from taxable income of corporations (proposed by some as part of tax reform) would eliminate the need for special treatment of cooperatives.

AGRICULTURAL POLICY

U.S. agricultural cooperatives, with few exceptions, have played only a minor role in the determination of U.S. agricultural policy. Dairy cooperatives are among only a few that have maintained a high profile in the agricultural policy formulation process. Policies such as price supports and acreage adjustment programs have had a major impact on farmers and their cooperatives. These policy decisions often have been based on the needs of farmers *as farmers* rather than of farmers *as active members* of cooperatives. Farmers have been represented only by general farm organizations or commodity organizations in the policy process.

Some argue that the traditionally neutral position of farm cooperatives regarding government programs has not served their members well. According to Knutson and Black (p. 58), "realistically, the interests of cooperative members would be most effectively reflected in the policy process if cooperatives, themselves, become more directly involved in farm policy making." For example, the 1983 acreage-reduction program using payment in kind proved to be a heavy blow to farmers' cooperatives. The volume of inputs and marketing of grain were dramatically reduced, and international grain marketing activities of cooperatives were seriously affected. It would, of course, be an error to ignore individual farmers' needs and to gear policy solely to the interests of cooperatives; however, to ignore the impact on farmer cooperatives when designing commodity programs may not serve the farmer well either.

Cooperatives are affected by a number of government programs in addition to price supports and production adjustments that seriously limit the building of a marketing program. A number of programs, such as contract law and trade practice regulation, define the rules of trade. Marketing orders may be used to promote orderly marketing. Properly designed and used, they can facilitate a cooperative's marketing programs. It may be necessary for cooperatives to take a more active role in the agricultural policy process to assure that farmers' interest in their cooperatives is heard.

SUMMARY

The pace of change in agriculture has accelerated. Cooperatives must change with the needs of their farmer-patrons or they will have no role in the agriculture of the twenty-first century. A number of factors will shape the survivors.

Emerging biological and informational technologies will affect farmers and their cooperatives. These technologies will probably favor larger farms and specialized operations. A major question is whether cooperatives will be able to provide the technical services needed for the medium-sized farms to survive in the new environment. The application of biotechnology may result in crop and animal lines designed for specific end uses. This would increase the need for close coordination of production and processing.

The increasing disparity in the size of farms challenges cooperatives to serve patrons with very different needs equitably. They will either adapt to serve the large operation appropriately (often with less service included) or forfeit the fastest-growing portion of the market.

Agricultural markets have become worldwide rather than national. This trend is not likely to reverse. Cooperatives have not generally gained a major role in export markets. Farmer commitment to export has been weak, and this fact may limit the potential for cooperatives in export markets.

Being a cooperative presents some special problems as well as opportunities. There will always be free riders because not all the benefits of a cooperative can be limited to those who support it. The nature of the distribution of benefits limits the ability of a cooperative to take full advantage of market power even if they could obtain it. Obtaining equity capital will require innovative programs and more attention to keeping patron capital investment proportional to the use of the cooperative. Cooperatives must develop programs to reward management for performance that benefits patrons. They will also have to guard against (1) providing more services than patrons are willing to pay for and (2) keeping uneconomical facilities or services too long.

Member relations are a continuing challenge for cooperatives. A more diverse patron group may mean that more attention to equity in the treatment of patrons is needed. Equal net margins may be more equitable than equal prices for all. New technologies that will require greater coordination will mean more centralized decisions and a new relationship with patrons in many cases.

Cooperatives should not take for granted the generally favorable public policy toward agricultural cooperatives. Unique tax treatment and special antitrust status must be defended and the conduct of cooperatives as economic entities must not violate the public trust. Some argue that cooperatives should take a more active role in the federal farm policy process.

DISCUSSION QUESTIONS

24-1. What major factors will affect the role and nature of agricultural cooperatives into the twenty-first century?

24-2. What is the most likely change in size and number of farms during the next 15 years?

24-3. What does it imply for cooperatives if, as the Office of Technology Assessment says, "biotechnologies will have the greatest impact because they will enable agricultural production to become more centralized and vertically integrated"?

24-4. Can existing cooperatives serve both very large and small farmers?

24-5. What is a major factor that has limited cooperatives' role in export marketing?

24-6. Explain why members of a group cannot be expected to behave in the best interest of the group.

24-7. Why might paying a manager a bonus based on cooperative net income not always be a good idea?

24-8. Why is it a greater problem for a cooperative to close a facility or stop a service than it is for an IOF?

24-9. Why not sell cooperative equity to the general public?

24-10. Which structure do you believe will serve the cooperatives of the future best—centralized or federated systems? Why?

REFERENCES

HARRINGTON, D. H., AND A. C. MANCHESTER, "Profile of the U.S. Farm Sector," *Agricultural Food Policy Review: Commodity Program Perspectives.* Washington, DC: USDA ERS AER 530, July 1985, p. 25.

KNUTSON, RONALD D., "Cooperative Principles and Practices: Future Needs," *Farmer Cooperatives for the Future,* ed. Lee F. Schrader and William D. Dobson, workshop, St. Louis, MO.; Dept. of Agr. Econ., Purdue Univ., 1985.

KNUTSON, RONALD D., AND W. E. BLACK, *Cooperative Involvement in Issues of Domestic Farm Policy.* Dept. of Agr. Econ. Info. Rep. 86-4, Texas A&M Univ., 1986.

OLSON, M., *The Logic of Collective Action: Public Goods and the Theory of Groups.* Cambridge, MA: Harvard University Press, 1971.

U.S. CONGRESS, OFFICE OF TECHNOLOGY ASSESSMENT, *Technology, Public Policy, and the Changing Structure of American Agriculture.* Washington, DC: U.S. Congress, Mar. 1986.

Glossary

acquisition: the process of one firm purchasing part or all of a previously independent firm. The facility acquired is absorbed and the acquiring firm continues its identity.

advances: a term generally used to denote partial payments made to patrons for products marketed.

adverse selection: a term meaning that the most likely participants are those with the poorest risk or highest cost of service.

agency contract: a marketing contract under which the cooperative does not take title to products delivered by members to the association but handles them on an agency basis.

allocated equity: equity that is assigned by amounts to individuals or organizations, typically in the form of retained patronage refunds and/or per-unit capital retains; investments by patrons for which they have received notification of the allocation. [*syn:* certificate of equity or of investment, revolving fund capital and others]

articles of incorporation: a legal document filed with the state under which a corporation is legally created. It consists of the powers, rights, and liabilities of a corporation granted by the state. It gives authority to proceed as an artificial person subject to the laws of the state where the incorporation took place.

association: an organization of persons with a common purpose united in a formal structure; often used as a synonym for "cooperative."

Banks for Cooperatives: institutions which are part of the Farm Credit System; they are owned by farmer cooperatives, subject to federal control; they make loans to farmer cooperatives.

bargaining cooperative: a cooperative whose sole or principal function is to bargain with buyers for price and other terms of trade on behalf of its members. It typically does not handle and may not take title or handle products.

base capital plan: a plan for providing equity; each member's capital obligation is determined each year by the member's share of total patronage for a base period. Underinvested patrons build equities and overinvested patrons' equities are redeemed in several ways. [*syn*: permanent capital, adjustable capital, adjusted balances, modified revolving capital, capital quota, capital investment, equity pool fund, capital requirement]

bylaw: a legally enforceable rule established by members and/or directors, which specifies operational practice and policy.

Capper-Volstead Act: a federal law establishing a limited antitrust exemption for cooperatives marketing products for farmers.

cash patronage refund: distributions of patronage refunds paid to patrons in cash.

centralized cooperative: a cooperative all of whose members are producers and hold direct membership in the central association.

certificate of membership: a document, given to persons upon payment of a fee, that sets forth the rights, privileges, and conditions of membership in a nonstock cooperative.

Clayton Act: an amendment to the Sherman Antitrust Act, passed in 1914, to, among other things, legalize the organization of agricultural or horticultural associations created for mutual help and not having capital stock or conducted for profit.

closed membership: a policy of no new members admitted to the cooperative or of severe restrictions to qualify for membership.

common stock: shares or stock certificates in a corporation which carry voting rights, unless otherwise indicated, and which may be eligible to receive dividends. There is no due date on such stock and it carries risks associated with investment in equity. Many cooperatives restrict ownership to one share per member in order to achieve the one-member, one-vote condition.

conglomerate integration: the process of adding to a firm business activities which are unrelated to existing lines of business regarding raw materials, manufacturing, marketing channels, and/or product.

consolidation: formation of a new firm from two or more previously independent firms which subsequently lose their identity. *See also* acquisition and merger.

consumer cooperative: a purchasing cooperative that sells primarily goods and services for final consumption.

cooperative: a user-owned and user-controlled business that distributes benefits on the basis of use. [*syn*: association, patron-owned]

corporation: an artificial person or entity created under the laws of a state to act as a single person and legally endowed with specific rights and duties. Powers, liabilities, and rights are separate and distinct from individuals owning the corporation.

democratic control: Members, by majority vote, determine the operation of their cooperative. Decisions can be made on the basis of one-member, one-vote or through proportional voting based on patronage.

director: one of a group of persons elected by the members to govern or control the affairs of the cooperative.

dividend: a distribution of current or accumulated net income paid according to invested capital and paid independent of current patronage.

equitable: fairness or absence of one group subsidizing another.

equitable financing: a situation in which patrons of a cooperative provide equity in proportion to their patronage.

equity: ownership or risk capital in the cooperative generally arising from direct investment, retained patronage refunds, per-unit capital retains, and nonmember business. Total assets less total liabilities. [*syn:* net worth, stockholders' or capital stock or members' and patron's equity, retained margins]

equity redemption: the payment in cash or other property for funds previously invested in the equity of the cooperative.

established prices: prices used to denote competitive prices for goods sold or purchased, or prices set by a board of directors. [*syn:* commercial market price]

exclusive agency bargaining: bargaining for terms of trade by an association that has authority to represent all farmers in a bargaining unit, whether members or not, and to establish rights and obligations of members, nonmembers, and handlers or processors.

exempt cooperative: *See* section 521 cooperative.

fair: *See* equitable.

favorable prices: relatively high prices for farm products and relatively low prices for supplies.

federated cooperative: a cooperative whose members are other cooperatives. Members retain their autonomous character.

free rider: a person or firm that receives benefits of an effort without paying the price or contributing to that effort.

horizontal integration: combining under a single management two more like units at the same stage of the production process.

integration: the act of bringing two or more units or functions under a single management.

investor-oriented firm (IOF): a business firm other than a cooperative, an investor-oriented business in contrast to a user-oriented business or a cooperative. [*syn:* noncooperative, private, for profit, proprietary, ordinary, standard, investor owned, other corporation]

IOF: *See* investor-oriented firm.

joint venture: an association of two or more businesses to carry out a specific economic activity or enterprise and sharing net income and risks but the identities of the participants remain independent.

marketing agencies-in-common: federated cooperatives that act as marketing agencies for their members.

marketing agreement: *See* marketing contract.

marketing board: a marketing agency granted wide powers (by governments in certain foreign countries) that direct the purchase and sales of a given commodity in both domestic and foreign markets.

marketing contract: a legal agreement between a member and a cooperative under which the member agrees to market salable products mentioned in the contract through the association, and the association agrees to market the products for the member. [*syn*: marketing agreement, member contract]

marketing cooperative: a cooperative that markets products for its patrons.

marketing order: a legally binding instrument managed by a federal or state agency that specifies how a particular farm product will be marketed; the primary purpose is to foster orderly marketing of the product.

member: a person that has met membership requirements of the cooperative and is entitled to voting privileges.

membership agreement: *See* marketing contract.

membership certificate: *See* certificate of membership.

membership fee: an amount paid for membership in a nonstock association or unincorporated cooperative.

merger: the act of two or more independent firms forming a single company with one firm surviving and the other(s) losing identity. *See also* acquisition and consolidation.

needed equity: the net worth required by a cooperative to sustain its operation for a specified period, such as the coming year, as determined by the board.

net income: total income from all sources less total expenses. [*syn*: net savings, margins, proceeds, earnings, or profit]

net savings: *See* net income.

noncash refund: *See* retained patronage refunds.

noncooperative: *See* investor oriented firm.

nonexempt cooperative: *See* nonsection 521 cooperative.

nonpatronage income: income not arising from business with or for patrons.

nonprofit association: membership associations, charities, and other organizations that are exempt from certain taxes and that generally cannot financially reward their owners or members.

nonqualified allocation: a noncash patronage refund or per-unit capital retain allocation, which is not deducted from the taxable income of the cooperative and on which the cooperative has a tax obligation. When a nonqualified allocation is later redeemed in cash, the cooperative deducts the allocation from its taxable income, and the patron recognizes the amount, with minor exceptions, as ordinary income. [*syn*: nonqualified refund]

nonsection 521 cooperative: a cooperative that does not qualify for exemption under section 521 and therefore must take all income except qualified allocations into account for computing federal income taxes.

nonstock cooperative: a cooperative formed without capital stock.

open membership: a policy of allowing membership to anyone meeting minimal qualification standards.

overinvested: the condition of patrons who have more than their share of equity, based on patronage, invested in a cooperative.

partnership: two or more persons jointly carrying on a business without being incorporated. Each is fully responsible for the debts, commitments, and obligations incurred by any one of the partners. In a limited partnership the obligations of the partners vary as prearranged.

patron: any person with whom or for whom the cooperative does business, and for whom the cooperative was organized to benefit, whether a member or a non-member of the cooperative, and regardless of legal form of organization.

patron consent: an agreement by a patron to include the full face value of a qualified written notice of allocation in the patron's taxable income.

patronage refund: net income of a cooperative allocated, under a prior existing obligation, to a patron in proportion to the value or quantity of the person's patronage, whether distributed in cash or invested in the cooperative. [*syn:* patronage dividends, distribution, saving, or rebate]

per-unit capital retain: equity invested in a cooperative by a patron based on the value or quantity of products marketed or purchased for the patron and withheld from the proceeds of products marketed or added to purchase price. [*syn:* per-unit retain allocation, capital retain]

percent-of-all-equities plan: a system of redeeming equity; a percentage of allocated equity, regardless of year of issue, is redeemed.

pooling: A method of handling products whereby lots of the same product from different producers are combined by grade and contributors receive average net payments. Typically, each grower's products lose their identity and are treated collectively as one lot by grades. All producers receive the same average price for the specific grade. Multiple pools determine proceeds on the basis of two or more grades, varieties, or periods.

producers: persons directly producing a crop, including persons having a share-crop interest which is also at *risk* and has the *chance of profits* the same as the actual producers of the crop. It includes owners or tenants that cultivate, operate, or manage farm for profit and bears the risk of production.

proxy voting: a written authorization empowering another person to vote for and act on behalf of a member. Proxy voting in cooperatives is prohibited by law in many states.

purchasing cooperative: *See* supply cooperative.

qualified allocation: a patronage refund or per-unit capital retain allocation that the cooperative can exclude from its taxable income and that the patron agrees to have taxed as if received in cash. At least 20% of a qualified patronage refund allocation must be paid in cash.

qualified check: A check, paid as part of a patronage refund, on which there is a statement that the endorsement and cashing of the check constitute the consent of the receiving patron to include the total amount of the patronage refund in the person's gross income.

qualified written notice of allocation: a written notice of allocation which the patron can redeem in cash at face value within 90 days after the date of notice, or a written notice which the patron has consented to include in his or her taxable income upon receipt in the same manner as cash. A written notice is not *qualified* unless it is distributed as part of a patronage, at least 20% of which is in cash or a qualified check as defined in the Internal Revenue Code. All other written notices of allocation are *nonqualified.*

regional cooperative: a cooperative serving members in a relatively large geographical area, usually involving a fairly large section of a state or several states.

reserve: *See* unallocated reserves.

retain: *See* per-unit capital retain.

retained patronage refunds: noncash allocations of net income, allocated to members but retained by the cooperative to increase member investment in the cooperative. Patrons are informed by written notices of allocation. These allocations are usually redeemed in cash at a later date.

revolving fund plan: A system of redeeming equity; the earliest investments of members are redeemed first. Equities may originate from retained patronage refunds, per-unit capital retains, and cash investments.

section 521 cooperative: a farmers' cooperative that has satisfactorily met the provisions of section 521 of the Internal Revenue Code. Such cooperatives are not required to pay federal income taxes on dividends paid on equity or on nonpatronage income paid to patrons on a patronage basis. [*syn*: tax-exempt cooperative]

service cooperative: a cooperative that primarily renders services such as housing, financing, insurance, artificial breeding, electricity, and telephone service.

Sherman Antitrust Act: a federal law prohibiting restraint of trade and monopoly practices.

sole proprietorship: an unincorporated business or firm owned by one person.

spatial monopolist: monopoly power obtained by locating at a distance from competitors when the size of the market is too small relative to economies of size to support more than one enterprise. Consumers are unwilling to travel to the next available source of supply.

special situation plan: a system of redeeming equity; no redemption is made until the situation of a member changes in a specified way such as death or age that initiates redemption of equity. [*syn*: special plan]

stock or share: a certificate showing investment in an incorporated firm and certifying ownership rights.

subchapter T: the portion of the Internal Revenue Code (sections 1381-1388) that covers the tax principles applying to any business operating on a cooperative basis.

supply cooperative: a cooperative that provides supplies or inputs for its patrons.

unallocated reserves: equity not allocated to individual members. Sources include tax-paid net income retained but not allocated, unclaimed checks, and appraisal surplus. [*syn*: capital reserve, retained margins, retained earnings, undistributed margins, earned surplus, tax-paid surplus, reserves for losses]

underinvested: the condition of patrons who have less than their share of equity based on patronage invested in the cooperative.

vertical integration: combining two or more sequential producing and/or marketing stages under a single management. "Vertical back" relates to source of supply and "vertical forward" relates to markets.

written notice of allocation: a written notice from a cooperative that discloses the amount allocated and the portion constituting patron refund and/or per-unit capital retains to the patron. A written notice of allocation may be qualified or nonqualified.

Subject Index

Name Index